THE GREAT RECONFIGURATION

This book is intended for researchers, policymakers, and practitioners interested in the dynamics and governance of low-carbon transitions. Drawing on the Multi-Level Perspective, it develops a whole system reconfiguration approach that explains how the incorporation of multiple innovations can cumulatively reconfigure existing systems. The book focuses on UK electricity, heat, and mobility systems, and it systematically analyses interactions between radical niche-innovations and existing (sub)systems across techno-economic, policy, and actor dimensions in the past three decades. Comparative analysis explains why the unfolding low-carbon transitions in these three systems vary in speed, scope, and depth. It evaluates to what degree these transitions qualify as *Great Reconfigurations* and assesses the future potential for, and barriers to, deeper low-carbon system transitions. Generalising across these systems, broader lessons are developed about the roles of incumbent firms, governance and politics, user engagement, wider public, and civil society organisations. This title is also available as Open Access on Cambridge Core.

FRANK W. GEELS is Professor of System Innovation and Sustainability at the University of Manchester and a world-leading scholar on socio-technical system transitions. He has published six books and more than 80 peer-reviewed articles, and he was selected in the list of Highly Cited Researchers. He is a lead author of the Working Group III contribution to the 2022 IPCC Sixth Assessment Report, a member of the Scientific Committee of the European Environment Agency, and a member of the World Economic Forum's Global Future Council on the New Agenda for Economic Growth and Recovery.

BRUNO TURNHEIM is a permanent research scientist at the French National Research Institute for Agriculture, Food and Environment (INRAE), based at the Interdisciplinary Laboratory for Science, Innovation and Society (LISIS), Gustave Eiffel University. His research focuses on socio-technical transitions, notably processes of destabilisation and experimentation. He currently leads the WAYS-OUT project on the governance of socio-technical destabilisation pathways and phase-out strategies in the context of climate governance.

'To protect the climate the world needs a massive transformation in its energy system. Policy makers and scholars need reliable guides from history for how those transformations can occur. Here is that guide — a book steeped in rich detail about how the world really works along with a fresh look at the big picture of how transformations come from interlocking technological and social processes.'

David Victor, University of California, San Diego

'One of the most analytically rich and robust assessments of low-carbon transitions I have ever seen...The comparative treatment of heat, electricity, and mobility is compelling and apt, the connections to broader issues of climate governance or system reconfiguration strong. Ground-breaking in its conceptualising but down-to-earth in its policy implications, I recommend this for students and professional researchers alike.'

Benjamin K. Sovacool, University of Sussex

'A rapid reconfiguration of large parts of the global economy is now essential for our survival. Geels and Turnheim look deeply into how such change can happen, what holds it back, and how we can speed it up. An essential guide.'

Simon Sharpe, Director of Economics for the UN Climate Action Champions

'. . .Within the field of sustainability transitions research, the multilevel perspective of Geels, Turnheim, and colleagues is arguably the most influential approach. *The Great Reconfiguration* shows how taking into account the complex relationships among actors, technologies, and governance can help to identify opportunities for change that would otherwise be overlooked. This book is the definitive statement of the new developments in the multilevel systems approach and its contribution to transition policy research.'

David J. Hess, Vanderbilt University

'. . .an impressive volume that offers remarkable insight into the progress made in decarbonising the electricity, heat, and mobility systems in the UK over the past three decades. The book's strength lies in its comprehensive analysis of system realignment within and across these three crucial domains. . .It will interest both researchers and practitioners and reveals the power and analytical sweep of contemporary transition scholarship.

James Meadowcroft, Carleton University

'This book clearly shows the power of sustainability transition research for identifying ways to mitigate the climate crisis. It advances a new and important agenda through its focus on how multiple innovations reconfigure several systems. It pairs an excellent theoretical framing with rich and well-structured empirical cases. A must read for anyone interested in understanding how sustainability transitions happen.'

Johan Schot, Utrecht University

'One of the most exciting additions to sustainability science over the last decade has come from the vibrant community of researchers exploring historical transitions in socio-technical systems. In *The Great Reconfiguration,* two leaders of that community offer a lucid summary and extension of the relevant theory, use that theory to explain the complex co-evolution of today's interlinked production-consumption systems, and conclude with practical guidance for the interventions to promote more sustainable development pathways to the future.'

William C. Clark, Harvard University

'Now that most people agree that climate change is a real problem, the big debate is about how to solve it, and whether this demands marginal reforms to the economy, a wholesale revolution of industrial and capitalist society, or something lying between these two extremes. Relying on extensive empirical research, Geels and Turnheim convincingly map out a viable middle path — the reconfiguration of key production-consumption systems — that is already beginning to achieve the emissions reductions we need, while being politically pragmatic and feasible.'

Tony Patt, ETH Zürich

THE GREAT RECONFIGURATION

A Socio-Technical Analysis of Low-Carbon Transitions in UK Electricity, Heat, and Mobility Systems

FRANK W. GEELS
University of Manchester

BRUNO TURNHEIM
French National Research Institute for Agriculture, Food and Environment (INRAE)

CAMBRIDGE
UNIVERSITY PRESS

CAMBRIDGE
UNIVERSITY PRESS

University Printing House, Cambridge CB2 8BS, United Kingdom

One Liberty Plaza, 20th Floor, New York, NY 10006, USA

477 Williamstown Road, Port Melbourne, VIC 3207, Australia

314–321, 3rd Floor, Plot 3, Splendor Forum, Jasola District Centre, New Delhi – 110025, India

103 Penang Road, #05-06/07, Visioncrest Commercial, Singapore 238467

Cambridge University Press is part of the University of Cambridge.

It furthers the University's mission by disseminating knowledge in the pursuit of education, learning, and research at the highest international levels of excellence.

www.cambridge.org
Information on this title: www.cambridge.org/9781009198240
DOI: 10.1017/9781009198233

When citing this work, please include a reference to the DOI 10.1017/9781009198233

First published 2022

A catalogue record for this publication is available from the British Library.

ISBN 978-1-009-19824-0 Paperback

Contents

Figures

Tables

Preface

At the time of writing (August 2021), the climate change problem featured high on societal and political agendas because of its increasingly visible effects (such as droughts, heat waves, and wildfires), civic activism and calls for action, the recent publication of the sixth assessment report by the Intergovernmental Panel on Climate Change (IPCC, 2021), and the 2021 United Nations Climate Change Conference in Glasgow.

In response to the growing sense of urgency, many countries have expressed commitments to net-zero targets by mid-century and a desire to bring about low-carbon system transitions. None of these countries, however, has yet presented a realistic and feasible implementation policy plan for reaching these targets, neither in terms of sector-specific transition pathways nor in terms of specific policies.

This problem is, at least partly, due to the dominance of integrated assessment models in climate policy debates. While these abstract formal models, which draw strongly on economic and engineering knowledge, have many strengths, they also have notable weaknesses, including oversimplified representations of social realities and system transition processes, limited attention to actors and behaviours (e.g., beliefs, meanings, strategies, politics), and narrow understandings of institutions, policies, and real-world implementation processes (Turnheim et al., 2015).

Although integrated assessment models will remain important for exploring potential long-term futures, acknowledgement of their limitations is increasingly leading to calls for deeper understandings of real-world drivers and barriers in the low-carbon transitions that are presently unfolding at varying speed, scope, and depth in different sectors and countries. The IPCC's 1.5°C report (IPCC, 2018), for instance, started analysing the multi-dimensional feasibility of different low-carbon innovations and system transition pathways, while the UK Committee on Climate Change (2021: 33) aims to provide more implementation-relevant knowledge by seeking to 'broaden our assessment of real-world progress, including public attitudes, corporate commitments, finance and the green recovery, as well as consumption emissions and the factors affecting them'.

Our book aims to contribute to these developments by deepening the social scientific knowledge of real-world system transition processes, which we think are relevant for developing realistic and feasible policy plans for reaching net-zero targets. While there are many books that analyse particular dimensions of low-carbon transitions, this book presents a reconfigurational perspective that aims to be integrative and interdisciplinary by addressing multiple relevant dimensions in transitions, including techno-economic, socio-political, cultural, and business dimensions. There are trade-offs between breadth and depth, however. So, although our integrative socio-technical system approach aims to generate more comprehensive and multi-dimensional understandings of low-carbon transitions, it may not quite satisfy disciplinary scholars who would desire a more in-depth analysis of a particular dimension such as politics or economics.

Another characteristic of our reconfigurational perspective is that it conceptualises socio-technical systems (in electricity, heat, and mobility) as heterogeneous entities and focuses on the endogenous processes, innovations, and activities that can change the elements of these entities and their interlinkages. Rather than seeing system transitions as singular disruptive processes, we thus conceptualise them as more dispersed processes that can cumulatively change the elements and architectures of existing systems. We empirically investigate if and to what degree unfolding low-carbon system reconfigurations are 'Great' (or not), by evaluating the depth and scope of changes in relevant dimensions of socio-technical system configurations. We also analyse the speed of change in unfolding transitions in relation to net zero targets and the urgency of reducing greenhouse gas emissions. Although our empirical analyses focus on electricity, heat, and mobility systems in the UK, the conceptual and conclusions chapters articulate general insights, findings, and propositions that will be of interest for non-UK academics and policymakers wanting to understand the policies and dynamics that drive real-world transition processes.

The intellectual background for the book's reconfigurational approach is socio-technical transitions research, which started two decades ago in the specialised field of innovation studies. Drawing on neo-Schumpeterian evolutionary economics and sociology of innovation, scholars in this field developed the Multi-Level Perspective, which conceptualises socio-technical transitions as multi-dimensional struggles between emerging niche-innovations and established systems, against the backdrop of exogenous context ('landscape') developments. The creation of the Sustainability Transitions Research Network in 2009 contributed to the deepening, broadening, and theoretical elaboration of socio-technical transitions research, as scholars from mainstream disciplines joined the rapidly growing community and made important conceptual and empirical contributions. This also nurtured important and ongoing dialogues with sociology, political science, management, and environmental sciences.

The cumulative result of these contributions was that socio-technical transitions research increasingly matured, leading scholars to venture beyond specialised outlets and to start publishing in general science journals such as *Nature*, *Science*, and *Proceedings of the National Academy of Sciences* (e.g., Geels et al., 2017, 2016a; Markard, 2018; Rosenbloom et al., 2020; Sovacool and Griffiths, 2020), which enhanced visibility and credibility. Policymakers and policy-oriented organisations also showed increasing interest in socio-technical transitions research, leading transition scholars to engage in translational efforts aimed at articulating policy-relevant insights and recommendations (EEA, 2019a; Geels, 2020a; OECD, 2015; Victor et al., 2019). Many policy-oriented organisations and NGOs also adopted the socio-technical transitions framework to guide their thinking and activities (EEA, 2019b, 2018, 2017, 2016; Global Alliance for the Future of Food, 2019; JRC, 2020; Leadbeater and Winhall, 2020).

This book benefits from the maturation of socio-technical transitions research, particularly from the hundreds of detailed UK case studies that have been published on single low-carbon innovations, particular actors, or particular dimensions. This wealth of empirical qualitative research enables this book to make a next step by integrating the detailed findings into more comprehensive analyses of whole systems reconfiguration that for each domain investigate between 6–8 niche-innovations and 2–4 systems or sub-systems. Although whole systems research is well-established in the engineering and modelling communities, this kind of integrated whole systems research is unprecedented in the socio-technical transitions community. Another novel contribution is the systematic comparison of unfolding low-carbon transitions between electricity, heat, and mobility systems. While comparative research between countries is becoming more common, comparisons between systems, and reflections on the influence of their morphological differences, break new ground in socio-technical transitions research.

This book also attempts to further strengthen and deepen conversations with mainstream climate research and policy debates. Because such inter- and trans-disciplinary conversations tend to benefit from some degree of simplification (Diercks, 2018; Turnheim et al., 2020), our book streamlines the socio-technical transitions approach in at least three ways, as Chapters 1 and 2 explain. First, regarding the tangible elements of socio-technical systems, we focus on techno-economic elements and flows, which we also map quantitatively using a wide variety of statistical databases. This simplification hopefully facilitates interactions with economic modellers and engineers. Second, our operationalisation and empirical research of rules and institutions focuses on formal policies and governance, which hopefully facilitates interactions with political scientists and policymakers. Third, with regard to actors and activities, we focus on firms, users, policymakers, and wider publics, including civil society organisations, which means we pay less attention to a wide range of other actors that are also relevant in socio-technical transitions.

This book has been a long time in the making. Initial ideas about socio-technical whole systems analysis, comparisons between systems, and efforts at interdisciplinary bridging were developed in the context of the PATHWAYS project ('Exploring transition pathways to sustainable, low carbon societies'). This three-year research project (2013–2016), which was funded by the European Commission's Seventh Framework Program, enabled fruitful interactions between integrated assessment modellers, socio-technical transition researchers, and project implementation specialists. It also produced a range of empirical analyses of socio-technical system transitions in various domains and countries, including the UK, Netherlands, Germany, Sweden, and Portugal, which demonstrated the feasibility and relevance of the initial ideas.

A one-year post-doc project in 2017–2018, sabbatical leave in 2018–2019, and Professorial Enhanced Research Leave in 2019–2020 enabled further development of the conceptual ideas and the empirical research that led to this book. We are grateful to the Alliance Manchester Business School and the Faculty of Humanities at the University of Manchester for the financial support that made this possible. We are also grateful for having received support under the French 'Programme d'Investissements d'Avenir' (ANR-19-MPGA-0010), which allowed us to publish the book in Open Access and so reach out to a wider readership.

Over the past year, we have worked hard to finish the book, which had to be done in between other projects and commitments, and in the context of the COVID-19 pandemic, which substantially affected our available work time since we both have care responsibilities for young children. We initially intended to also include the agri-food system in our analysis, because it differs in interesting ways from the electricity, heat, and mobility systems. We reluctantly decided to exclude it, however, because of time constraints.

We are very pleased that the book is finally finished and hope that researchers from multiple disciplines as well as policymakers will find something interesting and relevant in it. Last but not least, we want to thank our families for their support, understanding, and patience while we worked to complete the book.

1
Introduction

1.1 The Need for Low-Carbon System Transitions and a Reconfigurational Approach

Climate change is a grand societal challenge that in the coming decades will increasingly affect many aspects of society either through its impacts (e.g., droughts, floods, crop failures, fires, sea level rise, heat stress) or through mitigation efforts that attempt to transform energy, mobility, industrial, and agri-food systems in low-carbon directions (IPCC, 2018).

Recognition of the seriousness of these threats and the scale of the mitigation challenges has increased public attention to climate change since the mid-2000s (see Figure 1.1), fuelled by events such as Hurricane Katrina (2005), Al Gore's movie *An Inconvenient Truth* (2006), the Stern Review (2006), and the Fourth IPPC Assessment Report (2007). Public attention decreased after the 2007/8 financial crisis, but has increased again in recent years, along with highly publicised events such as the Paris Agreement (2015), protests by school children and civil society organisations (e.g., Extinction Rebellion, Climate Justice movement), and new framings such as 'climate emergency' since 2019.

In response, an increasing number of countries have adopted net-zero greenhouse gas (GHG) emission targets, and broadened and strengthened their low-carbon transition plans. Public attention remained high throughout the COVID-19 pandemic, creating pressures on policymakers, although there is an indication of a slight decrease in coverage during 2020, due to competing societal issues related to the pandemic.

It is now widely recognised that achieving net-zero targets will require system transitions in core societal domains. The Intergovernmental Panel on Climate Change (IPCC), for instance, calls for 'rapid and far-reaching transitions in energy, land, urban and infrastructure (including transport and buildings), and industrial systems. These systems transitions are unprecedented in terms of scale, but not

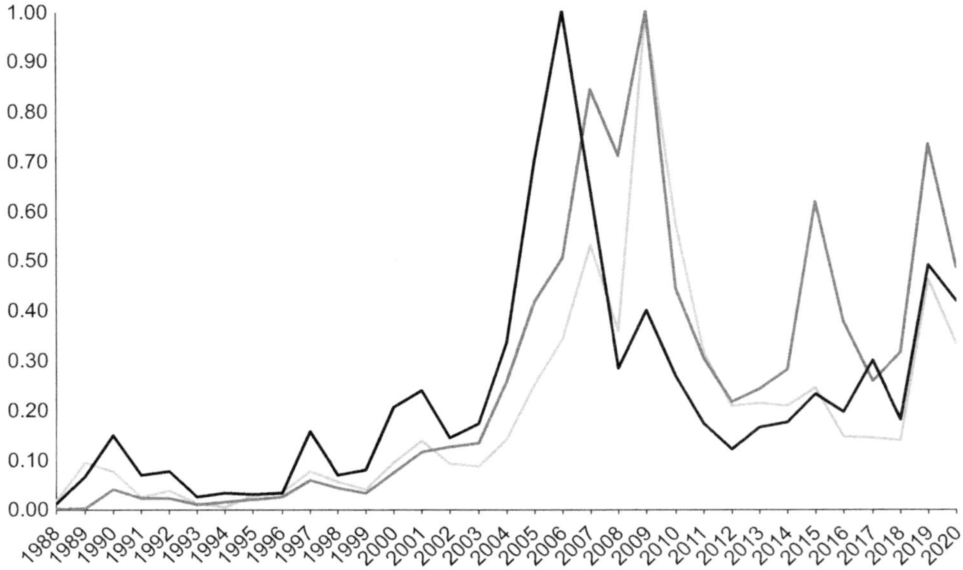

Figure 1.1 Yearly number of articles in selected UK national newspapers related to climate change (the graph is based on data from a keyword search in the digital archives of these newspapers, using the search string [Text, 'climate change' OR 'global warming' OR 'global heating' OR 'greenhouse effect' OR 'greenhouse gas' OR 'climate emergency' OR 'climate crisis' OR 'decarbonisation' OR 'decarbonization' OR 'low-carbon'] within the title and first hundred words of the articles. Duplicated articles were excluded. To facilitate visual comparison between different data sets, we normalised the time series to the year with the maximum number of counts)

necessarily in terms of speed, and imply deep emission reductions in all sectors' (IPCC, 2018: 21).

The European Commission's long-term climate strategy likewise acknowledges that:

> economic and societal transformations are required, engaging all sectors of the economy and society, to achieve the transition to net-zero greenhouse gas emissions by 2050. . . . This transition will radically transform our energy system, land and agricultural sector, modernise our industrial fabric and our transport system and cities, further affecting all activities of our society.
>
> *(EC, 2018: 5–6)*

The European Environment Agency (EEA) also assesses that addressing climate change (and other persistent environmental problems) 'will require fundamental transitions in core production-consumption systems such as those meeting European demand for food, energy, mobility and housing. Such transitions will necessarily entail profound changes in dominant institutions, practices, technologies, policies, lifestyles and thinking' (EEA, 2019a: 7).

While the need for low-carbon system transitions is now widely acknowledged, there is disagreement, however, in both public and academic debates, about *what system transitions are and how they come about.* Building on Geels et al. (2015), we distinguish three analytical approaches that resonate with different scientific theories and policy approaches: 1) reform, 2) revolution, and 3) reconfiguration.

The *reformist* approach, which can be found in engineering, modelling, and mainstream economics (Acemoglu et al., 2016; Dangerman and Schellnhuber, 2013; Hawken, 2017; Rockström et al., 2017; Stiglitz et al., 2017), conceptualises transitions as driven by the development and market adoption of low-carbon technologies that substantially lower the carbon-intensity of existing provisioning systems in energy, mobility, food, and industrial production. It sees research and development (R&D), subsidies (to stimulate new technologies), and carbon pricing instruments (to influence company investment decisions and consumer purchases) as the main policy instruments. This approach is 'reformist' because it assumes that low-carbon transitions only involve technical component substitution and do not affect other elements of transport, energy, and agri-food systems. Because the approach assumes that the technological substitutions mainly involve rational economic decisions by firms, investors, and users, it also does not pay much attention to political, social, or behavioural dimensions.

While the reformist position rightly emphasises the importance of low-carbon technologies, investments, and markets, it also has several limitations: 1) it has an over-simplistic 'linear model' understanding of innovation as 'pushed' by upstream R&D investments and 'pulled' by downstream market demand (Schot and Steinmueller, 2018), 2) it pays little attention to non-technological kinds of innovations such as social, business model, and grassroots innovations, 3) it pays too little attention to non-market actors such as wider publics, civil society organisations, industry associations, and other lobby organisations, 4) it also pays too little attention to various forms of agency and processes such as institutional change and power struggles, business activities and strategic games, cultural meanings, and demand-side dynamics in social practices.

The *revolutionary* approach, which involves multiple sub-streams that share deep critiques of the status quo and current policymaking, views low-carbon system transitions as involving the complete overhaul of socio-economic deep structures. Neo-Marxist (Schnaiberg, 1980) and critical political economy scholars (Newell, 2021; Paterson and P-Laberge, 2018), for example, highlight the need to overthrow or transform capitalism (particularly its focus on commodification, market competition, and capital accumulation) and neo-liberalism (particularly its faith in free markets). Scholars within revolutionary strands have proposed a 'new economics' (Schor, 2014) that includes 'de-growth' (Kallis, 2011), more emphasis on third sector and community-based enterprise (Jackson and Victor, 2011), a shift from GDP measures towards happiness (Gough, 2010), and returning the globalised financial system to 'its role as servant, not the master of the economy and ecosystems' (Pettifor, 2019). Cultural and moral critics also call for changes in consumer society and underpinning values, which should be transformed in the direction of frugality and sufficiency (Alcott, 2007; Princen, 2005) or towards 'meaningful' activities such as 'fine education, arts, healthcare, childcare and elderly services, [. . .] and community development' (Vergragt, 2013: 124). These cultural value changes have been characterised as The Great Mindshift (Göpel, 2016) or The Great Transition towards a new planetary civilisation, which involves a 'fundamental shift in the paradigm of development – indeed, in the very meaning of human progress. A Great Transition would make solidarity, fulfilment, and resilience the heart and soul of human endeavour' (Raskin, 2016: iii).

While the revolutionary position rightly draws attention to macro-economic and macro-cultural issues, it also has several limitations. First, these assessments of macro-level 'deep structures' are often reductionist because they try to bring complex realities back to single 'root causes'. Second, they tend to be rather abstract and distanced from concrete experiences of real-world actors. Many, though not all, critical analyses of capitalism or neo-liberalism are disempowering because their focus on an 'all-encompassing entity can easily come to appear as a kind of gigantic, all-powerful [. . .] force that causes everything else to happen' (Ferguson, 2010: 171), which is difficult to alter by situated actors. Third, macro-level analyses of capitalism lack the explanatory granularity to satisfactorily explain why some sectors, such as electricity, have made much more decarbonisation progress than other sectors, such as heat.

Fourth, despite their interest in fundamental change, some sub-streams in the revolutionary position are paradoxically static, restricting analysis to critiques of deep structures or advancing utopian visions of communitarian, local, and sustainable societies. These sub-streams thus offer little insight about change mechanisms or dynamic pathways that could bring about the desired system transitions. Other revolutionary sub-streams place high hopes on the transformative power of community initiatives, grassroots innovations, or social movement activism, but often fail to articulate how local initiatives bring about large-scale system change. Steward (2018: 100) in this regard notes that 'Such case studies [of community activism] are certainly impressive and inspiring. However, they do not demonstrate to academic critics that this is a route for a transition to a low-carbon society at a broader level'. O'Brien and Signa (2018: 40) similarly observe an analytical gap between local initiatives and large-scale transformation, noting that there are many studies of the former but few of the latter, 'There are (as yet) relatively few empirical examples of successful large-scale transformations of socio-ecological systems towards sustainability.' These strands in the revolutionary position therefore suffer from a 'lack of empirical grounding' (Feola, 2015: 377), which means that related interpretations of system transition remain more normative and political than analytical.

Because of the limitations of the reformist and revolutionary approaches, this book mobilises a *reconfiguration* approach, which builds on the general scientific notion that 'the whole is best understood from a systemic perspective and should be viewed as a constellation of interconnected elements' (Fiss et al., 2013: 2). Instead of privileging an ultimate cause, reconfigurational approaches imply a commitment to multidimensional analysis that traces endogenous interactions between multiple components and processes that together produce larger outcomes: 'What makes configurational thinking unique is its insistence on putting particular pieces together into larger wholes' (Abbott, 2001: 119). Reconfigurational approaches are particularly suited for analysing changes in large-scale systems made up of heterogeneous entities (see Section 3.1), and they often involve processual analyses that explain how outcomes are produced through co-evolving causal processes: 'This interest in combinations of causes dovetails with a focus on "how" things happen [...] to understand causally relevant conditions as intersections of forces and events' (Ragin, 2008: 109).

With regard to low-carbon system transitions, the reconfiguration approach has been particularly developed in socio-technical transitions research, which focuses on deep changes to socio-technical systems that fulfil societal functions such as mobility, thermal comfort, or sustenance. Drawing on sociology of innovation and evolutionary economics (Geels, 2020b, 2004), socio-technical systems are conceptualised as heterogeneous configurations of elements including technical artefacts, scientific knowledge, industry structures, markets, consumption patterns, infrastructure, policy, and cultural meanings (Figure 1.2).

In this book, we build on socio-technical transitions research and conceptualise low-carbon system transitions as involving substantial changes in both the elements of socio-technical systems and the architecture of their linkages (Geels et al., 2017). Taking innovation (in technologies, business models, social practices) as the analytical *entrance point*, the socio-technical transitions approach follows the emergence, diffusion, and societal embedding of innovations over time and analyses the relevant interactions between technical, social, cultural, political and economic processes and actors (Geels, 2005). Socio-technical transitions have the following characteristics (Köhler et al., 2019):

- *Multidimensionality and co-evolution*: Transitions are co-evolutionary processes involving interactions between multiple socio-technical dimensions. Because it

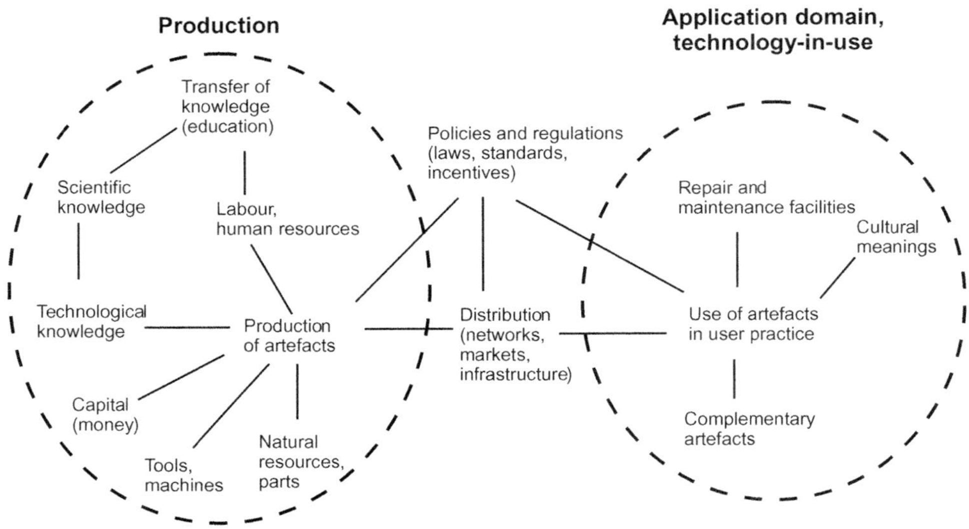

Figure 1.2 Basic elements of socio-technical systems (Geels, 2004: 900)

does not privilege one driver or dimension, the reconfiguration position is less reductionist than both the reformist position (which focuses on techno-economic processes) and the revolutionary position (which focuses on macro-economic or macro-cultural 'root causes').

- *Multi-actor process*: Transitions are enacted not only by firms and users (as in the reformist position) but also by social movements, wider publics, policymakers, industry associations, and other special-interest groups (Geels, 2004). These social groups have different interests and resources, and they engage in multiple activities (e.g., technological exploration and learning, public debates, political power struggle, investment, negotiation, coalition building) which make transitions very complicated processes that cannot be comprehensively understood by single theories or disciplines.
- *Goal-orientation*: Low-carbon transitions are particular kinds of socio-technical transitions because they aim to address environmental objectives rather than mere technical performance or economic objectives. This means that in addition to requiring deep and large-scale system changes, they involve the additional challenge of securing a particular directionality (Kemp and van Lente, 2011). Therefore, policymakers must play a central role in low-carbon transitions by adjusting institutions and policies, including regulations, standards, taxes, and subsidies.
- *Resistance, conflict, and struggle*: Since low-carbon transitions threaten the economic positions and business models of some of the largest and most powerful industries (e.g., oil, automotive, electric utilities, agri-food), such

incumbents will protect their vested interests, which may lead to conflict and struggle about the need for, and speed of, transitions and the types and stringency of policy instruments intended to advance them.

- *Long-term process*: Transitions are longitudinal processes that often take decades to unfold. One reason is that radical low-carbon innovations and practices usually take a long time to develop from their early emergence in small application niches to widespread diffusion. Another reason is that it takes time to destabilise and 'unlock' existing systems and overcome resistance from incumbent actors (Turnheim and Geels, 2012).
- *Non-linearity, uncertainty, and open-endedness*: Because there are *multiple* low-carbon innovations and initiatives in all socio-technical systems, it is difficult to predict in advance which of these will prevail. Since there are multiple possible transition pathways (Geels and Schot, 2007; Rosenbloom, 2017), future low-carbon transitions are open-ended. Uncertainty also stems from the non-linear character of innovation processes (which may experience failures, hype-disappointment cycles, or accelerated price/performance improvements), political processes (which may experience setbacks, reversals, or accelerations), and socio-cultural processes (which may experience changes in public agendas and sense of urgency).

These characteristics make socio-technical system transitions a special kind of phenomenon that requires a dedicated research approach. This book therefore aims to elaborate the socio-technical transitions approach, which has emerged in the past two decades in the innovation studies and sustainability transitions communities, and to introduce it into the mainstream climate mitigation debates. In doing so, it hopes to transcend the first two approaches that have long dominated mainstream environmental sustainability and climate mitigation debates, which has led to stale dichotomies between strong and weak ecological modernisation (Christoff, 1996) and strong and weak sustainable consumption (Fuchs and Lorek, 2005). Spaargaren and Cohen (2009: 257) criticised these traditional positions as overly limited, characterising them as 'the dark green romantic dismissal of modernity and the naïve endorsement of market driven liberal eco-technotopias'.

The book also aims to be relevant with regard to ongoing policy debates. The reformist approach is closely tied to the policy orthodoxy, which has long led climate policy debates to emphasise R&D subsidies and carbon pricing instruments to reorient financial investments (Energy Transitions Commission, 2017, 2016; IEA and IRENA, 2017; OECD, 2018; World Bank, 2015). While these generic economic policies are not irrelevant, the increasing emphasis on the *implementation* of low-carbon innovations and system transition is changing policy debates to focus more on specific innovations, the actors that develop and deploy them, and their real-world social and political feasibility (Meckling and Allan, 2020).

The European Commission (EC, 2019), for instance, acknowledges that 'New technologies, sustainable solutions and disruptive innovation are critical to achieve the objectives of the European Green Deal' (p. 18), that 'conventional approaches will not be sufficient' (p. 18), and that 'there is a need to rethink policies for clean energy supply across the economy, industry, production and consumption, large-scale infrastructure, transport, food and agriculture, construction, taxation and social benefits' (p. 4). The IPCC (2018) emphasises that climate mitigation policies should address six feasibility dimensions (technological, economic, socio-cultural, institutional, geophysical, environmental-ecological) that shape the real-world implementation and acceptance of low-carbon transitions. And the UK Committee on Climate Change (2021: 33) calls for deeper understandings of real-world implementation processes and actors to support policymaking:

> As Government makes the shift to focusing on implementation, the Committee's task must also evolve towards a focus on real-world progress and tougher scrutiny of Government plans. [. . .] The transition to Net Zero requires changes that go beyond the deployment-related metrics we have tended to track to date. We will seek to broaden our assessment of real-world progress, including public attitudes, corporate commitments, finance and the green recovery, as well as consumption emissions and the factors affecting them.

The socio-technical transitions approach to whole system reconfiguration, which this book will develop, aims to contribute to these recent policy debates.

1.2 The Multi-Level Perspective on Socio-Technical System Transitions

To further conceptualise socio-technical system transitions, we use the Multi-Level Perspective (MLP), which is a middle-range theory that combines insights from evolutionary economics, sociology of innovation, and institutional theory (Geels, 2020b, 2002). The MLP suggests that socio-technical transitions result from the interplay of developments at three analytical levels: socio-technical systems, niche-innovations, and exogenous socio-technical landscape developments (Geels, 2019, 2002; Rip and Kemp, 1998).

Before discussing these levels and their interactions, we articulate some foundational assumptions which build on Geels (2004) who distinguished three interrelated analytical dimensions: 1) tangible elements of socio-technical systems, 2) actors and social groups whose actions maintain, improve, repair, and change the system elements (through research, technology development activities, purchasing, debates, policymaking), and 3) rules and institutions (often called 'socio-technical regime') that shape actors' preferences, strategies, and actions.

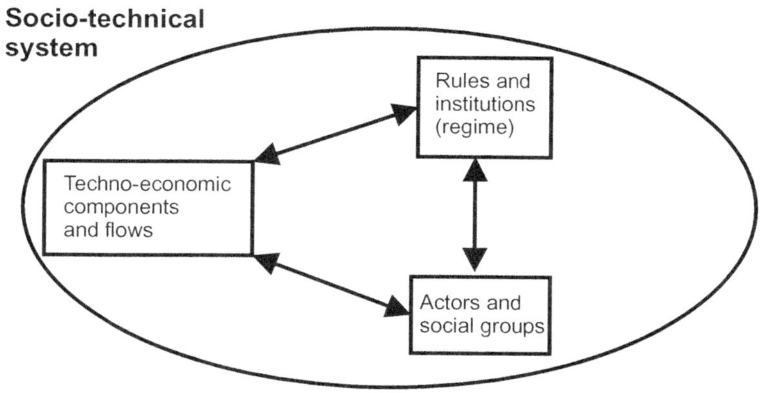

Figure 1.3 Three ontological dimensions of socio-technical systems (adapted from Geels, 2004: 903)

To facilitate interdisciplinary bridging conversations with technical, economic, and policy sciences (Cherp et al., 2018; Geels et al., 2016a; Turnheim et al., 2015) and increase relevance for policymakers, we make four simplifying adjustments in this conceptual scheme. First, rather than diffusely focusing on all tangible elements of socio-technical systems, we reformulate the first dimension of the distinction as listed in the preceding paragraph to focus more narrowly on material, technical, and economic elements and flows (e.g., artefacts, infrastructures, factories, flows of inputs and outputs). This resonates with, and gives more emphasis to, Geels' (2004: 904) observation that 'Technologies have a certain "hardness" or obduracy, which has to do with their material nature, but also with economic aspects. [. . .] This hardness also implies that artefacts cannot entirely be shaped at will.' This reconceptualisation also draws more analytical attention to the material and energy flows that sustain socio-technical systems.

As a second adjustment, we reconceptualise socio-technical systems as the entire configuration of the three analytical dimensions (Figure 1.3). This resolves the problem that the previous conceptualisation included three dimensions but had no concept to cover the whole configuration. This conceptualisation means that socio-technical systems have material, relational, and institutional dimensions.

As a third adjustment, we simplify the conceptualisation of rules and institutions, as Section 2.2.3 further explains. Previous conceptualisations (e.g., Fuenfschilling and Truffer, 2014; Geels, 2004), which build on neo-institutional theory (Powell and DiMaggio, 1991; Scott, 1995), distinguished three kinds of rules and institutions (regulative, normative, and cultural-cognitive) as enabling and constraining actors in different ways. *For the purpose of this book*, we simplify the conceptualisation to focus more narrowly on 'policies and governance structures', which is closer to the old institutional theory's understanding of institutions as rules of the game (Hirsch and Lounsbury, 1997; North, 1990). This conceptualisation of institutions is easier to operationalise and investigate, and intrinsically leads to a stronger focus on policy and politics, which hopefully appeals to readers with a policy interest.

Since we do not want to exclude norms and cultural-cognitive dimensions from the analysis, our fourth adjustment is to endogenise these dimensions in our conceptualisation of actors, which also includes intendedly rational strategic action, behavioural routines, and capabilities, as Section 2.2.2 further explains.

These theoretical assumptions underpin and inform the conceptualisation of the three analytical levels of the MLP and the dynamics of socio-technical transitions, to which we now turn.

Existing *socio-technical systems* are stabilised by various lock-in mechanisms that constrain incumbent actors and orient their activities towards incremental rather than radical change. These include a) techno-economic lock-in mechanisms such as sunk investments, material obduracy, low cost and high performance characteristic, b) social and cognitive lock-in mechanisms such as routines, heuristics (Nelson and Winter, 1982), shared mindsets, habits, and lifestyles (Barnes et al., 2004), and c) institutional and political lock-in mechanisms such as existing regulations and standards that favour existing systems and create an uneven playing field for emerging innovations (Walker, 2000) as well as institutional procedures that give incumbents more access to policy networks, where they can influence policymaking and protect the status quo (Geels, 2014; Kolk and Pinkse, 2007). These lock-in mechanisms stabilise existing systems, which is why system transitions do not happen easily.

Radical innovations, which are the seeds of transitions, emerge in small *niches* at the periphery of existing systems, through pioneering activities of entrepreneurs, start-ups, activists, or other relative outsiders (Kemp et al., 1998). Niches form ‘protected spaces’ that shelter radical (technical, grassroots, and business model) innovations from mainstream market selection pressure and nurture learning and development processes (Smith and Raven, 2012). The degree of radicality of niche-innovations depends on how much they deviate from the existing system on technical, social, business model, or infrastructural dimensions.

The struggles between niche-innovations and existing socio-technical systems are influenced by the *socio-technical landscape* (Rip and Kemp, 1998), which includes both slow-changing states or developments (e.g., demographics, cultural repertoires, societal concerns, geo-politics, macro-economic trends) and external shocks (e.g., wars, financial crises, accidents, oil price shocks, pandemics) (Van Driel and Schot, 2005).

Socio-technical transitions are non-linear processes that occur through the interplay between processes at niche, system, and landscape levels, which unfold

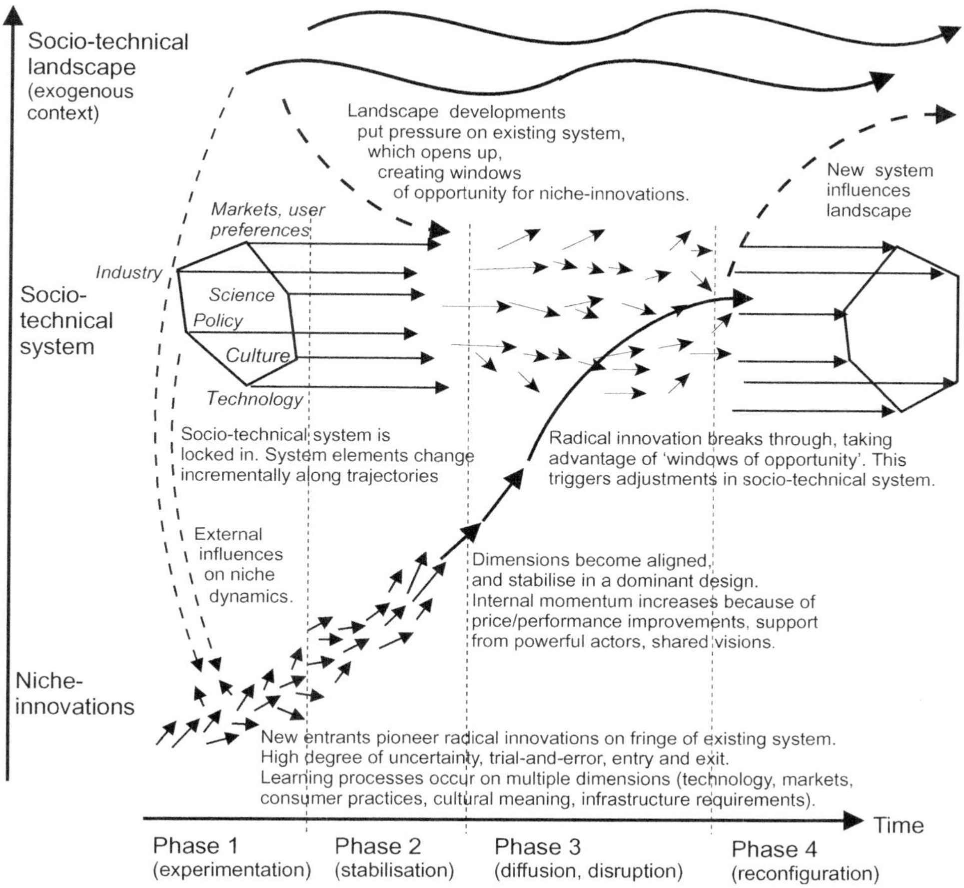

Figure 1.4 Multi-Level Perspective on socio-technical transitions (substantially adapted from Geels, 2002: 1263)

over time through four phases (Figure 1.4), which are further discussed in Chapter 2. In the first phase, radical innovations emerge in small niches. They gradually build up internal momentum in the second phase, but often face uphill struggles against entrenched systems. In the third phase, external landscape pressures and bottom-up niche pressures help destabilise the existing system, leading to highly visible struggles in business, socio-cultural, and political dimensions. In the fourth phase, diffusing innovations replace the existing system, trigger wider system reconfigurations, and become settled in a new status quo.

Instead of a single cause or driver, the MLP thus emphasises alignments between processes on multiple dimensions and at different levels which together culminate in system transitions. The MLP accepts the importance of technological and economic dimensions (e.g., firms, markets, investments), emphasised by the reformist approach, but is broader in at least two important ways. First, it also acknowledges the potential role of other *types* of innovation such as grassroots and social innovation (Geels, 2019) or business model innovation (Bidmon and Knab, 2018; Van Waes et al., 2018). Second, the understanding of innovation and diffusion *processes* includes not only techno-economic dimensions but also socio-cultural and political ones such as discursive framing struggles (Hermwille, 2016; Roberts and Geels, 2018; Rosenbloom et al., 2016), political coalitions and empowerment (Hess, 2014; Markard et al., 2016; Smith and Raven, 2012), and societal embedding (Kanger et al., 2019; Mylan et al., 2019).

The MLP's landscape concept accommodates the potential role of macro-level influences, but unlike the revolutionary position it does not a priori assume that these are the most important or 'ultimate' drivers of socio-technical transitions. Instead, the role of exogenous landscape pressures should be empirically investigated and analysed, including how these pressures are interpreted and mobilised by actors at niche and system levels.

1.3 Aims and Contributions of the Book

The book aims to make three scientific contributions to different debates.

1.3.1 The Great Reconfiguration

With regard to the different views on low-carbon systems transitions, summarised in Section 1.1, the book's first aim is to elaborate and demonstrate the usefulness and validity of the reconfiguration approach. By analysing unfolding low-carbon system transitions, the book demonstrates that this approach better resonates with the empirical evidence and provides a more comprehensive, differentiated, and policy-relevant understanding than the reformist and revolutionary approaches.

One reason for choosing *The Great Reconfiguration* as the book's title is thus to highlight that reconfiguration is a fruitful approach for analysing transitions in socio-technical systems. Another reason is that the title relates it to and differentiates it from Polanyi's (1944) book *The Great Transformation: The Political and Economic Origins of Our Time*, which scholars from the revolutionary position often use as a source of inspiration. Polanyi analysed how fundamental changes in mentalities (e.g., liberalism) and formal institutions (e.g., property rights) in the early nineteenth century helped to create capitalist market societies by dis-embedding economic activity from moral, customary, and religious constraints, which resulted in the commodification of labour, land, and money and the unleashed pursuit of individual monetary advantage. Polanyi also showed how the negative consequences of free market capitalism led to counter-movements in the late nineteenth century, which resulted in some degree of re-embedding of capitalism in the first half of the twentieth century (e.g., anti-trust and banking regulations, welfare state policies).

While we agree with the revolutionary approach that addressing climate change requires deep and fundamental change, we argue that a focus on reconfiguring socio-technical systems offers greater analytical traction and stronger socio-political appeal than the overhaul of capitalism or calls for frugality and sufficiency. We also take inspiration from Polanyi's notion of the double movement, which suggests that major transitions always involve conflicts and struggles between an initial movement (propelled by actors advocating change) and a subsequent countermovement (by actors with vested interests in the status quo or actors suffering unintended negative consequences). And we refer to Polanyi, because his work demonstrates the importance of longitudinal and historically informed diagnoses of the present, which is something we aim to emulate for low-carbon transitions.

Another reason for choosing the book's title is that the word 'Great' implies that low-carbon transitions require a particular kind of reconfigurational change, which has substantial scope and depth. Substantial *scope* refers to the breadth of change, meaning that many socio-technical system elements undergo change, that is, not just techno-economic elements but also actors and institutions (Figure 1.3). Substantial *depth* means, first, that techno-economic changes involve not just incremental but also radical innovations and architectural change; second, that actors do not just change their activities and resource allocations but also their goals, strategies and interpretations; and, third, that institutional change is not just about new settings of existing policy instruments but also involves new types of policy instruments and new governance paradigms (Hall, 1993).

1.3.2 Socio-Technical Transitions and the Multi-Level Perspective

With regard to debates about socio-technical transitions and the Multi-Level Perspective (MLP), the book's second aim is to make important adjustments in the MLP to address relevant criticisms and to conceptualise our Great Reconfiguration approach. One relevant criticism of the MLP is that many empirical studies of low-carbon transitions have focused on the emergence and diffusion of niche-innovations (Berkhout et al., 2004), which tends to lead to a singular bottom-up (or 'point source') view of change (Geels, 2018a). Our book therefore shifts the analytical emphasis towards *existing systems* and how these can be reconfigured in interaction with emerging niche-innovations (see Geels, 2018b; McMeekin et al., 2019 for initial explorations).

Another relevant criticism is that empirical studies of socio-technical transitions have focused too much on supply-side technological development and too little on users and social practices (Hargreaves et al., 2013; Shove and Walker, 2010). This criticism is somewhat misplaced because user practices have from the start been conceptualised as part of socio-technical systems (see Figure 1.2). Nevertheless, to address this criticism and analyse the reconfiguration of entire socio-technical systems, our book will explicitly investigate the perceptions and actions of users and households.

A third criticism is that the socio-technical transitions literature has focused more on actors and institutions as causal factors than on material and economic dimensions (Cherp et al., 2018; Svensson and Nikoleris, 2018). The book's Great Reconfiguration approach therefore explicitly analyses techno-economic dimensions as well as actors and institutions, both for existing systems and niche-innovations.

A fourth criticism is that the portrayal of socio-technical transitions as involving a single niche-innovation struggling against a single system (including in schematic representations such as Figure 1.4) is too simple (Andersen et al., 2020; Papachristos et al., 2013; Verbong et al., 2008). Our conceptualisation of whole system reconfiguration will therefore emphasise the role of *multiple* niche-innovations and *multiple* (sub)systems, which can interact in various ways.

The various elaborations of the MLP change the transition imagery from singular 'bottom-up' disruption towards a more dispersed 'whole system' reconfiguration process, resulting from multiple change mechanisms including incremental improvements in existing system components, replacement of system components by niche-innovations, incorporation of niche-innovations into existing systems (as add-ons), changing size of systems, or changing linkages between system components. This creates the possibility of a variety of low-carbon transitions pathways (Geels and Schot, 2007), which can be explored empirically.

While there is an understandable tendency in the socio-technical transitions literature to 'zoom' in and focus on particular actors or dimensions (Köhler et al., 2019), our book aims to demonstrate the importance and fruitfulness of 'zooming out', especially for addressing whole system transitions, which has arguably been the original focus of socio-technical transitions research (Elzen et al., 2004; Geels, 2005).

1.3.3 Comparative Analysis of System Reconfiguration in UK Low-Carbon Transition

The book's third aim is to empirically analyse ongoing low-carbon transitions in three major systems (electricity, heat, mobility) within a single country, and to systematically compare reconfiguration patterns across these systems. We deliberately chose this single country research design because it means that the wider socio-economic, cultural, and political contexts of the three systems are the same. This enables systematic comparative analysis of the three systems, because differences in speed, scope, and depth of low-carbon transitions will relate more to system-specific innovations, actors, and policies than to wider contexts. This research design therefore enables insights about similarities and differences between the systems, as well as reflections about a national style of governance and management of low-carbon transitions.

Methodologically, the book is inscribed within a comparative turn in transitions studies, but it deviates from previous comparisons, which mostly compare one system or technology in multiple countries. Our research design required us to develop a unified operational approach that can be applied to different systems (Chapter 3), and the formulation of a conceptual framework focussed on key dimensions of whole systems reconfiguration (Chapter 2). This ambitious exercise also required us to revisit foundational conceptual debates and offer suggestions for conceptual elaboration (Chapter 2).

The book's empirical analyses apply our conceptual framework, demonstrating the usefulness of a socio-technical system reconfiguration approach. We have chosen to focus our single country analysis on the UK, which has reduced its domestic greenhouse gas (GHG) emissions by 41% between 1990 and 2019, while the economy grew by 78% (CCC, 2020). These reductions made the UK one of the leading countries in climate mitigation (Le Quéré et al., 2019), especially when compared to other major developed economies (Figure 1.5).[1]

Since low-carbon transitions are beginning to unfold in the UK, it is important and interesting to not only analyse the low-carbon innovations that directly shape GHG emissions but also the underlying techno-economic, actor, and policy changes that explain the scope, depth, and speed of associated system reconfigurations. Previous studies of particular innovations and domains have shown that UK decarbonisation journeys were not linear technological deployment processes but are instead full of struggles, setbacks, and tensions (Kivimaa and Martiskainen, 2018; Kuzemko, 2016; Lowes et al., 2020). A socio-technical system reconfiguration approach is hence highly relevant to better understand the UK case.

The UK case is also interesting because GHG emission reductions varied substantially between different systems, as Figure 1.6 shows. Our book focuses on the electricity, heat, and mobility systems, which are among the largest GHG producers and together generated 59% of UK domestic emissions in 2019. The associated socio-technical systems range from production to infrastructure to end-use social practices, which makes them very suited for our whole system

[1] Although the UK consists of four devolved nations (England, Wales, Scotland, Northern Ireland), it is beyond the book's scope to systematically disaggregate our analysis to the nation level. Some of the statistical data, however, focus on Great Britain (England, Wales, Scotland) rather than the UK.

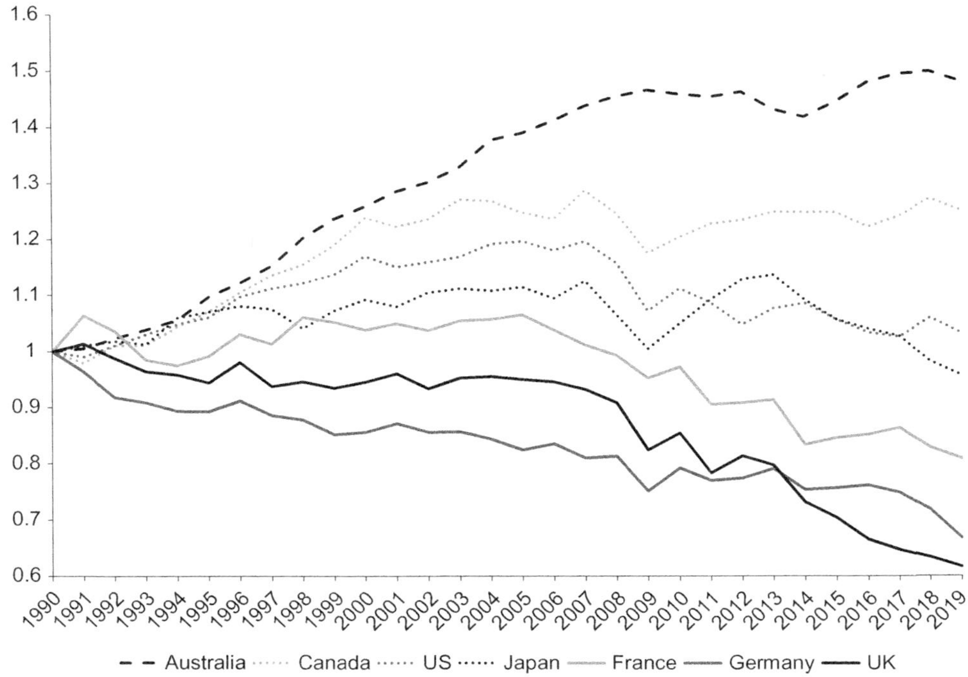

Figure 1.5 Indexed CO_2 developments of major developed economies, 1990–2019 (1990=1) (constructed using data from 'Our World in Data' database; https://ourworldindata.org/)

reconfiguration approach. This is not the case for industry, the second largest GHG producer in 2019, which includes upstream production (e.g., steel, cement, chemicals) but not end-use social practices. Although we excluded industry from our analyses, future research could fruitfully investigate the low-carbon innovations, actors, and policies in this domain, while also addressing specificities such as capital intensity and low-cost competition on international commodity markets.

The three selected systems have contrasting performances in GHG emission reduction (Figure 1.6).

- GHG emissions of the electricity system decreased by 71% between 1990 and 2019, from 204.2 $MtCO_2e$ to 58.5 $MtCO_2e$, experiencing steep declines in the 1990s, increases between 1997 and 2006, and another steep decline since 2006.
- GHG emissions in the mobility system declined by 4.6% between 1990 and 2019, experiencing steady increases until 2007, a sharp decline after the 2008 financial crisis, some increase between 2013 and 2017, and another decline since then.
- Emissions in the heat system decreased by 15.4% between 1990 and 2019, experiencing steady declines between 1996 and 2014, but increasing emission trends since then (Figure 1.5).

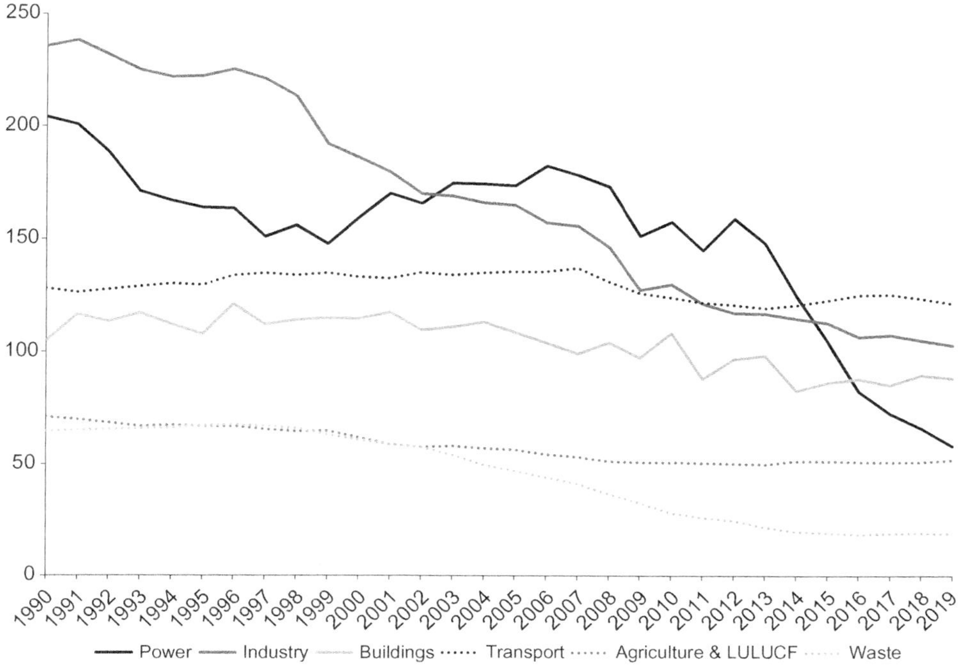

Figure 1.6 UK domestic GHG emissions in metric tons of carbon dioxide equivalent ($MtCO_2$), 1990–2019 (constructed using final UK greenhouse gas emissions national statistics 1990–2019, BEIS, 2021)

So, while a low-carbon transition is clearly unfolding in the electricity system, this is not (yet) the case in the heat and mobility systems. These different performances in GHG emission reductions are the empirical puzzle that the book aims to explain. Using the socio-technical system reconfiguration approach we address the following research questions:

(1) Which innovations and system changes contributed directly to the varying GHG emission performances in the three systems?
(2) What are the underlying techno-economic, actor, and policy reconfigurations, and what do these changes imply for the scope and depth of socio-technical system reconfiguration?
(3) Are the unfolding low-carbon transitions moving in the direction of a Great Reconfiguration, characterised by high scope and depth of system changes?
(4) What explains the different speed between unfolding low-carbon system reconfigurations?

To answer these questions, our book makes in-depth, multi-dimensional analyses of the electricity, heat, and mobility systems, with particular attention to understanding the dynamic interaction of sources of change (e.g., technological, political, and societal innovation) and sources of stability (e.g., structural lock-ins and their enactment in various strategies). For each system, we analyse the lock-in mechanisms, degree of resistance and reorientation of incumbent actors, and gradual developments in the *existing* system as well as multiple radical niche-innovations that have emerged in the past two or three decades. To understand and assess the momentum of and potential for low-carbon transitions in the present, we thus follow Polanyi's lead and analyse the *longitudinal* (multidecadal) developments that have led to present situations in the different systems. The analysis of both existing systems and emerging niche-innovation examines *techno-economic* developments (e.g., material system elements, technical improvements, market shares, costs), *actors* and activities (focused on firms, users, policymakers, and wider publics), and *institutions* (addressing both formal policies and informal governance styles). The book combines quantitative and qualitive analyses to make comprehensive and comparative assessments of unfolding reconfigurations in three socio-technical systems.

Since this kind of analysis of low-carbon system transitions does not yet exist in the literature, neither for the UK nor for other countries, our book makes a major empirical contribution to climate mitigation debates. To be sure, many other kinds of analyses of UK low-carbon transitions do exist, but these mostly focus on particular dimensions, innovations, or actors. Economists, engineers, and modellers, for instance, have made many *techno-economic* analyses of recent and future UK low-carbon transitions, which focus on the performance and costs of various technologies that are competing in markets, which are shaped by policies (Hardt et al., 2018; Skea et al., 2019; Staffell, 2017; Wilson and Staffell, 2018). Modellers have also made techno-economic whole-system analyses, which investigate how multiple technologies may interact to change systems in low-carbon directions (Kannan and Strachan, 2009; Strachan et al., 2009). Since its creation in 2008, the UK Committee on Climate Change (CCC) has also made very informative annual progress reports to Parliament, which factually describe policy, emission, and technical trends, but do not address actors, activities, struggles, or informal institutions, which are beyond its (current) remit. We will use findings from these techno-economic analyses in our book but go beyond them by also investigating the actors and institutions in relation to existing systems and low-carbon niche-innovations.

Socio-technical transition scholars have also made many analyses of unfolding UK low-carbon transitions, but these mainly focus on particular niche-innovations such as solar-PV (Smith et al., 2013), offshore wind (Kern et al., 2015), electric vehicles (Mazur et al., 2015; Skeete, 2019), or low-energy houses (Kivimaa and Martiskainen, 2018; O'Neill and Gibbs, 2014). Although these in-depth studies analyse the roles of actors and institutions (e.g., learning, network building, struggles, visions, lobbying), they are limited by their focus on single innovations. But because there are now hundreds of socio-technical analyses of particular UK low-carbon innovations, we can make a next step by using their finding as inputs for the socio-technical whole system analyses we undertake in this book.

1.4 Structure of the Book

The book is structured as follows. Chapter 2 makes conceptual adjustments in the MLP to enhance its analytical traction for investigating whole system reconfiguration. On the one hand, it elaborates the techno-economic, actor, and institutional dimensions in the innovation journey of niche-innovations through the four phases. On the other hand, it elaborates how existing systems can be reconfigured through the incorporation of niche-innovations, the reorientation of existing actors, and adjustments in policies and governance styles. Chapter 3 discusses methodological issues such as the need for a particular explanatory style to investigate multi-dimensional longitudinal change processes in large-scale heterogeneous entities such as socio-technical systems. Chapter 3 also provides an operational analytical template for the empirical analyses in subsequent chapters and discusses the data-sources we used.

Chapters 4, 5, and 6 use the conceptual framework and analytical template to make socio-technical analyses of the unfolding low-carbon transitions in UK electricity, heat, and mobility systems. These empirical chapters, which form the bulk of the book, analyse longitudinal multidecadal developments in both existing systems and multiple niche-innovations, which are summarised in Table 1.1. While some of these niche-innovations are also addressed in techno-economic analyses and CCC reports (e.g., onshore wind, offshore wind, bio-power, solar-PV, heat pumps, electric vehicles), others are not (e.g., smart meters, smart grids, demand-side response, tele-working, car sharing, intermodal transport, and self-driving cars), which means our reconfiguration analysis is also broader in terms of the types of innovations that are considered.

Importantly, we link our analysis of niche-innovations to our analysis of system-level developments, which enables an assessment of opportunities for niche breakthrough and system reconfiguration. Each empirical chapter ends with conclusions about reconfiguration patterns along three socio-technical dimensions: techno-economic reconfigurations, actor reconfigurations, and policy reconfigurations.

Chapter 4 analyses the electricity system, which we divide into three sub-systems (generation, grid, consumption) that are distinct in terms of technologies, actors, and institutions. These sub-systems are closely integrated because electricity generation and consumption need to be closely matched to avoid

Table 1.1. *The existing (sub)systems and emerging niche-innovations that will be analysed*

	Electricity	Heat	Land-based passenger mobility
Systems or sub-systems	- Power generation sub-system - Grid infrastructure sub-system - Electricity consumption sub-system	- Heat supply and heat generation system - Buildings system (which shapes heat demand)	- Auto-mobility system - Railway system - Bus system - Cycling system
Niche-innovations	- Onshore wind - Offshore wind - Bio-power - Solar PV - Energy-efficient lighting, including CFL and LEDs - Smart meters - Smart grids - Flexibility-enhancing options: battery storage and demand-side response	- Heat pumps - Biomass heating - Solar thermal heating - Heat networks - Gas grid repurposing to hydrogen or biomethane - Passive house designs - Whole-house retrofits	- Electric vehicles - Biofuels - Tele-working - Car sharing - Intermodal transport (including smart cards and Mobility-as-a-Service) - Self-driving personal cars

blackouts. The electricity system is organised along an energy carrier, which can nowadays be used to fulfil multiple societal functions such as lighting, freezing/cooling, hygiene/washing, cooking, and entertainment. In that sense, it differs somewhat from the other two systems, which are linked to single societal functions such as heat and mobility.

Chapter 5 analyses mobility systems, which can be divided into systems for land, water, and air, and for passengers and freight. The book focuses on land-based passenger mobility systems, because these account for the largest part (46% in 2017) of transport-related GHG emissions.[2] We will analyse four specific mobility systems (auto-mobility, railways, buses, cycling), which are distinct and separate in terms of technologies, actors, and institutions (although there are some overlaps such as shared road infrastructures for cars, buses, and bicycles).

Chapter 6 analyses the heat system, which involves two closely related but separate systems. The dominant UK heating (supply) system is organised around a gas supply infrastructure and domestic gas boilers, which generate heat in buildings. The buildings system, which consists of all the building components and

[2] In 2017, domestic and international aviation accounted for 22%, vans and heavy goods vehicles for 24%, and domestic and international shipping for 8% (DfT, 2019c).

supply chains, shapes heat demand, which in the UK is quite high since many houses are draughty and poorly insulated.

Each chapter also analyses multiple low-carbon niche-innovations (see Table 1.1), which have emerged and developed since the 1990s and have contributed in varying degrees to unfolding low-carbon system reconfigurations. The niche-innovations relate to different parts of the (sub)systems, offering the potential for system reconfigurations with substantial scope and depth.

The analyses of (sub)systems and niche-innovations addresses both techno-economic developments (using many quantitative time-series), actors and activities (which are often more qualitative), and institutions (addressing both formal policies and informal governance styles). All analyses are longitudinal, going back to the post-war decades for the (sub)systems to trace their emergence, stabilisation, and gradual reorientation. Analyses of niche-innovations vary in longitudinal scope depending on specificities of their emergence and diffusion: while these analyses start in the 1990s for some renewable electricity technologies, they start in the 2000s or 2010s for other niche-innovations.

Unlike many other studies of low-carbon transitions, we do not assume that climate mitigation is the only, or most important, concern of various actors. The relative importance of climate mitigation versus issues such as cost, convenience, comfort, energy poverty, energy security, mobility access, safety, congestion, jobs, or business opportunity is an empirical question. Our analysis of actors and institutions will therefore address climate change and other salient concerns, which can both change over time.

The concluding Chapter 7 answers the research questions and provides a comparative analysis of unfolding low-carbon transitions in the three focal systems. It also inductively draws conclusions about cross-cutting topics with salient differences and similarities between the three systems, including: the roles of incumbent firms, governance style and politics, users, wider publics and civil society organisations, and exogenous 'landscape' developments and shocks. Chapter 7 ends by discussing future low-carbon transitions, articulating policy recommendations, and offering suggestions for future research.

2

Conceptualising Socio-Technical System Reconfiguration

This chapter presents our conceptual elaborations, which, together with the Multi-Level Perspective (MLP) described in Section 1.2, will guide the empirical analyses of system reconfigurations in Chapters 4, 5, and 6. It consists of two complementary parts. The first part, which stays close to the existing literature, conceptualises socio-technical transitions by following the innovation journey of niche-innovations. It builds on the MLP but elaborates four different phases. For each phase, we describe the main processes and mechanisms with regard to three dimensions: 1) techno-economic developments, 2) actors and social networks, and 3) policies and governance. The systematic conceptual discussion of these three dimensions over time has not been done before and thus constitutes a contribution to the socio-technical transitions literature. In particular, it more systematically introduces techno-economic developments in the MLP (Cherp et al., 2018).

The second part, which aims to make more substantial contributions to the literature, conceptualises system reconfiguration by describing change processes in the *existing* socio-technical system. Focusing on techno-economic and material components, we build on McMeekin et al. (2019) in distinguishing four change processes: 1) 'modular incrementalism', which incrementally improves existing components, 2) 'modular substitution', which involves replacement of particular components, 3) 'architectural stretching', which extends or elaborates linkages between existing components, and 4) 'architectural reshaping', which involves component replacement and changing linkages between components. Focusing on actors and social networks, we describe reorientation processes of incumbent actors, who can shift their attention, support, or resources from the existing system to emerging niche-innovations. Focusing on policies, we describe how existing policy and governance frameworks can be adjusted to accommodate new issues such as climate change.

We propose and will empirically demonstrate that a comprehensive analysis of low-carbon system reconfiguration should use both analytical lenses, which trace the emergence of low-carbon niche-innovations as well as reorientation processes in existing systems.

2.1 The Emergence and Diffusion of Radical Innovation in Socio-Technical Transitions

Radical innovations, which deviate substantially from the existing system on one or more dimensions, are important in low-carbon system reconfiguration, because incremental improvements will not be enough to deliver the large reductions in GHG emissions needed to address climate change. It therefore makes sense to analyse the innovation journey of niche-innovations with reference to ideal-typical phases, although this heuristic also has limitations, as we will discuss further. As indicated in Figure 1.4, the existing socio-technical transitions literature often divides long-term transitions into four phases, which are characterised by different processes and mechanisms. Combining insights from the sociology of innovation, evolutionary economics, and innovation management, we describe the main processes for each phase, organised along three analytical dimensions (techno-economic, actors, institutions).

2.1.1 First Phase: Experimentation in Protected Spaces

In the first phase, radical innovations emerge in small niches, often outside or on the fringe of the existing system (Schot and Geels, 2008). Through subsidised R&D or pilot projects, niches provide ‘protected spaces’ that shelter radical innovations from mainstream market selection pressure and nurture learning and development processes (Smith and Raven, 2012).

Techno-Economic: The technological (or social or business model) innovation is just emerging in this phase, and there are many uncertainties about technological characteristics, user preferences, policy, infrastructure requirements, and cultural meanings (Kemp et al., 1998). Multiple design variations may co-exist, which exacerbates uncertainties. Performance tends to be low and costs high, so the innovations struggle to survive economically, and often depend on financial support from policymakers (e.g., subsidies for R&D or demonstration projects) or specialised investors (e.g., venture capital, business angels). The fluidity and divergence of niche-innovations is represented with small diverging arrows in the bottom-left corner of Figure 1.4. Markets may not readily exist for radical innovations. There may be uncertainty about who the users are, what their preferences are, and what the final functionality of the new technology will be. Pervasive uncertainties complicate the use of cost-benefit calculations in this phase. In fact, over-reliance on financial assessment tools may act as 'innovation killers' in this phase (Christensen et al., 2008).

Policies and Governance: There are no stable design rules, guidelines, standards, policies, or governance structures in this early phase, given that radical innovations do not initially 'fit' with prevailing regulatory and selection environments. If there is policy support, this tends to be small and relatively uncommitting, often in the form of seed money for demonstration projects or subsidies for R&D.

Actors and Social Networks: Radical innovations are often pioneered by inventors, entrepreneurs, start-ups, activists, or other relative outsiders (Van De Poel, 2000), although incumbent firms can also develop novelties in their R&D laboratories. The social networks of niche-innovators and their wider support coalitions are small and unstable in this phase and characterised by high degrees of entry and exit. The Strategic Niche Management (SNM) literature (Kemp et al., 1998; Schot and Geels, 2008) distinguishes three actor-related processes that drive the emergence of niche-innovations: 1) learning processes through experiments or on-the-ground demonstration projects, aimed at reducing uncertainties and gradually improving the innovation, 2) the enlargement of *social networks* and the enrolment of more actors to expand the social and resource base of niche-innovations, 3) the articulation of *expectations* or *visions* to provide direction to the innovation activities and attract attention and funding from external actors (Borup et al., 2006). Champions of radical innovations may over-inflate their promises (of the innovation's performance, market potential, or transformative effects), which can lead to hype-disappointment cycles (van Lente et al., 2013) if the initial 'buzz', media-attention or investment influx are followed by setbacks, problems, and delays.

Network building, learning processes, and the articulation of credible and appealing visions may take a long time: 'There may be long periods when only a few pioneers advocate change without much attention, before a tipping point comes which leads to a swarm of competing alternatives, that is then followed by a period of winnowing out, and then the consolidation of a much smaller number of models that turn out to be viable' (NESTA, 2013). Radical innovations are risky, and many new entrants, innovations, and promises fail to survive the lengthy first phase, because of a lack of financial and organisational means (Olleros, 1986).

In the first phase, niche-innovations do not (yet) form a threat to the existing system, which is entrenched in many ways (institutionally, organisationally, economically, culturally). The existing system is not inert but changes incrementally due to small improvements in existing technologies, policy instruments, markets, and cultural meanings, which produce predictable trajectories (represented as stable lines in Figure 1.4).

2.1.2 Second Phase: Stabilisation in Small Market Niches

Techno-Economic: In the second phase, radical innovations break out of protected spaces and establish a foothold in one or more market niches. This provides a more reliable flow of resources, which stabilises the innovation and makes it more attractive for other actors. Learning processes focus on improving functionality and performance rather than cost: 'Performance dominates cost in initial market niches' (Wilson and Grubler, 2011: 168).

Policies and Governance: Learning processes also gradually lead to the stabilisation of a dominant design (Anderson and Tushman, 1990), which becomes institutionalised in design guidelines, product specifications, best practice formulations, and standards. The innovation thus develops a trajectory of its own because of the stabilisation of rules and social networks (represented in Figure 1.4 with converging arrows in the second phase). Policy support often becomes stronger in this phase and may take the form of investment subsidies for firms, purchase subsidies for consumers, public procurement, or feed-in-tariffs, which help to create and expand market niches.

Actors and Social Networks: Social networks and alliances become bigger in the second phase as a dedicated community (of firms, engineers, policymakers, users) emerges. The participation of more actors (including powerful incumbents) increases the legitimacy of the innovation and brings more resources into niches (Schot and Geels, 2008). Social interactions, learning, and articulation processes begin to reduce uncertainties, and future visions become more precise and more broadly accepted. Dedicated professional groups (e.g., engineering communities, branch organisations) emerge, which help to codify the new body of knowledge (Geels and Deuten, 2006) and lobby for more policy support. Technological stabilisation and emerging economic opportunities increase the willingness of firms, policymakers, and financial actors to invest. The involvement of more mainstream actors may reduce the radical scope and visions compared to the intentions of initial innovators, which can be seen as the 'price of success' (Smith et al., 2014).

Innovation may also happen on the user side, as people 'domesticate' radical innovations and transform them from unfamiliar things to familiar objects embedded in the routines and practices of everyday life (Lie and Sørensen, 1996). The articulation of positive cultural visions may help to legitimate innovations and attract further support.

Innovations may, however, also be opposed by social groups who experience negative side-effects or by citizens who feel insufficiently consulted in decision-making. Such opposition may result in controversy and social acceptance problems that can hinder further progression of the innovation, as happened with nuclear energy, genetically modified food, and onshore wind turbines in some countries (Batel, 2018; Geels et al., 2007).

Innovations may remain stuck in market niches for a long time, especially when they face a mismatch with the existing system (e.g., infrastructure, user preferences, institutional barriers). Niche advocates may try to alter wider contexts through political lobbying and institutional entrepreneurship (Smith and Raven, 2012), but incumbent regime actors may actively resist these changes (Geels, 2014).

2.1.3 Third Phase: Diffusion and Struggles against the Existing System

Techno-Economic: In the third phase, the innovation diffuses into mainstream markets, where it competes head-on with the existing technology in terms of techno-economic performance and wider socio-technical system in terms of 'institutional fit'. Diffusion often follows a pattern of 'niche-accumulation' (Geels, 2002), with an innovation emerging in a technological niche, then moving to a small market niche or application domain, and subsequently into larger mainstream markets (Figure 2.1).

Important techno-economic drivers of diffusion are cost reductions and performance improvements, which Arthur (1988: 591) related to five positive feedback mechanisms of increasing returns to adoption: 1) scale economies in production, which allow the price per unit to go down, 2) learning-by-using: the more a technology is used, the more is learned about it, the more it is improved; 3) technological interrelatedness: the more a technology is used, the more complementary technologies are developed; 4) network externalities: the more a technology is used by other users, the greater its usefulness and performance; 5) informational increasing returns: the more a technology is used, the more information becomes available and is shared among users.

The entry in mainstream markets leads to economic competition between new and existing technologies, the outcome of which depends not only on price/performance characteristics but also on the institutions that shape markets and economic frame conditions.

Policies and Governance: Diffusion into mainstream markets is often accompanied by adjustments in regulations and policies so that these become more supportive of radical innovations (EEA, 2019a). New product regulations

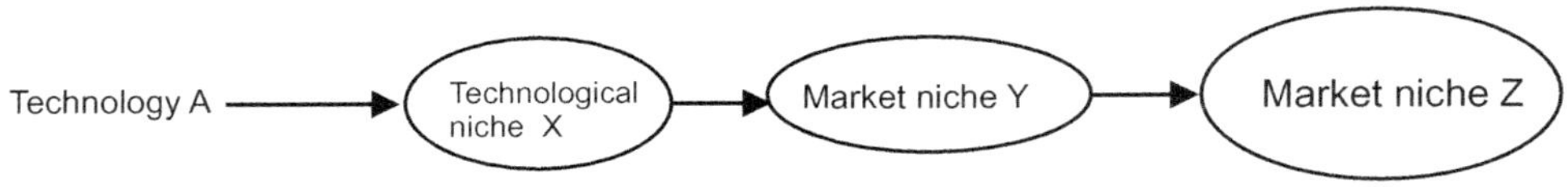

Figure 2.1 Diffusion as a process of niche-accumulation (adapted from Levinthal (1998: 243))

(e.g., energy efficiency standards for new cars, appliances, or houses) and performance regulations (such as renewable energy obligations for utilities or electric vehicle sales targets for automakers) can drive company engagement. Capital grants or interest-free loans can also stimulate investments and uptake by firms. And purchase subsidies and information campaigns may advance user adoption of innovations (Brand et al., 2013). These policy instruments are often embedded in and supported by new policy goals, visions, and strategies.

Actors and Social Networks: The number of actors increases rapidly in the diffusion phase due to interactions and positive feedbacks (Kanger et al., 2019; Mylan et al., 2019) such as the following: a) decreasing costs stimulate adoption by more users, which increases visibility and markets, b) growing markets and improving technologies attract more firms, which may lead to 'swarming effects' (Schumpeter, 1927) that increase investments and industry size, c) increasing investments further improve technologies and lower costs, while increasing industry size enhances lobbying power, which may result in more favourable policies that stimulate adoption or company investment, d) positive user experiences and the emergence of new industries and jobs may validate positive cultural discourses, which can alter user preferences and legitimate further policy support (Roberts and Geels, 2018).

To protect their vested interests, incumbent actors may try to resist or delay the transition. Widespread diffusion is therefore often characterised by highly visible struggles and conflicts between actors associated with niche-innovations and existing systems. On the business dimension, there may be struggles between new entrants and incumbents, which may follow different patterns: 1) new entrants may outcompete and replace existing firms (Christensen, 1997); 2) incumbent firms may defend the existing system by improving the existing technology, hindering the new innovation (through pricing strategies or political tactics), or buying up the new firms (to eliminate the risk and/or acquire new capabilities), 3) incumbent firms may diversify and reorient themselves towards new technologies. Car manufacturers, for instance, can diversify towards electric vehicles (Penna and Geels, 2015), while electric utilities can diversify towards renewables (Geels et al., 2016b). This means that incumbent actors *can* play constructive roles in transitions, even when they initially tend to resist.

On the political dimension, there may be conflicts and power struggles about the *settings* of policy instruments (e.g., adjustments in subsidies, taxes, and regulations) and the *kinds* of instruments (e.g., market-based, regulatory, informational). Political struggles are also about which problems appear on agendas, how they are framed, and what degree of urgency is attached to them (Kern, 2011). These struggles involve traditional policy actors (e.g., bureaucrats, Ministers, advisory committees, political parties, Parliament) but also wider interest groups, which often have differential degrees of access to policy networks (Lockwood et al., 2017). Successful transitions are deeply political processes, because they usually require major changes in policy instruments and in market metrics or measurement tools (Meadowcroft, 2009). Incumbent actors tend to resist such changes, whereas niche-actors push for them. Policy change therefore often requires changes in power relations, for example, strengthening of change coalitions and weakening of incumbent networks.

Transitions also involve cultural and discursive struggles about the framing of problems and solutions (Geels and Verhees, 2011). It matters, for instance, if the problem of climate change is framed as a 'market failure' (which is likely to lead to market instruments such as a carbon pricing) or as a 'planetary boundary' (which may lead to stronger policies with greater urgency). It also matters how particular solutions are framed and given meaning. For instance, are wind turbines primarily seen as renewable energy producers or as ugly artefacts that kills birds? Are nuclear power plants seen as low-carbon energy producers or as existential threats? Different social groups may have different views and interpretations, which often leads to contestation. These cultural dimensions are also important with regard to social acceptance of solutions and the legitimacy of policy efforts (Markard et al., 2016).

There is no guarantee that niche-advocates inevitably win the various struggles. Niche-innovations may fail to build up sufficient endogenous momentum or suffer setbacks. Incumbent actors may successfully counter-mobilise and thwart or stall niche-innovations. Or the existing system remains deeply locked-in and proves difficult to dislodge.

The MLP therefore suggests that broad diffusion often involves not only *endogenous* drivers of niche-innovations but also *external* landscape developments that create pressure on the existing system, which may lead to tensions and cracks (represented by diverging arrows in Figure 1.4) that, in turn, may create windows of opportunity for niche-innovations. Problems and tensions that may destabilise existing systems include the following (Turnheim and Geels, 2012): a) performance problems that cannot be met with the available technology, b) changes in markets and mainstream user preferences, c) changing cultural discourses that delegitimise existing technologies (Roberts, 2017), d) changes in policy agendas that lead to stricter regulations, e) competition and strategic games that lead incumbent firms to diversify away from existing technologies and towards niche-innovations (Penna and Geels, 2015).

2.1.4 Fourth Phase: Reconfiguration

Techno-Economic: In the fourth phase, new technologies replace existing ones, which thus decline. This replacement is accompanied by further system reconfiguration, including the creation and expansion of new infrastructures and industrial supply chains, which create forward and backward economic linkages (Hughes, 1994). The post-war establishment of the auto-mobile system, for instance, involved not only an expanding car industry but also stronger linkages with rubber, steel, and glass industries as well as roadbuilding, oil, and servicing industries.

Policies and Governance: The new system becomes anchored in safety regulations, (technological) performance requirements, tax and subsidy rules, and professional standards. New government departments and regulatory agencies may be created to oversee and inspect the system. And new teaching curricula may be developed to train new staff. New policies may be needed to mitigate negative unintended consequences that may be generated by the expanding system. Expanding automobility, for instance, generated more traffic accidents and air pollution, which gave rise to new safety and environmental regulations.

Actors and Social Networks: Social networks gradually expand and stabilise in relation to the new system. The majority of users switch to new technologies and social practices, which stabilises new habits of use and views of normality (Shove, 2003). Successful firms expand their factories, market shares, and supply chains, while 'old' firms decline and shrink. The 'losers' in transitions (e.g., firms, employees, regions) may need to be helped or compensated to mitigate disruptive effects and limit potential resistance (Mayer, 2018; Vögele et al., 2018).

2.1.5 Limitations and Directions for Elaboration

This 'innovation journey' approach is widely used in socio-technical transitions research, especially for analyses that focus on the emergence of disruptive niche-innovations such as solar-PV (Smith et al., 2013), wind turbines (Jolly et al., 2016), electric vehicles (Dijk et al., 2016; Mazur et al., 2018; Sovacool, 2017; Sprei, 2018), community energy (Tom Hargreaves et al., 2013), hydrogen and fuel cell vehicles (Upham et al., 2018). Despite its usefulness it also has several limitations for understanding system transitions, as critics have pointed out.

One limitation is that this approach has a bottom-up bias, representing a 'point source' approach to transitions (Geels, 2018a), which see emerging innovations as the driving force. Berkhout et al. (2004: 62) also criticised this approach for being 'unilinear in that they tend unduly to emphasise processes of regime change which begin within niches and work up'. To address this limitation, and because of our interest in whole system transition, this book shifts the analytical focus towards *existing socio-technical systems*, which we conceptualise as both locked-in and dynamically evolving, both through endogenous processes and pressures from niche-innovations.

Another limitation is that the 'innovation journey' approach tends to focus on singular radical niche-innovations that emerge and overthrow the existing system. This approach is clearly too simple for understanding low-carbon system transitions, since there are multiple niche-innovations in electricity, heat, and mobility systems. We will therefore build on research that has investigated *multiple* niche-innovations (Dijk et al., 2013; Markard and Hoffmann, 2016; Marletto, 2014; Raven, 2007; Sandén and Hillman, 2011; Verbong et al., 2008) and how these may compete, complement, or build on each other.

A third limitation is that the 'innovation journey' approach and the standard MLP representation (also in Figure 1.4) assume that niche-innovations struggle against a *single* existing system, which has been criticised as too limited (Andersen et al., 2020; Rosenbloom, 2020). Some research has tried to address this limitation by investigating interactions between *multiple* systems in transitions (Geels, 2007a; Konrad et al., 2008; Marletto, 2014; Papachristos et al., 2013; Raven and Verbong, 2007). We will build on this research in our investigation of whole electricity systems (which we divided into generation, grid, and consumption subsystems), passenger mobility (for which we distinguish auto-mobility, train, bus, and cycling systems), and heat (which we relate to heating and buildings systems).

A fourth limitation is that the focus on radical niche-innovations may lead transition scholars to underestimate the potential of incremental change in reducing GHG emissions: 'Preoccupation with disruptiveness [...] risks marginalizing and overlooking [...] mundane, incremental and continuity-based innovation, and possibilities for adapting existing systems' (Winskel, 2018: 235). An exclusive focus on radical niche-innovations may also lead to simplistic or normative views of transitions: 'One sometimes gets the idea that the change that really matters is truly dramatic change, the overturning of big systems. [...] Yet we should take care here. Our concern should be solving societal problems not tilting at "systems"' (Meadowcroft, 2009: 337). To address this limitation, our conceptual elaborations and our empirical investigations will explicitly accommodate incremental change and more substantial adjustments in *existing* systems.

A fifth limitation is that existing socio-technical systems and incumbent actors are sometimes presented as inert monolithic entities, which overlooks the possibility of internal tensions and endogenous change (Jørgensen, 2012; Turnheim and Sovacool, 2020). This limitation relates to the first one, discussed previously: if one focuses on niche-innovations as drivers of transitions, then existing systems easily become seen as static barriers to be overcome. To address this limitation, we will elaborate the notion of the 'semi-coherence' (Geels, 2002) of existing systems, and, on the one hand, acknowledge disagreements, tensions, and ongoing dynamics in existing systems but, on the other hand, accommodate the relative stability and lumpiness of techno-economic elements, social networks, and institutions.

2.2 Reconfiguration of *Existing* Socio-Technical Systems

To address the limitations of the bottom-up 'innovation journey' approach to transitions, this section provides a complementary perspective on reconfiguration dynamics in *existing* socio-technical systems. Although these systems and associated actors are stabilised by lock-in mechanisms, they are not inert and can change over time. Drawing on several different literatures, we here elaborate the MLP by developing new conceptualisations of reconfiguration in techno-economic elements, actors, and rules and institutions.

2.2.1 Techno-Economic Reconfiguration

Socio-technical systems have techno-economic and material dimensions, which include artefacts, material goods, infrastructures, factories, and flows of inputs and outputs through supply chains. These include energy flows, which can be represented with Sankey diagrams (Cullen and Allwood, 2010), which show how energy inputs (e.g., oil, gas, coal, nuclear, renewables, biomass) feed into conversion devices (e.g., engines, motors, burners) to heat or power passive systems (e.g., vehicles, appliances, or buildings) to deliver final services (such as passenger transport, thermal comfort, sustenance, or hygiene). But they also include material consumption (e.g., steel, coal, plastics, concrete) and many technical components such as signalling systems for railways, transmission and distribution grids for electricity, road infrastructures and gas pipelines, and electronic information provision systems at bus stops, which are not represented in Sankey diagrams because of their focus on energy flows and conversion.

Existing socio-technical systems are stabilised by economic lock-in mechanisms such as sunk investments, economies of scale (Arthur, 1989), and competitive economic characteristics (e.g., low price, high performance) that make it difficult to dislodge them in mainstream markets (Arthur, 1989; Dosi, 1982). Artefacts, infrastructures, and factories also have material 'hardness' and momentum (Hughes, 1994) because of obduracy (Hommels, 2005) or complementarities between components and sub-systems (Rycroft and Kash, 2002). Despite these lock-in mechanisms, material system elements are not static and inert but usually evolve gradually through incremental technical changes and economic alterations in relative size, output, or market share (e.g., increasing car sales, road infrastructure length, or coal-fired electricity production). Deeper techno-economic reconfiguration is possible, as we will discuss after introducing some system coupling ideas.

From a technical systems perspective, it is useful to distinguish between components and the architecture of linkages between components. Components can be tightly or loosely coupled (Simon, 1973). In tightly coupled systems, a change in one component is likely to trigger or require changes in other components. Loose coupling means that components are organised as independent modules, which can operate (relatively) independently of the detail of other components; they are only connected through functional inputs and outputs. Loose coupling permits modular innovations, which are improvements or replacements within one component without requiring synchronous changes in other components that make up the system. Modular innovation thus enables distribution of labour, specialisation, and flexible innovation (Baldwin and Clark, 1997; Robertson, 1992).

Building on these ideas, Henderson and Clark (1990) proposed that innovation in *existing* systems can be directed at components, architectures, or a combination of both, leading them to propose a typology with four kinds of innovation in technological products. We follow McMeekin et al.'s (2019) elaboration, which makes the typology more suitable to socio-technical systems by slightly changing the conceptual labels to 'modular incrementalism', 'modular substitution', 'architectural stretching', and 'architectural reshaping', and by differentiating the 'modular substitution' category to accommodate three different kinds of (partial) replacement:

- System-to-system switching results from competition between *existing* systems or dominant technologies, leading the former to decline and the latter to increase in size or output (Raven and Verbong, 2007). A modal switch from auto-mobility to railways is one example; a shift from coal- to gas-fired power is another.
- Niche-to-system hybridisation means that niche-innovations are added to and incorporated in existing systems (Berkers and Geels, 2011; Geels, 2002; Raven, 2007), leading to partial replacement of unsustainable components. Possible examples include the co-firing of biomass and coal, the blending of biofuels in petrol, hybrid electric vehicles.
- Niche-to-system replacement means that niche-innovations substitute particular (sub)system components. Examples include solar-PV or wind turbines replacing coal-fired power plants in the electricity generation sub-system; electric vehicles replacing diesel or petrol cars; heat pumps replacing gas boilers.

Based on these considerations, we propose Table 2.1, which we will use in the empirical chapters to analyse techno-economic reconfiguration of existing (sub) systems. One implication is that whole system reconfiguration involves multiple change mechanisms, which can all contribute to GHG emission reduction. The empirical chapters will try to identify the relative importance of different kinds of changes in unfolding low-carbon transitions in UK electricity, heat, and mobility systems. Another implication is that this conceptualisation changes the transition imagery from singular 'bottom-up' disruption towards a greater variety of reconfiguration pathways, including more gradual and dispersed reconfiguration.

We will not only use Table 2.1 to map and categorise different low-carbon innovations but also to diagnose temporal developments such as spillovers, knock-on effects, or innovation cascades between innovation categories

Table 2.1. *Analytical framework to map techno-economic reconfigurations of existing systems (adapted from McMeekin et al. (2019: 1226))*

	Core elements reinforced	Core elements substituted		
Architecture unchanged (linkages between components)	*Modular incrementalism*	*Modular substitution*		
		System-system switching	Niche-system hybridisation	Niche-system replacement
Architecture changed	*Architectural stretching*	*Architectural reshaping*		

(Berkers and Geels, 2011). For example, modular substitutions in the electricity generation sub-system (e.g., solar-PV and wind replacing coal) can trigger subsequent innovations in the electricity grid and consumption sub-systems (e.g., battery storage, back-up capacity, smart grids, or demand-side response), which may lead to architectural reshaping.

We will also use Table 2.1 to assess the *speed* and *scope* of techno-economic low-carbon transitions in electricity, heat, and mobility systems. We will assess *speed* by empirically analysing the deployment and market diffusion of different low-carbon innovations. We will assess *scope* by diagnosing the relative speeds and degrees of activity across the four different innovation categories. Substantial scope means that there is much activity and deployment in all four categories, whereas limited scope implies that activity is more concentrated on particular kinds of innovations.

2.2.2 *Reorientation of Mainstream Actors*

The material elements of existing socio-technical systems do not function autonomously but are the outcome of activities of incumbent and mainstream actors such as manufacturing firms, suppliers, policymakers, users, and civil society groups (Geels, 2004). Because incumbent actors repeatedly interact with each other, their networks and relations can be characterised as organisational fields (Geels, 2020b). 'The concept of field identifies an arena (a system of actors, actions, and relations) whose participants take one another into account as they carry out interrelated activities' (McAdam and Scott, 2005: 10). Field boundaries are not fixed and 'the players that populate the field and the nature of their play can change over time' (Davis and Marquis, 2005: 337).

There is always something at stake in organisational fields and field actors occupy varying positions, which are differentially advantageous in terms of economic resource flows, political power, or socio-cultural influence. While some interactions are routinised, these differences in position give rise to strategic jockeying for position and ongoing struggles in existing organisational fields (Hoffman, 1999), both between incumbents and between incumbents and challengers. Fligstein and McAdam (2012: 3) suggest that 'constant low-level contention and incremental change are the norm in fields'.

For socio-technical systems, this implies that some degree of flux and incremental change is normal, leading to gradual adjustments and incremental trajectories in technology, markets, user practices, policies, and infrastructures (Geels, 2004). Larger changes and reorientations are more challenging and rare, however, because incumbent and mainstream actors are embedded and locked-in in various ways. To better understand these lock-ins and possible reorientations, we develop a multi-dimensional view on actors that draws on the configurational approach in organisation theory and generalises its logic to other actors.

Configurational approaches in organisation theory understand organisations holistically as a constellation of interconnected elements or dimensions that cluster together in particular patterns or archetypes, which have a substantial degree of stability and lock-in (Doty et al., 1993; Fiss et al., 2013; Ketchen et al., 1993; Meyer et al., 1993; Miller, 1996, 1990). Although there are ongoing debates about the specification of the organisational elements, many scholars (Gavetti and Rivkin, 2007; Greenwood and Hinings, 1996, 1993; Tushman and Romanelli, 1985) have proposed that organisational configurations have a hierarchical structure that, at least, involve the following dimensions: 1) core beliefs, goals, and interpretations (which are *cultural-cognitive categories* that provide overall or strategic direction and make sense of the organisation in its contexts), 2) capabilities, knowledge, and skills (which provide organisations with *abilities and resources* to perform certain tasks and innovate), 3) routines, habits, and operating procedures (which provide *behavioural templates* for standard operations and tasks).

Following Geels (2021), we propose that this logic of multiple hierarchically structured dimensions can be generalised from business organisations to other kinds of mainstream actors in socio-technical systems, although there are obvious differences between them, and the logic is perhaps somewhat tenous for wider publics, which are more dispersed and less task-oriented. This generalisation provides a multi-dimensional conceptualisation of important actors in socio-technical systems (firms, policymakers, users, wider publics), which acknowledges coherence and lock-in but also allows for reconfiguration (because the specified elements are not fixed 'hard cores'). Building on Geels' (2021) mobilisation of different disciplinary literatures, we aim to instantiate and exemplify this logic for different social groups but refrain from a deeper theoretical discussion, as this would hamper the argumentative flow. We first discuss the different dimensions for important actor groups and then briefly address change and depth of reorientation.

- For business actors, the following theories instantiate the three dimensions: 1) cognitive and interpretive management theories perceive core beliefs, mission, or mindset as the most foundational dimensions that guide the strategies and actions of firms in particular directions (Gavetti, 2005; Phillips, 1994); 2) the resource-based view of the firm (Barney, 1991) and evolutionary economics (Nelson and Winter, 1982) see technical knowledge and capabilities as core enablers and strategic assets of firms; 3) the behavioural theory of the firm suggests that day-to-day activities are guided by routines, rules of thumb, and standard-operating procedures (Cyert and March, 1963).
- For policy actors, the following theories instantiate the three dimensions: 1) policy paradigm theory suggests that policy goals, worldviews, problem framing, and governance styles guide policymaking in particular directions (Hall, 1993); 2) historical institutionalism argues that institutional arrangements (e.g., access rules, agenda-setting procedures, political 'rules of the game'), policy networks, and knowledge enable and structure policymaking processes and political struggles (Thelen, 1999); 3) the theory of (disjointed) incrementalism suggests that most day-to-day policymaking consists of small adjustments in the settings of existing policy instruments, because civil servants use heuristics and routines in the cycle of policy formulation, implementation, evaluation, and adjustment (Lindblom, 1979; Weiss and Woodhouse, 1992).
- For users, the following theories instantiate the three dimensions: 1) convention theory highlights the role of socio-cultural frames, value systems, and 'orders of worth' in user evaluations (Boltanski and Thévenot, 2006; Wilkinson, 2011), while sociologists proposed the term 'doxa' to refer to society's taken-for-granted, unquestioned truths that underpin understandings of normality and normal behaviour (Bourdieu, 1977); 2) social practice theory proposes that people purchase and use products in the course of engaging in daily life practices, which are stabilised clusters of activities involving multiple elements, including competencies, meanings, and materials (Shove et al., 2012; Warde, 2005); 3) social psychological theories suggest that day-to-day user behaviour is shaped by habits, routines, and heuristics, which do not require conscious thought and thus help people save time and energy (Maréchal, 2010).
- For public debates, the following discourse theories instantiate the three dimensions: 1) cultural sociology (Alexander, 2003) and macro-cultural discourse theory (Lawrence and Phillips, 2004) suggest that cultural deep structures such as symbolic categories, cultural repertoires or ideographs such as 'freedom', 'democracy', 'progress', or 'sustainability' (McGee, 1980) undergird and steer public debates in particular directions; 2) critical discourse analysis draws attention to more specific discourses, which are 'ensembles of ideas, concepts, and categories through which meaning is given to social and physical phenomena' in

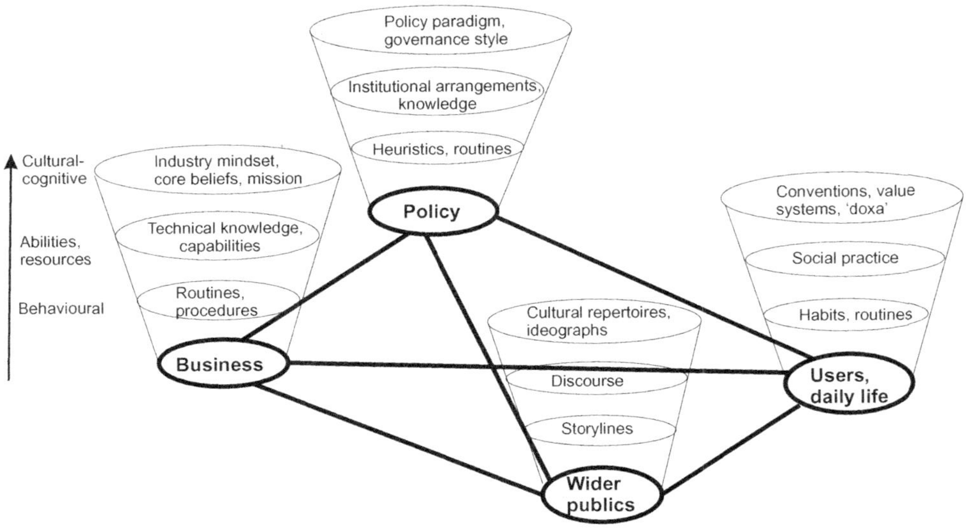

Figure 2.2 Configurational dimensions of incumbent actor groups (Geels, 2021)

a particular domain of practice (Hajer and Versteeg, 2005: 175) and the interests and power relations behind these discourses (Fairclough, 1995); 3) argumentative and rhetorical discourse theory focuses on the storylines, metaphors, slogans, and catchphrases that participants use in specific debates, which they incrementally adjust in response to each other's claims and arguments (Heracleous and Marshak, 2004).

Figure 2.2 schematically summarises the main elements at the three configurational dimensions for each actor group. One advantage of this novel conceptualisation is that it accommodates multiple kinds of agency, including boundedly rational strategic action through which actors aim to achieve goals and advance their interests (which need to be interpreted and can change over time), routine-based action, learning and knowledge development, and strategic sense-making.

Another advantage is that this conceptualisation makes it relatively straightforward to understand different *depths* of actor reconfiguration as involving changes in different dimensions: substantial reconfiguration involves changes in cultural-cognitive beliefs, repertoires, conventions, or policy paradigms; moderate-depth reconfigurations involve changes in technical capabilities, social practices, discourses, or institutional arrangements; and limited-depth reconfiguration involves changes in routines, habits, and standard-operating procedures, which happen relatively frequently, generating incremental change. Deeper changes are more transformative but also rarer and more difficult.

A third advantage is that this conceptualisation allows for differentiation within organisational fields and the possibility that different actor groups reconfigure to different degrees and depths (which we will empirically analyse for different systems). A fourth advantage is that this conceptualisation of actors endogenises cultural-cognitive dimensions, which we, for the purpose of this book, exclude from our conceptualisation of rules and institutions (further discussed next). It thus makes these cultural-cognitive dimensions less free-floating and less consensually shared between all actor groups than in some neo-institutional theories.

This novel conceptualisation thus makes it possible to further develop the increasing recognition that transitions do not only result from new entrants overthrowing incumbents (which follows a heroic 'David versus Goliath' plotline) but can also be enacted by incumbent actors who reorient from existing systems towards niche-innovations (Bergek et al., 2013; Berggren et al., 2015; Penna and Geels, 2015; Turnheim and Geels, 2019).

Such reorientation is not easy, because it involves overcoming various lock-in mechanisms. Especially for core beliefs and capabilities, actor reorientation therefore usually requires: 1) increasing external pressures, 2) decreasing beliefs in the viability of the status quo, and 3) perceptions of attractive opportunities towards which actors can reorient (Geels and Penna, 2015; Turnheim and Geels, 2013). The development of a comprehensive reorientation framework for all actor groups is beyond the book's ambition. For our empirical analyses of incumbent actor reorientation in electricity, heat, and mobility systems (in Chapters 4, 5, and 6), we mainly aim to describe and map the *depth* of reorientation by different actor groups and identify some salient drivers. We will use Figure 2.2 as a heuristic framework for this empirical mapping. We do not assume beforehand that all incumbent actor groups are reorienting in equal depth in unfolding low-carbon transitions. In fact, we will investigate reorientation depths for different actor groups and the transition patterns that result from this.

As to the directionality of change, our empirical analyses also do not assume that climate change is the only issue to motivate actor reorientation. In fact, we assume and will empirically show that incumbent actors in UK electricity, heat, and mobility systems are historically more concerned about other issues such as congestion, road safety, parking, domestic manufacturing and jobs, prices, energy security, energy poverty, and housing shortages. So, to comprehensively understand the depth of low-carbon reorientation of incumbent actors, we will not only analyse their climate mitigation strategies and activities but also other issues that concern them and the associated activities, which may hamper their engagement with climate mitigation.

2.2.3 Policy Reconfiguration

Actors and activities in organisational fields are shaped (but not determined) by rules and institutions (Powell and DiMaggio, 1991). Building on neo-institutional theory (Powell and DiMaggio, 1991; Scott, 1995; Thornton et al., 2012), previous conceptualisations of rules and institutions in socio-technical transitions research distinguished regulative, normative, and cultural-cognitive institutions (Geels, 2004) or used the institutional logics concept (Fuenfschilling and Truffer, 2014; Smink et al., 2015). But the three types of institutions are easier to distinguish conceptually than to investigate empirically, and the institutional logics concept has been criticised as too all-encompassing and too consensual:

> The use of the term "institutional logics" tends to imply way too much consensus in the field about what is going on and why and way too little concern over actors' positons, the creation of rules in the field that favour the more powerful over the less powerful, and the general use of power in strategic action fields. [...] We see fields as rarely organized around a truly consensual "taken for granted" reality. [...] In contrast, for us, there is constant jockeying going on in fields as a result of their contentious nature. (Fligstein and McAdam, 2012: 12)

For the purpose of this book, we therefore draw more on 'old' institutional theories that focus on formal-regulative institutions such as laws, regulations, standards, financial incentives, and subsidy schemes, which act as legally sanctioned 'rules of the game', as well as governance structures and the role of the state (Hirsch and Lounsbury, 1997; North, 1990). To not exclude cultural-cognitive dimensions from our overall analysis, we have accommodated them in our conceptualisation of incumbent actor groups, as described previously.

Our focus on formal policies and governance style implies that we see policymakers as having special responsibilities and resources in governing organisational fields and socio-technical systems (see also Scott et al., 2000). This also resonates with the empirical reality that electricity, heat, and mobility systems are highly regulated systems with salient roles for policymakers, ministries, and regulatory agencies. It further resonates with the fact climate change is an externality for actors in these socio-technical systems and that public policies will therefore be essential to drive their low-carbon reconfiguration.

Policy reconfiguration for low-carbon transitions is a struggle because existing electricity, heat, and mobility systems already have many policies and governance arrangements in place. Many of the existing policies and arrangements are tailored to incumbent interests and oriented towards non-climate goals such as congestion, road safety, domestic car manufacturing in (auto)mobility, energy security or prices in electricity, and energy poverty and boiler safety in heating. Low-carbon transitions thus require the new issue of climate change to be integrated into existing policies and governance structures. To further conceptualise this struggle, we build on the literatures of environmental policy integration (EPI), climate policy integration (CPI), and policy mixes.

The basic idea behind EPI, which goes back to the 1990s, is that environmental problems cannot be fully addressed by environmental ministries but that this requires involvement from the sectors that drive and cause the environmental problems. EPI thus involves the mainstreaming or 'incorporation of environmental objectives in non-environmental policy sectors, such as agriculture and transport, rather than pursuing environmental protection through specialised environmental policies and legislation by environmental institutions. In this way, EPI aims to target the underlying driving forces rather than symptoms of environmental degradation' (Persson and Runhaar, 2018: 141).

EPI is not easy, however, because it involves the integration of 'a traditionally less prioritized policy objective, typically supported by less powerful actors, into "mainstream" sector objectives, typically supported by well-organized interests. This sets the stage for resistance from incumbents which arguably requires significant political will to be overcome' (Nilsson and Persson, 2017: 37).

Policymakers in electricity, heat, or mobility systems may (initially) find environmental or climate change issues less important than existing sector-specific goals. Because they are locked-in by prevailing routines, institutional arrangements, and policy paradigms (as discussed in Section 2.2.2), policymakers in these systems may be reluctant to accommodate the new issue, as Jordan and Lenschow (2010: 153) explain:

> At the administrative level, contention arises from distinct cultures and routines in the bureaucratic segments of an administration and from the 'rational' inclination of each part to protect its competences, resources and ways of doing from the intervention of other parts. At the end of the day, greater policy integration does often require political leadership from above.

Although the *principle* of EPI enjoys broad policy support, actual implementation has remained limited because of these contentions. In their state-of-the-art review, Jordan and Lenschow (2010: 153) conclude that: 'With the exception of a very few cases, EPI is pursued as an "add-on" rather than as a process that challenges the underlying rationale for spending public money on unsustainable practices. [. . .] While governments have undoubtedly extended their repertoire of instruments, they have done so in a largely piecemeal fashion'. Nilsson and Persson (2017: 36) similarly conclude that EPI initiatives 'rarely result in having significant impact on policy but tend to get stuck at the level of policy statements'.

Drawing on the limited set of positive cases, Persson and Runhaar (2018) identified several external success factors for EPI that create pressure on policymakers: 1) public awareness and support for addressing an environmental issue, 2) stakeholder and interest group support, 3) support by other governmental actors, and 4) compatibility with pre-existing sectoral policy frames. They also identified several internal success factors: 1) political will, 2) overlap with sectoral objectives, 3) perceived urgency of the issue to be integrated, 4) overarching policy frameworks, and 5) organisational provisions for intersectoral cooperation, leadership, and resources.

Climate policy integration (CPI), which is a subset of EPI, has been relatively more successful (Di Gregorio et al., 2017; Dupont, 2016), as indicated by substantial increases in legislative activity within particular sectors, although Schmidt and Fleig (2018) find more CPI activity in energy than in transport. This greater success of CPI is partly due to the strengthening of several of the external and internal success factors that were discussed earlier. But it is also due to three specific differences identified by Adelle and Russel (2013: 9): 1) CPI is narrower and more tangible than the broader and vaguer concept of EPI, 2) CPI outputs are more easily measured (e.g., as GHG emissions or diffusion of green technology) and easier to communicate in a media friendly way than the less appealing administrative processes associated with EPI, and 3) CPI is less about expansive integration across all policy sectors and more about engaging a narrower set of sectors to work together to meet specific goals.

Building on the EPI and CPI literatures, we conclude that institutional low-carbon reconfiguration is possible but challenging because the integration of climate change objectives into existing policies and governance styles is likely to encounter various hurdles that require internal and external pressures to be overcome.

The policy mix literature offers further relevant insights for analysing institutional reconfiguration. This literature distinguishes between two analytical levels: 1) policy strategies, including policy goals and governing styles, and 2) policy instruments (Rogge and Reichardt, 2016). It understands policy mixes as 'complex arrangements of multiple goals and means which, in many cases, have developed incrementally over many years' (Kern and Howlett, 2009: 395). It also proposes useful concepts to analyse the interplay between multiple goals and instruments in a policy mix (Rogge and Reichardt, 2016; Schmidt and Sewerin, 2019), including: 1) the *coherence* of policy goals (i.e., the degree to which they contradict each other), 2) the *consistency* of policy instruments (i.e., the degree to which they reinforce or undermine each other in achieving policy goals), and 3) *comprehensiveness* of policy instruments (i.e., the balance of *types* of instruments). With regard to the latter, Schmidt and Sewerin (2019: 3) observe that: 'An unbalanced policy mix that relies on one or a few instrument types is less likely to address all issues and reach all relevant actors. Consequently, it is less likely to be effective'. Relevant instrument types for low-carbon transitions typically include financial incentives, regulatory instruments, information instruments, and direct government investments (Grubb, 2014).

Policy mixes in particular domains are rarely designed from scratch. Instead, they 'more often than not evolve through layering of potentially incoherent policy goals and inconsistent instruments over time' (Kern et al., 2019: 2).

Combining these ideas with the EPI/CPI literatures, we propose that climate change goals and policy instruments are initially likely to be layered on top of existing system-specific policy goals and instruments. This layering often results in an incoherent policy mix, which will be relatively ineffective if climate goals or policies remain singular and isolated 'add-ons' to the existing policy mix (Howlett and Rayner, 2013). To advance low-carbon transitions, it will be necessary to make deeper reconfigurations in policy goals, governance styles, and instruments so that the system-specific policy mix becomes more coherent, consistent, and comprehensive. The *depth* of institutional low-carbon reconfiguration thus depends on the elevated importance of climate mitigation goals and alignment with system-specific policy goals, changes in governance style (e.g., from hands-off to interventionist approaches; from piecemeal to systemic approaches), and increases in the number, stringency, and types of climate mitigation instruments.

Our empirical analyses will map the varying depths of institutional low-carbon reconfiguration in electricity, heat, and mobility systems. They do not aim to systematically explain these differences, although external and internal causes may be inductively alluded to.

3

Methodology

While Chapters 1 and 2 discussed the book's rationale, novelty, scope, and conceptual approach, this chapter addresses methodological challenges and operational issues.

Section 3.1 engages with broader methodological issues for research of longitudinal large-scale change processes. Drawing on the wider social sciences, it argues for a particular explanatory style since mainstream methods have limitations for this research topic. Section 3.2 justifies the analytical demarcation choices (e.g., selection rationales, system boundaries), describes how we operationalise the conceptual framework and underlying categories, and discusses data sources. Section 3.3 addresses the evaluation of system reconfiguration patterns across various dimensions.

3.1 Broader Methodological Issues

Before addressing operational issues, we discuss some broader methodological challenges that relate to special characteristics of the research topic (Geels and Schot, 2010; Köhler et al., 2019). First, socio-technical system transitions are longitudinal processes that unfold over time, often involving non-linear and complex interactions. Second, socio-technical systems are large-scale, heterogeneous entities with multiple co-evolving elements. Third, constitutive elements (like actor properties, techno-economic elements, and rules) are themselves transformed as the transition process unfolds.

These characteristics pose challenges for mainstream explanatory formats since 'contemporary social scientists are strongly predisposed to focus on aspects of causal processes and outcomes that unfold very rapidly' (Pierson, 2004: 13). Mainstream analytical formats (such as regression analysis or structural equation modelling) approach the world as consisting of variables or factors, with independent variables ('causes') having effects on the dependent variable ('outcome'). This *push-type* causality (Mohr, 1982), which has come to dominate many social sciences (Abbott, 2004), assumes that causes 'operate at equal speed and in the same way across all cases. [...] Explanations should emphasize immediate causation. [...] It is not necessary to know the particular twists and turns of an entity's history to explain it' (van de Ven, 2007: 152–153). While mainstream explanatory formats may be suitable for narrowly defined research topics with *linear causality*, which involves 'a straightforward, direct chain of events that characterizes simple phenomena' (George and Bennett, 2004: 212), they have major limitations for explaining longitudinal transformation processes in heterogeneous macro-entities.

Analysing those kinds of entities, like socio-technical system transitions, requires a different explanatory style, as social scientists with longitudinal processual interests have emphasised. For political science, for instance, Pierson (2004: 1–2) notes that:

> Contemporary social scientists typically take a "snapshot" view of political life, but there is often a strong case to be made for shifting from snapshots to moving pictures. This means systematically situating particular moments (including the present) in a temporal sequence of events and processes stretching over extended periods. ... The systematic examination of processes unfolding over time warrants a central position in the social sciences.

Hall (2003: 387) similarly observes that:

> Comparative politics has moved away from ontologies that assume causal variables with strong, consistent, and independent effects across space and time toward ones that acknowledge more extensive endogeneity, path dependence and the ubiquity of complex interaction effects. ... Accordingly, parsimony is no longer seen as a key feature of explanation in political science.

For business and management, Langley et al. (2013: 1) diagnose a similar trend:

> A growing number of management scholars have been researching process questions. ... Process studies address questions about how and why things emerge, develop, grow, or terminate over time. ... Process research, thus, focuses empirically on evolving phenomena, and it draws on theorizing that explicitly incorporates progressions of activities as elements of explanation and understanding.

More broadly, Tilly (2008: 9) suggests that: 'In social science as a whole, a substantial intellectual movement has formed to adopt mechanism- and process-based explanations'.

Situating ourselves in this broader epistemological turn, the book's research is oriented by the following guiding principles, which also resonate with the Multi-Level Perspective.

3.1.1 Conjunctural and Configurational Causality and Explanation

Our explanations of socio-technical system transitions are based on conjunctural or configurational causalities, in which outcomes result from multiple interacting processes.

> Heterogeneous entities are composed of disparate components. … complex social happenings are almost invariably composed of multiple causal processes and components rather than existing as unitary systems. The phenomena of a great social whole … should be conceptualised as the sum of a large number of disparate processes with intertwining linkages and often highly dissimilar tempos. (Little, 2016: 8–9)

Conjunctural or configurational causality

> characterizes a specific mode of explanation … in which researchers consider how multiple factors combine to form larger combinations, complexes, and causal packages. One reason this configurational analysis figures so prominently is because the large-scale outcomes … are themselves often aggregated combinations of multiple events and processes. (Thelen and Mahoney, 2015: 7)

This explanatory style is common in qualitative research and suitable for answering 'how' questions: 'Qualitative researchers especially tend to think in terms of combinations and configurations because of their interest in context and in understanding social phenomena holistically. … This interest in combinations of causes dovetails with a focus on 'how' things happen' (Ragin, 2008: 109).

Instead of linear causal chains, longitudinal socio-technical transitions should thus be explained through processual alignments and co-evolution: 'Most historical sociologists reject the notion of a single master process, acknowledging multiple processes that overlap and intersect one another. Explaining a particular outcome or pattern of development thus involves a particular logic of explanation: situating events or outcomes in terms of their location in intersecting trajectories with independent temporalities' (Aminzade, 1992: 466).

3.1.2 Causal Reconstruction and Process Tracing

Our explanations of socio-technical transitions aim to trace and reconstruct the interaction of co-evolving causal processes: 'The majority of macro-phenomena … cannot be explained by applying one particular mechanism model. Instead, the causal reconstruction of macro-phenomena … involves a chain of different mechanisms that jointly generate the outcome' (Mayntz, 2004: 254). Héretier (2008: 75) adds that the causal reconstruction approach 'is appropriate when the number of cases is small, the explanatory factors are highly dependent on each other and … outcomes are the result of complex interaction effects and various forms of multicausality'. Mayntz (2004: 238) further explains that:

> Causal reconstruction does not look for statistical relationships among variables but seeks to *explain* a given social phenomenon – a given event, structure, or development – by identifying the processes through which it is generated. Causal reconstruction may lead to a (more or less complex) historical narrative, but in its theoretically more ambitious version, causal reconstruction aims at generalizations – generalizations involving processes, not correlations.

As a general methodological orientation, we use process tracing, which is presented as 'more appropriate than other methods in the study of phenomena characterized by complex causality or multiple causal pathways' (Falleti, 2016: 456). Process tracing studies are concerned with explanations 'that indicate how the process unfolds over time' (Poole et al., 2000: 12). The analytical focus is hence set on understanding *how* and *under which conditions* things change. A process is understood as a developmental event sequence (Langley, 1999; Sminia, 2009), in which processual outcomes (e.g., the occurrence of a particular transition pattern) are explained as the result of sequences of events and related to the identification of underlying generative mechanisms.

There are different types of event-sequences with varying relevance for socio-technical transitions: 1) *self-reinforcing sequences* are 'characterized by the formation and long-term reproduction of [established] patterns' (Mahoney, 2000: 508); they typically stem from cumulative causation and increasing return mechanisms, which may be particularly relevant with regard to techno-economic developments (e.g., cumulative cost reductions or performance improvements), and 2) *reactive sequences* in which each step depends on prior steps and involves reaction and counterreaction mechanisms, that is, 'backlash processes that *transform* and perhaps *reverse* early events' (Mahoney, 2000: 526); this may be particularly relevant with regard to actors (who jockey for position or are involved in struggles) and policies and rules (which may be introduced, removed, or adjusted, depending on effects, learning, and (counter-)lobbies).

3.1.3 Longitudinal Case Studies

A particularly suited methodological strategy for process tracing is the rich longitudinal case study. It involves reconstructing sequences of events, cumulative trends, and evolving contexts with attention to their temporal ordering and the steps along the way.

The temporal delineation of cases is not entirely unproblematic and a usual cause for academic debate. However, a pragmatic approach is to rely on transparent quantitative indicators of the phenomenon at hand, such as the diffusion rates of particular technologies, the economic weight of a particular sector, or the salience of a particular societal problem in the public debate. Major changes in such indicators, observable through inflection points in their general trends, provide good pointers for where to start and end a particular case or how to periodize if a phase-approach is taken.

George and Bennett (2004) distinguish different kinds of process tracing case studies, including detailed narrative, use of hypotheses, analytic explanation, and more general explanation. The case studies in our book aim for analytic explanation, which 'converts a historical narrative into an *analytical* causal explanation couched in explicitly theoretical forms' (p. 211). The analytical narrative is grounded in a conceptual understanding of the processes and mechanisms at play and structured around theoretically informed categories and dimensions. Such dimensions indicate what kinds of entities, events, and relationships to follow over time, as well what kinds of developments may be indicative of continuity or change. In Section 3.2.2, we explicitly return to the analytical dimensions considered in the empirical chapters of this book by describing how we operationalize the main elements of our conceptual framework.

While historical case studies are fully retrospective accounts that benefit from hindsight, this book is concerned with analysing transitions 'in-the-making'. This has important implications concerning our approach. First, we do not adopt a strict case study methodology but are, rather, focused on evaluating changes over time. In doing so, however, we do adopt a processual approach to explanation and develop a rich narrative approach to make sense of developments over time and mobilise longitudinal data. Second, while we follow historical developments, we do not claim to be historians nor mobilise empirical strategies, such as first-hand archival data, that historians would. Third, we follow reconfiguration processes up to the present, which means that our evaluation remains open-ended: system developments are ongoing, the observed transitions cannot be fully circumscribed and delineated (i.e., they have no end), and are marked by significant uncertainties.

3.1.4 Comparative Research

Most transitions research to date has rested on single case studies, which tend to favour depth, richness, and accuracy over more generic insights. Single case studies remain the privileged method of enquiry in transitions studies, and recent developments have been oriented towards the exploration of non-standard cases (e.g., cases displaying significant variations from ideal-typical patterns), different contexts (e.g., Global South countries), neglected systems (e.g., agri-food, health), different scales (e.g., cities, regions, transnational), or the role of specific actors (e.g., civil society, the State). Cumulatively, this amounts to an extensive collective evidence base of rich and detailed transitions cases (historical and contemporary), which broadly share the same epistemological commitments and often mobilise one of the key transition frameworks.

This extensive collective evidence base now makes it possible to investigate more laterally across cases and develop novel comparative strategies: 'The single-case research design remains prominent in transitions research, also as new regions, new actors, new technologies and new societal domains are explored. In turn, the increasing wealth of case materials creates demands and opportunities for methodological approaches that reach for generic insights across cases' (Köhler et al., 2019: 18).

Recently, researchers have sought to develop comparative approaches to analyse contrasting transition patterns. Such efforts have to date primarily focused on comparing national trajectories within a single sector such as in energy (Geels et al., 2016b; Johnstone and Stirling, 2020), agri-food (Darnhofer et al., 2019), mobility (Mazur et al., 2015; Nykvist and Whitmarsh, 2008) or related policy debates (Lovio and Kivimaa, 2012; Upham et al., 2013). Others have compared cases within the same country and sector but across different time periods, such as gas transitions (Arapostathis et al., 2013; Pearson and Arapostathis, 2017) or transitions away from coal (Turnheim and Geels, 2012). The notion of transition pathway (Rosenbloom, 2017; Turnheim et al., 2015) has been particularly fruitful to support such comparative work, notably through the development of transition typologies (Geels and Schot, 2007; Smith et al., 2005) to make sense of a variety of patterns in case observations, and as means to contrast them. Keeping one dimension constant (e.g., sector or country) simplifies comparability and the identification of key variations.

Other comparative attempts, such as meta-analyses of transition patterns (Martínez Arranz, 2017; Raven et al., 2016, 2008; Sovacool, 2016; Wiseman et al., 2013) are broader in scope, as they are oriented towards the comparison of transitions across sectors, geographical context, and time periods. While the objective is to draw out regularities and deviations across essential features of transitions (e.g., speed, scope, primary drivers and mechanisms), it also implies significant trade-offs, such as oversimplification, losing sight of complex causation, or simply comparing incommensurate phenomena due to inconsistencies in analytical units (Grubler et al., 2016).

So, while there is an 'unmistakable drive towards systematic comparison and theory-building from cases' (Köhler et al., 2019: 19) in transitions studies, such research is only emerging, and is not a substitute for in-depth case studies. A major challenge concerns the importance of maintaining an appropriate degree of richness and attention to local particularism while seeking more generic insights through comparison. For this reason, the book format seems particularly appropriate to minimise ensuing trade-offs involved in comparisons.

In this book, we have privileged intersectoral comparisons within a given national setting (the UK) and a given temporal frame (1990–present), with the aim of identifying regularities and deviations across the examined sectors. A major advantage of such an approach is that all sectors will share common features at an aggregated level (e.g., national climate policy and decarbonisation frame, approach to industrial policy, or consumer culture) despite important changes over time, which enables us to examine sector-specific variations and explain major differences. The comparative approach put forward in this book rests on 1) the consistent deployment of an analytical grid operationalised at a rather fine level of detail, which enables a systematic comparison, and 2) the deployment of broader categories at the level of pathways, which enables the comparative interpretation of reconfiguration trajectories (e.g., their speed, scope, depth).

3.2 Analysing Longitudinal Socio-Technical Developments

Our core methodological commitment in this book is concerned with documenting, tracing, and analysing longitudinal socio-technical developments in different systems, in a way that enables cross-comparisons. Our comparative aim leads us to consider a systematic and transparent way to map systems and co-evolutionary developments over time, as well as to adopt consistent boundaries across all systems. This section details related choices and operationalisation concerning 1) analytical scope, 2) socio-technical dimensions and levels, and 3) data sources.

3.2.1 Analytical Scope and Selection Rationales

The core focus of this book is to analyse low-carbon reconfigurations in the UK. For this reason, we are concerned with systems characterised by significant GHG emissions, but also for which official emissions inventories are sufficiently detailed and consistent over time. Electricity, heat, and mobility are carbon-intensive sectors that have been the focus of dedicated emission reductions efforts for some time and for which techno-economic assessments of decarbonisation are available. Our system delineation for these sectors deviates nonetheless from climate assessments' conventional focus on power, transport, and buildings.

The empirical analyses in this book cover the period 1990 to the present, because 1990 is a common reference point in climate assessments and related policy discussions, with most emission reduction objectives referring to pre-1990 emissions as a baseline. This also means that detailed longitudinal emissions inventories are available for this period. More substantially, because we are interested in low-carbon transitions, 1990 offers a clear reference point for the mainstreaming of climate considerations as a public issue. From 1990, decarbonisation started to become a major issue for policymaking related to carbon-intensive sectors, with international commitments agreed at the United

Nations Conference on Environment and Development in 1992, followed by the ratification of the Kyoto Protocol in 1997. Of course, climate concerns and emission reductions pre-date the 1990s, so we do not limit ourselves too strictly to this reference point and have included developments pre-dating 1990 when they have significant influence over system trajectories, or when they help explain some of the lock-ins at play that constrain reconfigurations. The most recent years of our evaluation have been marked by significant events with potentially long-lasting implications (Brexit, COVID-19, Recovery packages). These implications are, however, riddled with uncertainties that call for interpretive caution concerning their actual decarbonisation effect and are largely vehiculated by promissory discourse. For this reason, we return to these issues in our conclusion.

The national scale has been the predominant scale at which transitions dynamics have been analysed in the literature. Reasons for this include the importance of national boundaries for technical infrastructure (e.g., roads, gas and electricity grids), institutions and policies (e.g., building regulations), strategic innovation programmes, or user attitudes. There are also pragmatic reasons related to the availability of longitudinal quantitative data (on emissions, sector investments, markets), which are often collected by national agencies. This *de facto* focus on the national scale has been challenged within transitions studies, given that many relevant developments are also constituted at local scales (e.g., local innovation projects and regional industrial dynamics), overflow across national boundaries, and are strongly influenced by transnational dynamics and actors (e.g., global corporations and supply-chains, international policies, social movements, or consumer cultures). Accordingly, transitions studies is increasingly considering multi-scalar processes (Coenen et al., 2012; Raven et al., 2012) and deploying approaches on different spaces. In this book, however, we have privileged the national scale because of our interest in decarbonisation policies and strategies, our dialogue with techno-economic assessments of national decarbonisation strategies, as well as comparative considerations. We nonetheless attend, where relevant, to the influence of supranational policies (notably European standards and regulations) and to local considerations (e.g., London's local transport system). The statistical databases used in this book vary in their national coverage, focusing sometimes on the UK and sometimes on Great Britain (which thus excludes Northern Ireland).

We focus on three carbon-intensive sectors, as justified by their historical contribution to GHG emissions. However, our interest in socio-technical dynamics and production-consumption systems leads us to depart slightly from system delineations as put forward in climate assessments (which distinguish power, buildings, and transport systems) and sectorial policy (which tend to take a supply-side orientation). First, we define socio-technical systems as primarily oriented towards a societal function (e.g., heat, mobility), which enables us to consider dominant configurations but also alternative or parallel systems contributing to said function. The electricity system differs from the other two systems because it can fulfil multiple societal functions such as lighting, freezing/cooling, hygiene/washing, cooking, and entertainment. Second, because of our interest in whole system reconfigurations, we include production and consumption elements, broadly ascribing to a supply-chain understanding. Furthermore, because we are interested in how production and end users interact in whole systems, we have privileged systems boundaries including private consumers as end-users, which leads us to focus on residential users of electricity and heat and individual passenger mobility, hence excluding commercial and industrial uses. While we adopt a consistent analytical delineation across the three examined systems, sectorial particularisms lead us to also adapt this framework accordingly. For this reason, the first section of each empirical chapter describes and maps the systems considered, and elaborates distinctions between sub-systems and parallel systems, where relevant, according to the aforementioned considerations.

Because our framework is derived from the MLP (see Chapter 2), we analyse existing systems separately from emerging niche-innovations, as well as considering their interactions over time. For the selection of relevant niche-innovations, we focus on the most significant innovations according to our three analytical dimensions (i.e., those with emerging markets, institutional and policy backing, as well as growing actor coalitions), but have also sought variations in types of innovations. We have considered non-technological innovations (e.g., demand-side response, teleworking, and car sharing) and innovations for which technological and market maturity might not be that well developed but for which policy backing and discursive promises are significant (e.g., greening of the gas grid or self-driving cars).

3.2.2 Documenting Socio-Technical Dimensions

Our analysis of socio-technical development follows the adapted MLP framework described in Chapter 2, which rests on a commitment to tracing processes longitudinally, a distinction between different structuration levels (niche, regime, landscape) and an understanding that socio-technical change is the outcome of interaction mechanisms between these levels, and an analytical distinction between three key dimensions (techno-economic, actors, policies and governance). We here describe how we translate these commitments and analytical notions into operational categories to enable the systematic evaluation of low-carbon transitions. Starting with existing systems and established configurations, we explain how the three key dimensions can be deployed to map systems and longitudinal changes. We then briefly discuss how this can be replicated at the level of emerging niche-innovations. Our empirical analyses do not separately discuss exogenous landscape developments, because this would increase the complexity by adding another analytical layer. Instead, landscape developments, such as oil price spikes, the financial-economic crisis, Brexit, and COVID-19, are discussed when they are immediately relevant for system and niche developments. The conclusions chapter will revisit the role of landscape developments on low-carbon transitions in the various systems.

Techno-Economic Dimensions

Because of our interest in a constructive dialogue with different system assessment approaches and our engagement with climate policy debates, for which techno-economic dimensions tend to be foregrounded (e.g., in carbon-intensive system modelling, in sectorial decarbonisation assessments, in industrial policy), this dimension has a special status in our analysis. The techno-economic dimension allows us to identify and map systems and serves as a distinct analytical dimension.

Each empirical chapter begins with a descriptive mapping of the key techno-economic features and system architecture, following the main material flows, mediating infrastructures, and technical components. To do this, we focus on the dominant techno-economic architecture over the entire period, and schematically depict the connection from production all the way to end uses through a supply-chain logic. This system mapping allows us to identify the main sub-systems involved (for electricity), the orthogonal systems (for heat), and the parallel systems (for passenger mobility), which are then mobilised to structure the analyses of system developments – described in separate sub-sections.

Because of our interest in decarbonisation, these schematic system maps are followed by a description of the main GHG emission reduction trends at play. These emissions reduction patterns provide an overall trend to be explained by our reconfiguration analysis.

For each (sub)system, we then describe techno-economic aspects and developments following a broadly chronological order. The following questions guide our analyses of techno-economic dimensions along material, market, and innovation (especially low-carbon improvements) aspects:

- What are the main technical components, infrastructures, material flows, and transformation activities involved? How are they linked in systems? Where multiple alternatives exist (e.g., fuel inputs, appliances, modes), what is their relative weight and relationship? How have these material components and their linkages changed over time?

- What are the main market developments within the considered systems and subsystems (e.g., input costs, infrastructure and maintenance, accessibility)? How have these changed over time?
- What are the main technical problems addressed through innovation? What are the main low-carbon innovations contributing to improving existing systems? How have these changed over time?

Actors

Several relevant social groups take active part in making systems function the way they do, maintaining them the way they are, framing and regulating their use, using them and deriving a service from them, and engaging in critical debate about relevant issues and desirable priorities for their further development. For each subsystem, we describe actors and their activities over time following a broadly chronological order. For the existing (sub)systems, we mostly discuss operational activities of established actors but also include actors that are not operationally related but nevertheless exert pressure for change, such as wider publics and civil society actors. Because of our analytical scope, we have limited the actors that we include in our analyses to the following, relatively consensual, categories: firms, policymakers, users, and wider publics.[1]

The following questions guide our analyses of actors and networks for each of the categories above:

- For all actor categories: What actors are involved? What actions do they perform with respect to using, maintaining, or altering systems?
- For firms: What are the distinctive features of firms involved in this system (e.g., role, specialisation, market and policy influence, historical presence)? How is the industry structured? What are the main strategies enacted by firms? What are the low-carbon strategies enacted by firms, and how important are these compared to other concerns? Have these strategies changed over time?
- For policymakers: What are the main policy objectives and priorities? What is the importance of climate and decarbonisation relative to other sectorial policy priorities? What are the main policy activities (e.g., debates, negotiations, struggles)? How have they changed over time?
- For users: What are the main users and user practices in this system? What is the level of user engagement with appliances and system choice? What are the main user priorities (e.g., cost, accessibility, comfort, convenience, experience)? What are the main attitudes concerning low-carbon alternatives? How have these changed over time?

[1] Under this label, we include civil society actors and societal issues influencing public opinion.

- For wider publics: What are the main societal issues being discussed and debated relative to this system? What is the relative importance of climate and decarbonisation in public debates? What are the main civil society groups seeking to influence this system?

Policies and Governance

For each (sub)system, we describe the main policies implemented and discuss whether features of a governance style are identifiable. Our interest in low-carbon transitions leads us to give a special status to related decarbonisation policies, but these are considered in the context of broader strategic interventions in the sector, particularly if these are supporting or hindering decarbonisation.

Concerning policies and their implementation, we follow a chronological order and discuss the policy instruments implemented to orient and accelerate the decarbonisation of existing systems, which includes the deployment of low-carbon innovations. These instruments include those typically discussed within environmental policy, such as formal regulatory instruments (e.g., rules, regulations, standards), market-based instruments (e.g., financial (dis-)incentives), and informational instruments (e.g., awareness raising, labelling), as well as additional instruments more specific to innovation and industrial policy (e.g., RD&D programmes, infrastructure investments, supply-chain development, training, and skilling).

For these policies, we document the objectives, degree of ambition, and implementation and seek to explain why particular interventions have led to significant achievements while others have been interpreted as failures or having insignificant positive outcomes. So, we are particularly attentive to the conditions of implementation, that is, understanding not only what works but also why and under which conditions.

Additionally, our interest in the governance of low-carbon system reconfigurations over time led us to pay attention to the interaction of multiple goals and instruments in policy mixes. Accordingly, we have identified three issues as critical to effective policy mix design and implementation in Chapter 2, namely the coherence of goals, the consistency of instruments, and the comprehensiveness of instruments.

Last, we seek to situate the observed succession of policy interventions within broader policy paradigms and related governance styles, which we understand as identifiable approaches to framing and handling policy problems.

Accordingly, the following questions guide our analyses of policy and governance issues:

- For policy instruments: Which policy instruments have been deployed and implemented? What are the objectives and rationales? Have they been successful in supporting low-carbon changes? Have they been criticised, and if so, why? Have they endured and strengthened over time?

- Concerning policy mixes: Are there multiple interacting policies driving changes in the system? Were their objectives *coherent* or working at cross-purposes? Are their intended and actual effects *consistent* between instruments (e.g., reinforcing or undermining) and over time (e.g., cumulative, interrupted, or reversals)? Are the policy instruments *comprehensive* in their scope, types, and in terms of the balance of targeted dimensions?
- Concerning governance styles: How can the main governance style be characterised? What are the underlying rationales justifying the needs and means of intervention? Is there more emphasis on market-based logics or regulations and standards? Is the governance approach targeting whole system reconfigurations or more limited piecemeal improvements? What is the degree of interventionism and prescription (versus reliance on more voluntary approaches)? What is the degree of government leadership and coordination concerning the steering of long-term changes and/or technological choice prescription? Are certain actors and networks privileged?

3.2.3 Data Sources and Practical Considerations

The intended scope and depth of this book calls for the mobilisation of a wide range of data sources and literatures. Concerning the empirical analyses, we rely on a combination of longitudinal quantitative data and qualitative interpretations of various developments, as well as primary and secondary data sources. The literatures and data sources broadly cover three aspects and related requirements.

Documenting longitudinal system changes requires reliance on descriptive quantitative time-series concerning GHG emissions and techno-economic variables (e.g., sector size, economic output trends, investment trends). For these, we rely mainly on official GHG inventories and sector-level economic statistics, due to their extensive scope. We do not enter methodological debates about the reliability of government statistics. Our main concern is to document relative changes, to enable inter-sector comparisons.

For actors, policies, and governance styles, we require more in-depth kinds of data and analyses. For this, we rely primarily on secondary literatures from energy, buildings, and transport studies, which cover specific dimensions (e.g., on actors and strategies, policies, struggles, learning) or combinations thereof, at various points in time. A significant interpretive challenge concerns making sense of disciplinarily disparate sources to identify the main issues and trends, and to reflect the variety of relevant social groups involved.

Documenting innovation and transition dynamics requires sources that are more focussed on temporal and processual dimensions (e.g., emergence and momentum, inertia and lock-in, transformative change). For this, we rely primarily on innovation and transitions literatures, which cover a range of analyses of individual niche-innovations or system-level dynamics. A significant interpretive challenge concerns making sense of multiple dynamics to turn them into multi-dimensional yet concise narratives about the focal system and its possible developments. Fortunately, we do not start from scratch. In the past decade, socio-technical transition scholars have made many in-depth analyses of specific low-carbon innovations, on which we can build our aggregate analysis. For our system analyses, we can also build on (socio-political and techno-economic) interpretations from domain-specific experts, many of whom have started to use the MLP as analytical frame, which facilitates our task. The interpretation of these empirical 'building blocks' inevitably introduces subjectivity, but it also allows for creativity, which is necessary to make sense of the many heterogeneous data.

Together, these different aspects and the range of sources mobilised is unprecedented to our knowledge. For instance, existing reports on sector decarbonisation such as those produced by the Committee on Climate Change usually do not cover actors and institutional aspects and adopt a rather narrow interpretation of innovation trajectories (e.g., costs, efficiencies, and barriers).

3.3 Evaluating Reconfigurations in the Making

The bulk of the empirical chapters are concerned with documenting and tracing socio-technical developments over time, by distinguishing relevant analytical dimensions and by following both existing systems and niche-innovations. While we have already discussed how these developmental narratives combine description and interpretation, these are followed by conclusions that make a more interpretive move: the evaluation of reconfigurations in-the-making.

The objective of our evaluation of system reconfigurations is threefold. First, we are interested in explaining a concrete phenomenon: progress with low-carbon performance. We seek to explain this by reference to concrete innovations and the underlying systemic changes, which we analyse through reconfigurations. Second, because of our commitment to the MLP, we seek to evaluate reconfigurations from a specific perspective, which emphasises the dynamic interactions between niches and systems. This leads us to explore reconfigurations as the combined result of a) system lock-in and unlocking, b) processes of niche-innovation (emergence, diffusion), and c) emergent reconfiguration patterns. Third, because we are interested in the unfolding of system reconfigurations over time and into the future, we aim to provide evaluative interpretations concerning prospective trajectories.

In terms of concrete operationalisation, we first provide an overall evaluation of the role of low-carbon innovations in driving GHG emission reductions. To do this, we seek to attribute the relative importance of examined innovations in system-level decarbonisation, as well as to provide a general interpretation of the core patterns at play (e.g., incremental innovation, diffusion in sheltered niches, substitution). Following this, we examine reconfiguration patterns systematically within each of our analytical dimensions (techno-economic, actors, policy), and qualify whether the changes observed are limited, moderate, or substantial.

For techno-economic reconfiguration, our main concern is to qualify the different kinds of reconfiguration patterns implied by the interaction of the examined niche-innovations with system dynamics. To do so, we first position the examined niche-innovations within Table 2.1, by considering the degree of departure from existing techno-economic configuration's core elements (reinforced or substituted) and its architecture (unchanged or changed). By examining the relative importance and momentum of different types of niche-innovations over time, we are able to describe systemic preferences for a particular techno-economic reconfiguration trajectory, as well as to make informed evaluation about possible developments, particularly in terms of more radical and transformative change.

Concerning actor reconfigurations, we seek to trace the main changes in the goals, agendas, interests, and strategies of actors and networks involved. We are particularly interested in the degree of actor reconfiguration and its evolution over time in each sub-system, that is, their propensity to change/stability and their actual change/stability. To do this, we first provide a synthetic evaluation of major changes concerning established actors and new actor coalitions. Following this, we delve into a more detailed and systematic evaluation of the kinds and degrees of change for each actor category, notably by distinguishing a) actor changes in support of low-carbon transitions (for both incumbents and new entrants), and b) developments indicative of actor lock-ins, resistance to change, or competing issues inhibiting low-carbon transitions. Compounded together, these indicative evaluations build-up into a rich picture of the momentum for change and the degree of inertia (lock-in) as related to various actors.

Concerning institutional reconfigurations, our focus is primarily set on policy interventions and governance style. To evaluate changes in policy instrumentation, we examine the degree of change in the kinds of instruments, their articulation into policy mixes, and their effectiveness in supporting low-carbon developments. We are primarily interested in tracing patterns, that is, regularities, changes, or deviations over time. Similarly, when examining changes in governance styles, we are particularly interested in tracing the key changes over time, which allow us to make informed evaluations about possible future developments.

Last, and this represents yet a higher level of aggregation in pattern tracing, we seek to derive some general interpretation of the scope, depth, and speed of reconfigurations.

4

Electricity System

4.1 Introduction[1]

The electricity system has several special characteristics. First, electricity is an invisible and undifferentiated energy carrier, which does not directly fulfil end-use functions. Instead, it powers electric appliances that provide functionalities such as lighting, heating, freezing/cooling, washing, drying, cooking, entertainment, computing, or powering electric motors that perform mechanical work in households, the commercial sector, and industry. Second, while production and consumption of electricity have become geographically separated, they are integrated in real-time because precise balancing is needed to prevent blackouts. This geographical separation implies the need for electricity grids to transport power from (often centralised) sites of production to (often decentralised) sites of consumption (e.g., households, offices, factories). There are high-voltage transmission grids, which transport power over long distances, and low-voltage distribution grids, which disseminate power locally.

In terms of system architecture, these considerations imply that the electricity system can be distinguished into three sub-systems: production, grids, and consumption. The system architecture has a horizontal 'hour-glass' shape: multiple upstream inputs (e.g., coal, gas, nuclear material, biomass, wind, sunshine) are transformed into a single homogenous product (electricity), which is transported by a fine-grained, integrated power grid to sites of consumption where its use in electric appliances enables multiple end-use functions (Figure 4.1). This means that grids not only *connect* production and consumption but also act as a *buffer* in the sense that users hardly notice 'upstream' changes in power generation (which is an important difference with the mobility and heat systems).

[1] Parts of this chapter draw on McMeekin et al. (2019), but substantially update, elaborate, and refocus the analysis.

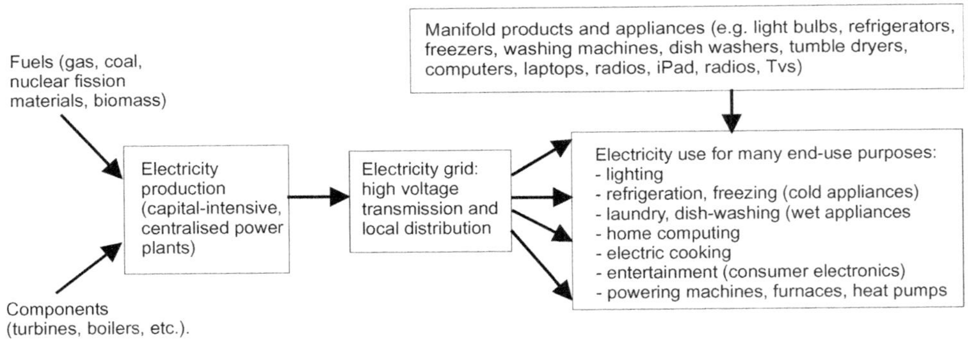

Figure 4.1 Schematic representation of the material elements and flows in the electricity system

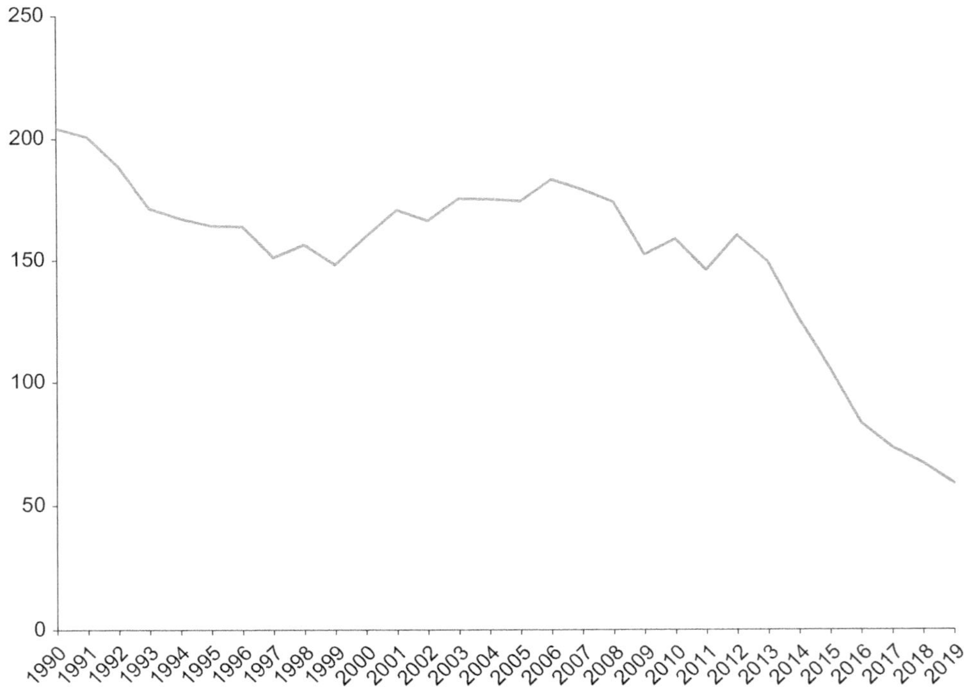

Figure 4.2 Greenhouse gas emissions from power stations in $MtCO_2$, 1990–2019 (constructed using data from National Statistics: final UK greenhouse gas emissions)

Direct greenhouse gas emissions, which are generated mostly in electricity generation, decreased by 71% between 1990 and 2019 (Figure 4.2), which suggests a low-carbon transition is well under way. The chapter aims to provide an interpretive assessment of underlying changes in various sub-systems. To that end, Sections 4.2, 4.3, and 4.4 respectively investigate the main developments in electricity generation (including coal, gas, and nuclear power), grid, and consumption sub-systems (including washing; cooling; information, communication, and entertainment; and lighting). For each sub-system, we first analyse techno-economic developments and then actors and institutions. Section 4.5 analyses nine niche-innovations across electricity generation (onshore wind, offshore wind, biomass, solar-PV), consumption (energy-efficient lighting, smart meters), and grids (smart grid, battery storage, demand-side response). Section 4.6 draws conclusions about the speed and depth of low-carbon system reconfiguration.

4.2 Electricity Generation Sub-system

4.2.1 Techno-Economic Developments

The electricity generation sub-system traditionally consists of large, centralised base-load units (coal, nuclear, or large gas turbines), complemented with flexible units for peak-load generation (e.g., smaller gas turbines). Power generation is a complex, engineering-heavy, and capital-intensive activity, linked upstream to specialised supply-chains for different fuels (e.g., coal, gas, uranium, oil) and equipment manufacturing, installation, and maintenance (e.g., thermal/nuclear reactors, turbines, boilers).

The direct economic relevance of the electricity supply industry decreased since the 1980s, rebounded since the late 2000s, and is still substantial in 2019, generating 98,000 jobs and contributing 1.15% to the Gross Value Added of the UK economy (Figure 4.3). The indirect economic relevance is broader, when taking into account supply chains and the importance of reliable electricity provision to the wider economy.

The relative importance of different fuel inputs to power generation has changed substantially since 1980 (Figure 4.4):

- nuclear power expanded in the 1980s and 1990s and then gradually contracted;
- the use of natural gas increased rapidly after the 1990 privatisation policies;
- renewable power generation has increased gradually since 1990, reaching 39% in 2019;
- coal use declined very substantially, reducing its relative contribution to power production from 69% in 1990 to 2% in 2019. This decline occurred in two steps. The first step was the market-driven 'dash for gas' in the 1990s. The second step was related to climate and energy policies, which stimulated the expansion of renewables and made coal more expensive (e.g., the Carbon Floor Price instrument).

Figure 4.4 also shows that electricity supply peaked in 2005 and subsequently decreased by 18%, owing to decreasing electricity demand, which will be discussed further in Section 4.3.

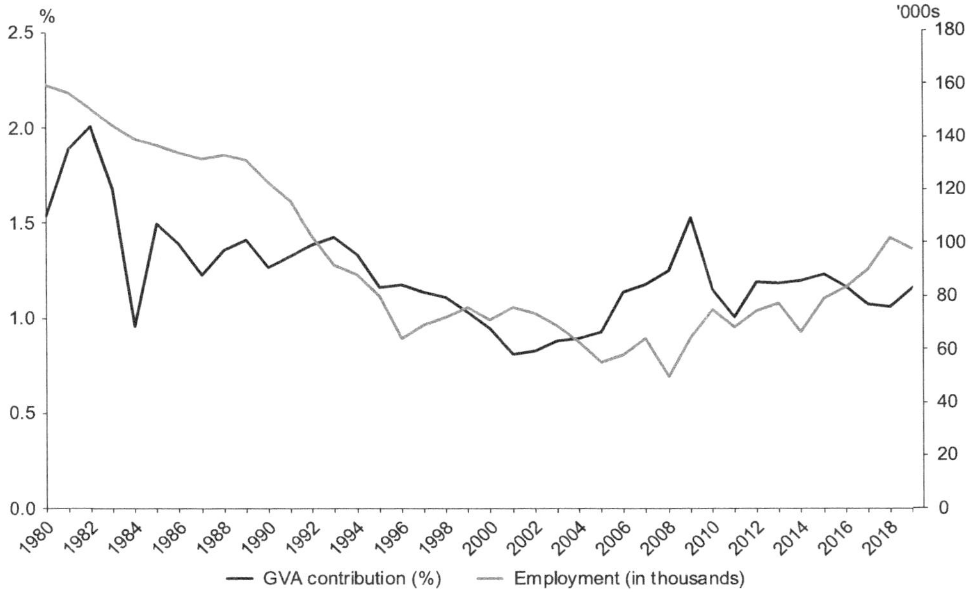

Figure 4.3 Relative contribution of UK electricity industry to Gross Value Added (in %, left-hand axis) and number of jobs (in thousands, right-hand axis), 1980–2019 (constructed using data from UK Energy in Brief dataset, Tables 1 and 2)

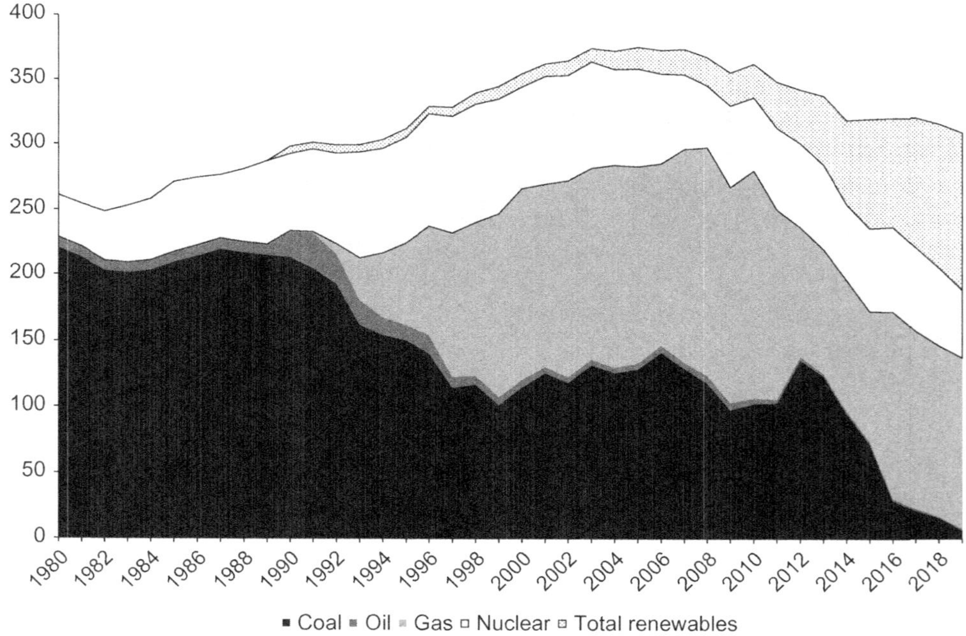

Figure 4.4 Electricity supplied by fuel type in TWh, 1980–2019 (constructed using data from Digest of UK Energy Statistics; Electricity Statistics; Electricity fuel use, generation and supply; Table 5.6)

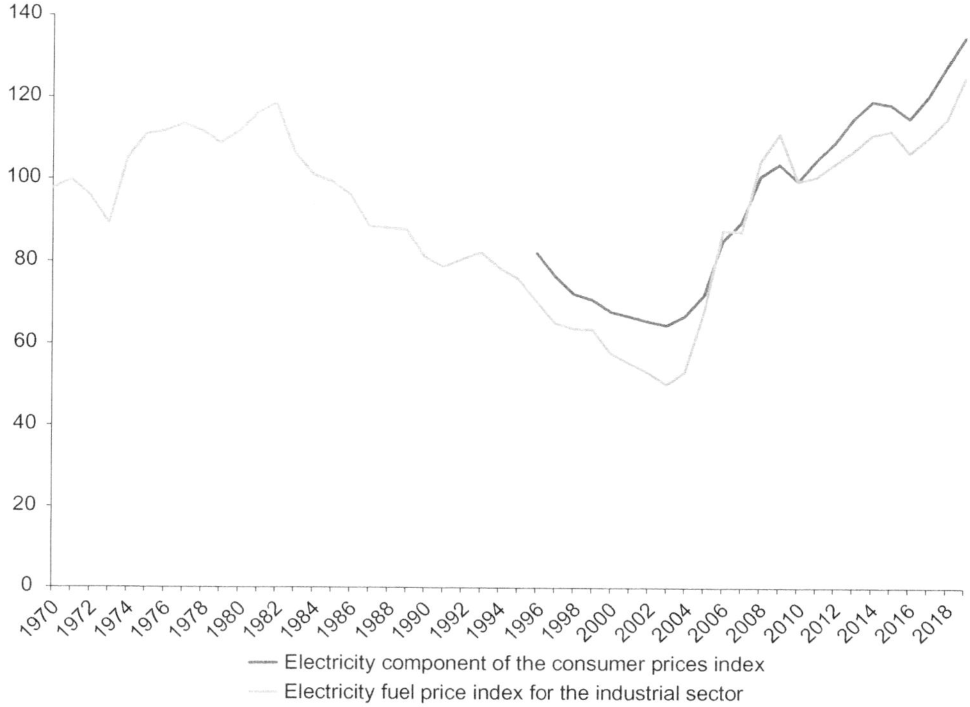

Figure 4.5 UK electricity price index in real terms for industrial and domestic sectors, 1970–2019 (2010=100) (constructed using data from Statistics at BEIS; Historical Electricity Data series)

Electricity prices decreased in the 1980s and 1990s (Figure 4.5) because of decreasing fuel input prices such as coal and gas (Figure 4.6) and because privatisation in 1990 increased competition between power producers, which led to lower prices. Since the early 2000s, however, electricity prices have started increasing (Figure 4.5), first because of rising gas prices (which are linked to oil prices) and second because of new investments, related both to decarbonisation and to replacement of old coal and nuclear power plants.

In terms of fuel supplies, the long-term decline in the domestic production of coal was accelerated by the 1984 coal miners' strike (Turnheim and Geels, 2012). Mine closures continued, leading to major declines in coal-related employment (Figure 4.7). Competition from cheaper coal imports and declining coal demand from power producers in the 1990s further reduced domestic coal production. The increasing reliance on imported coal helps explain why electricity-related coal phase-out since the mid-2010s received relatively little industrial opposition.

UK gas production increased very rapidly in the 1990s, owing to the exploitation of new North Sea gas fields (Figure 4.8). Between 1997 and 2003,

Figure 4.6 Average prices of fuel inputs (in pence per kWh in real terms) purchased by the major UK power producers, 1993–2019 (constructed using data from Statistics at BEIS; Industrial energy price statistics; Table 3.2.1.)

the UK even became a net exporter of natural gas. Since then, however, domestic gas production has decreased, while gas imports (from Norway, the Netherlands, and Belgium) have increased, reaching 49% in 2019 (Figure 4.8). The use of gas in the electricity sector grew rapidly in the 1990s (Figure 4.9). Other important gas users are households and services[2] (which mostly use gas for heating) and industries (which use gas for a range of purposes).

4.2.2 Actors

Firms: Following privatisation (1990) and liberalisation (1998), the UK electricity supply industry consolidated into the 'Big Six' electricity companies (EDF, E.ON, SSE, British Gas, Scottish Power, N-Power). Their strategies came to focus on price competition, sweating assets, decreased R&D spending, and fuel flexibility in response to fuel price fluctuations (Pearson and Watson, 2012). In the absence of clear product differentiation, competition mainly occurs on costs, and to a lesser extent on consumer relations and green profiles.

[2] Services includes public administration, agriculture, and commercial & miscellaneous.

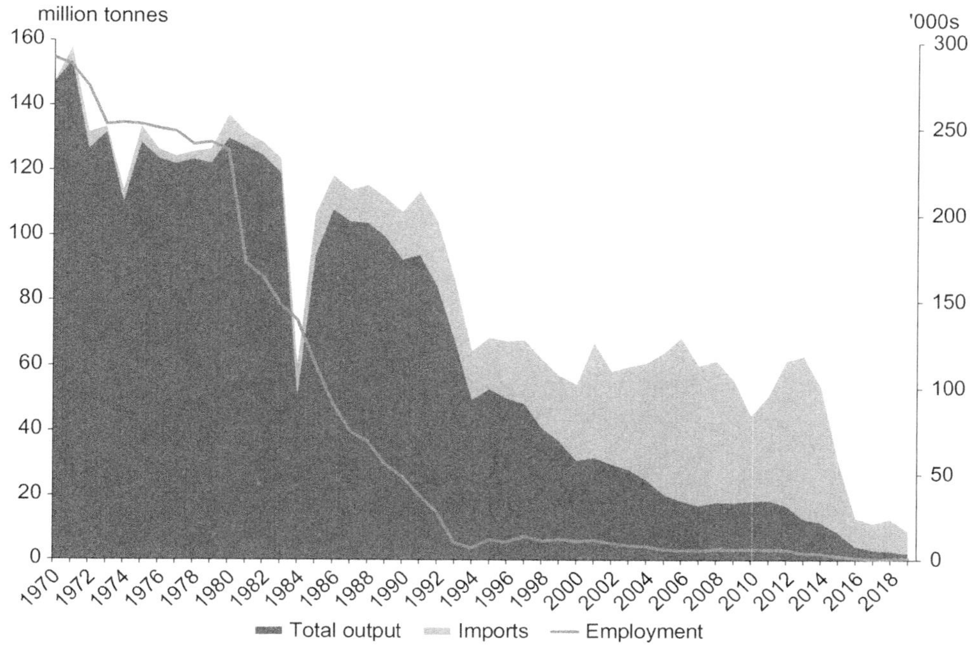

Figure 4.7 UK coal production and imports in million tonnes (left-hand axis) and employment in thousands (right-hand axis), 1970–2019 (constructed using data from Statistics at BEIS; Historical Coal Data Series)

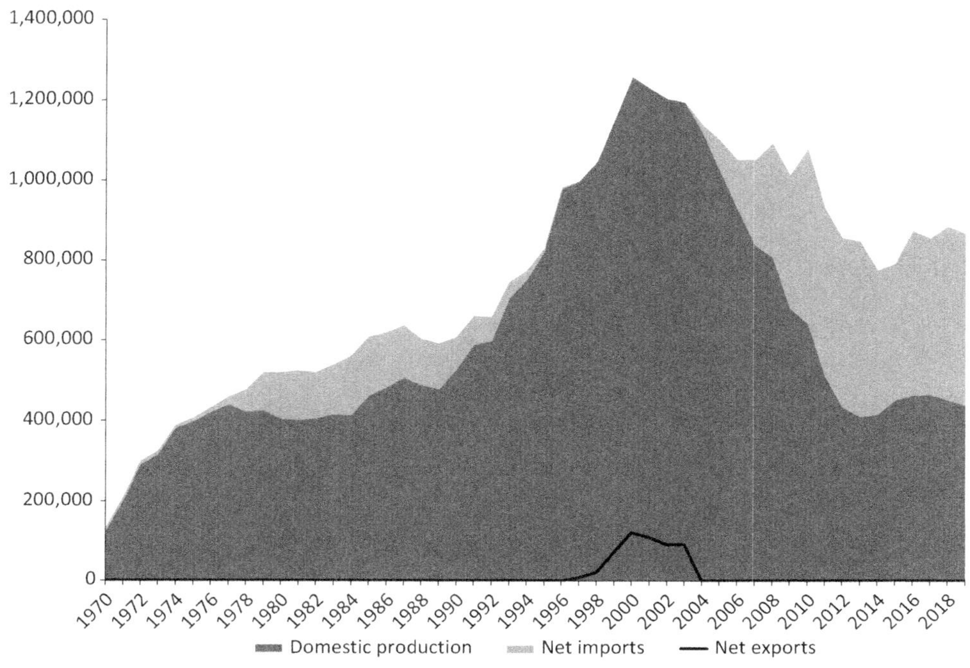

Figure 4.8 UK natural gas production, net imports, and net exports in GWh, 1970–2019 (constructed using data from Statistics at BEIS; Historical Gas Data Series)

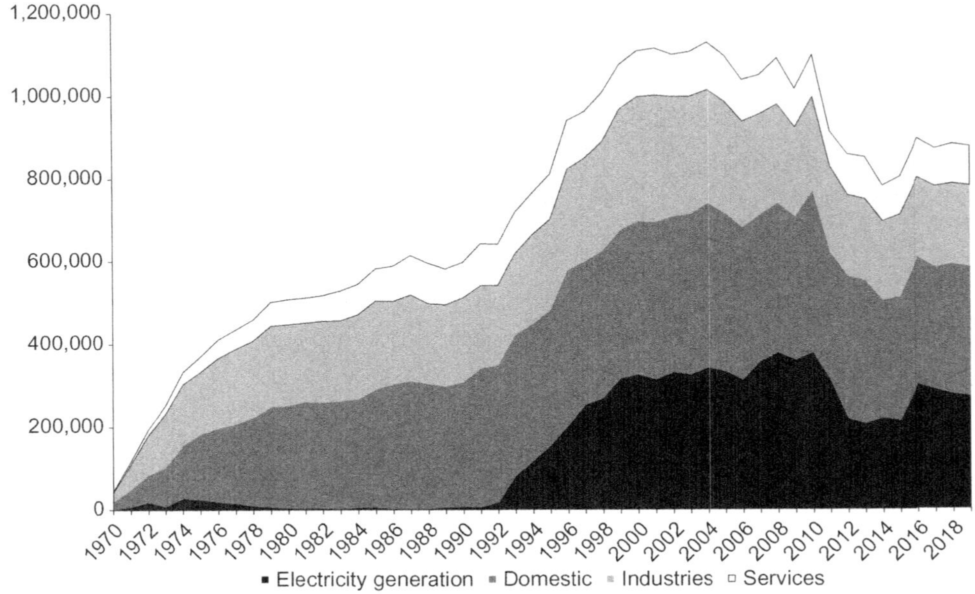

Figure 4.9 UK natural gas consumption by main user categories in GWh, 1970–2019 (constructed using data from Statistics at BEIS; Historical Gas Data Series)

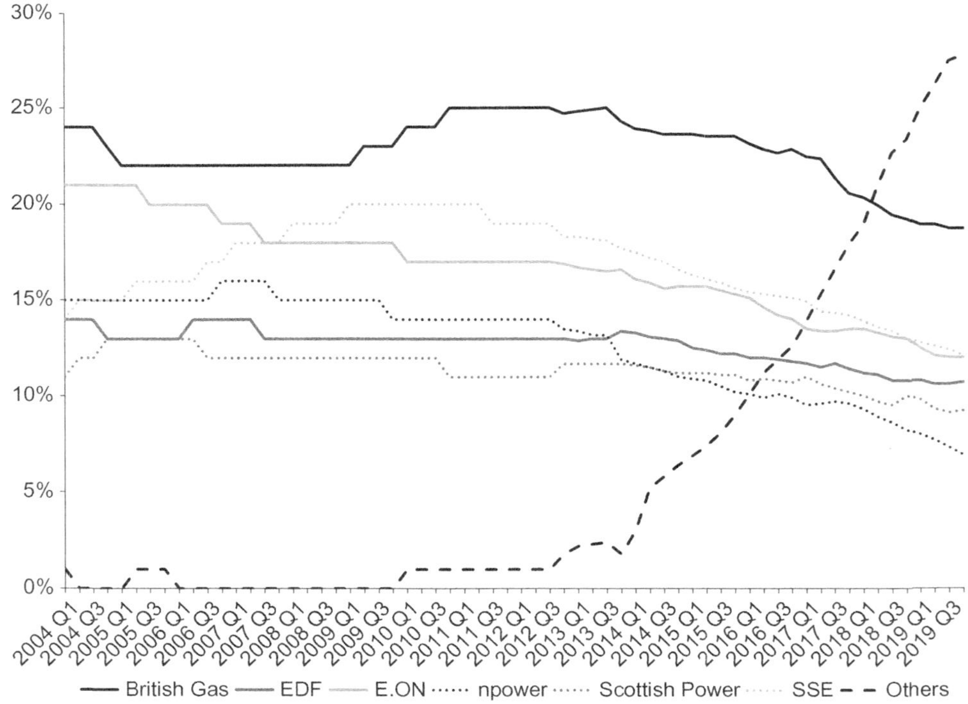

Figure 4.10 UK market share evolution of energy companies, 2004–2019 (constructed using data from Ofgem, available at www.ofgem.gov.uk/data-portal/electricity-supply-market-shares-company-domestic-gb)

The Big Six are vertically integrated (owning both electricity generation and retail) and dominate the market. But since 2014, new entrants (such as First Utility, Ovo Energy, Sainsbury's Energy) have begun to gain market share (Figure 4.10), which has increased competition. Organisations with new business models (e.g., community energy, transition towns) have remained small in the UK[3] because of unfavourable rules and institutions (Mirzania et al., 2019): 'key features of socio-technical regime for electricity provision continue to favor large corporations and major facilities' (Strachan et al., 2015: 106).

Technology strategies in recent decades have been shaped by public policies and economic considerations. In the 1990s, utilities switched from coal to *natural gas* (Figure 4.4). This 'dash for gas' was stimulated by various factors (Pearson and Watson, 2012): a) utility preferences for power generation units with short lead times, low capital costs, and quick returns on investment, which aligned well with combined cycle gas turbines (CCGT); b) price/performance improvements in CCGT; c) new North Sea gas finds and cheap international gas; d) environmental benefits of gas compared to coal.

Coal-fired power generation in the 1990s also faced pressures from the European Large Combustion Plants Directive (LCPD), which prescribed substantial reductions in SO2 emissions. Existing coal-fired plants therefore had to either invest heavily in flue gas desulphurisation, which would further erode their competitive position, or close by 2015 (Turnheim and Geels, 2012). Coal-fired plants also faced increasing pressure from climate change, which the newly elected (1997) Labour government saw as an important issue.

Nuclear power also faced difficulties in the 1990s because of nuclear waste storage scandals and because preparation for privatisation revealed its poor economic performance (Verhees, 2012). Privatisation was therefore delayed until 1996, when the government sold the nuclear plants to British Energy. But declining electricity prices created financial problems for British Energy, which had to be bailed out in 2002 (Hewlett, 2005).[4] EDF bought British Energy and its nuclear power plants in 2009 for £12.5 billion.

Rising gas prices in the early 2000s (Figure 4.6) changed the fortunes of *coal*, leading utilities to burn more coal in existing plants between 2000 and 2006 (Figure 4.4). Concerns about energy security (due to increasing reliance on gas imports) also increased the appeal of coal. Promises of 'clean coal' (based on flue gas desulfurisation, supercritical pulverised coal technologies, coal gasification, and Carbon

[3] Although there were more than 5,000 UK community energy groups in 2014, their cumulative renewable electricity generation capacity (60 MW) was small (DECC, 2014e), compared to 82.662 MW total capacity.

[4] The government also took responsibility for nuclear waste management and decommissioning costs of around £3 billion (Hewlett, 2005).

Capture and Storage) also intended to address environmental concern, including climate change (Geels et al., 2016b).

By 2008, utilities were seeking approval to build new coal-fired power plants, totalling over 11 GW, to replace plants that would be phased out by 2015/16 (Turnheim, 2012). Because these plans would threaten the ambitions of the 2008 Climate Change Act, policymakers announced that no new coal-fired plants would be permitted unless they incorporated Carbon Capture and Storage (CCS), which did not materialise. Since then, coal use has decreased rapidly (Isoaho and Markard, 2020), as some coal-fired plants converted to burning biomass (Drax, Ironbridge) while others closed (Kingsnorth, Cockenzie, Tilbury Didcot, Uskmouth, Ferrybridge) because of LCPD-legislation, end-of-life considerations, and the 2013 Carbon Floor Price policy, which made coal more expensive compared to other options. Decline accelerated after 2015 (Figure 4.4), when the government announced a coal phase-out by 2025 (Isoaho and Markard, 2020).

Increasing concerns about climate change and energy security (particularly increasing dependence on imported gas) also led to government plans for the expansion of *nuclear power* (see later). But utilities showed lukewarm interest, because waste processing liabilities, decommissioning costs, and unclear future electricity prices created uncertainties about the viability of nuclear investments, especially since the government had repeatedly ruled out subsidies. In 2013, the energy company Centrica abandoned new construction plans, leaving only EDF in negotiations with the UK government about a 3.2 GW plant at Hinkley Point C (Thomas, 2016). To enable the deal, the government broke its non-subsidy pledge and in 2016 agreed to pay EDF a guaranteed strike price (£92.50 per MWh, twice the wholesale price) for 35 years. Since then, however, the project has encountered problems in securing finance for the £18 billion investment and delays in scheduled opening dates (to 2025). Half of the UK's nuclear power plants are scheduled to retire in 2025 for end-of-life reasons, creating a potential capacity gap. Policymakers are therefore in discussion with possible suppliers about other new nuclear power plants (further discussed later).

In response to attractive government incentives, electricity generators also reoriented towards large-scale renewables such as biomass combustion in converted coal-plants (e.g., Drax, Ironbridge), onshore and offshore wind parks (Geels et al., 2016b), which is further discussed in Section 4.5.

Policymakers: The government's privatisation (1990) and liberalisation (1998) of the electricity supply industry were motivated by desires to unleash market forces and drive down electricity prices. The Labour government, elected in 1997, made climate change into an additional issue for energy policy besides low costs. The 2003 White Paper *Our Energy Future: Creating a Low-Carbon Economy* highlighted the need for a 60% reduction of GHG emissions by 2050 and committed to a target of 10% renewable electricity by 2010. Rising oil and gas prices in the early 2000s (Figure 4.6) and the 2005 Russia–Ukraine gas dispute (in which Russia closed gap supplies to pressure Ukraine) increased concerns about energy security and increasing reliance on gas imports (Figure 4.8). UK electricity policy was therefore increasingly framed in terms of an 'energy trilemma', which aimed to simultaneously address three goals: low cost, energy security, and climate mitigation (Kern et al., 2014a).

Public attention to climate change increased rapidly in the mid-2000s (Figure 1.1), making it an attractive issue for high-level politicians to compete on (Carter and Jacobs, 2014). This competition resulted in cross-party consensus about the importance of climate change. This consensus and pressure from environmental NGOs culminated in the ambitious 2008 Climate Change Act that legally committed the UK to 80% GHG reduction by 2050 and 34% reduction by 2020 (Lockwood, 2013). Nuclear power, Carbon Capture and Storage (CCS), and offshore wind (which will be discussed in Section 4.5) were seen as crucial low-carbon technologies.

The government's 2008 *White Paper on Nuclear Energy* announced plans to construct eight new nuclear power reactors by 2025, which would address both climate change and energy security issues. Since the mid-2010s, the government has spend much political capital on the first project (Hinkley Point C), which encountered delays, cost increases, and financing problems, and has been criticised for very costly support policies (Thomas, 2016).

CCS was seen as a crucial technology because it would enable continued gas and coal-fired power plants, while reducing GHG emissions. The government's 2007 *White Paper on Energy* (Meeting the Energy Challenge) wanted to 'make the UK a world leader in this globally important technology' (p. 15) and therefore launched a competition for a £1 billion subsidy to build a CCS demonstration plant. Nine proposals were received and four were selected for further development and negotiation. The BP and Peel Consortiums subsequently withdrew, leaving E.ON and Scottish Power/Shell/National Grid in the negotiation. But E.ON pulled out in 2009 and Scottish Power in 2011, because both deemed the economic risks too high (Kern et al., 2016). Despite these setbacks, the government's 2011 *Carbon Plan* repeated CCS aspirations, foreseeing up to 10 GW of CCS plants by 2030. A second £1 billion competition was launched in 2012, leading to prolonged discussions about the Peterhead project (led by Shell and SSE) and the White Rose project (led by Drax). But the government cancelled the competition in 2016, owing to concerns about the future costs for consumers.

Meanwhile several *political* countertrends gathered pace after the 2007/8 financial-economic crisis and the election of a new Conservative-Liberal Democrat government in 2010 altered political priorities. Public attention to climate change diminished, leading politicians to realise that they were ahead of their voters (Lockwood, 2013). Especially the right wing of the Conservative party became more vocal, questioning the climate change science and criticising subsidies for renewables. The financial-economic crisis also enhanced concerns about jobs, competitiveness, and energy prices. The Treasury used these concerns to regain influence over climate policy through the Levy Control Framework (Carter and Jacobs, 2014), issuing warnings that green policies should not hinder the economy. In 2013, cost concerns escalated into a full-scale political row over rising energy bills, with the Prime Minister reportedly telling aides to 'get rid of all the green crap' (*The Guardian*, 23 November 2013). These concerns led the government to scrap, delay, or water down various green policies (discussed later).

In this context, the government was also keen to emulate the US shale gas revolution, which it hoped would create jobs, reduce gas prices, and reduce dependence on natural gas imports. In 2012, it therefore lifted restrictions on fracking. In an official statement on 19 July 2013, the Chancellor promised tax breaks for shale gas companies, arguing that shale gas 'has the potential to create thousands of jobs and keep energy bills low for millions of people'. Since then, shale gas has developed slowly because of public protests, mixed results from underground explorations (because UK soils were found to be fractured and difficult to exploit), and correlations between fracking and local earthquakes that heightened public concerns (Williams and Sovacool, 2019). In November 2019, the government therefore announced a moratorium on further shale gas development.

In 2015, the newly elected Conservative Government announced an 'energy policy reset', which led to major reductions in financial support for renewables and CCS. To protect its green credentials, the government also announced in 2015 that it intended to phase-out unabated coal-fired power plants by 2025. This phase-out unfolded quicker than anticipated (Figure 4.4) and substantially decreased GHG emissions from the power sector (Figure 4.2). Since the mid-2010s, the government has also started negotiations about other new nuclear power plants, but several of these (Wylfa, Moorside, and Oldbury) have since stalled because project developers (Hitachi, Toshiba) pulled out in 2020 due to problems in securing funding. Negotiations about other plants (Sizewell, Bradwell) are ongoing but challenging because decreasing costs of renewables erode the business case for nuclear power plants (and the political will to subsidise them), creating major business and political uncertainties. Despite the various implementation and delivery problems, the 2020 *Energy White Paper* states that the government aims 'to bring at least one large-scale-nuclear project to the point of Final Investment Decision (FID) by the end of this Parliament', which is 2024 (HM Government, 2020a: 48).

In the context of 'climate emergency' debates, the UK government committed in 2019 to a net-zero target by 2050. For the electricity sector specifically, the 2020 *Energy White Paper* aimed for 'an overwhelmingly decarbonised power system in the 2030s'. Nuclear power and offshore wind (further discussed later) are thought to play important roles in this decarbonisation process, while the government also aims for 'at least one' CCS plant for power production to be operational by 2030 (HM Government, 2020a).

Users: UK consumers have limited direct involvement in power generation. Electricity production by households (through rooftop solar-PV) has remained relatively small in the UK (see Section 4.5.4). Nevertheless, households indirectly shape upstream power generation investments since they ultimately pay for them, either through their electricity bills (which allows utilities to pass costs onto consumers) or through general taxation (which pays for government subsidies to power generators). Additionally, consumer switching between electricity suppliers accelerated after 2013, providing space for new electricity suppliers (Figure 4.10).

Wider Publics: Increasing public attention for climate change and the 2006 Big Ask campaign by environmental NGOs prepared the ground for the 2008 Climate Change Act and stimulated the development of low-carbon electricity plans and policies. Public attention for climate change declined as the 2007/8 financial crisis and austerity increased public concerns about jobs, growth, and energy costs (Lockwood, 2013). Cost concerns underpinned various public debates about specific issues. Complaints about rising electricity bills culminated in a political row in 2013 and subsequent efforts to reduce public spending on renewables. There were also heated debates about pricing strategies of utilities, which were accused of being slow to pass on decreases in fuel input prices to consumers. They were also accused of too rapidly increasing standard variable tariffs, which would disproportionally affect lower-income people on these tariffs who were traditionally less likely to switch suppliers. Additionally, there were public debates about dysfunctional markets and insufficient competition between utilities that underpinned the market power abuse. This concern led to an energy market investigation by the Competition and Markets Authority, which in 2016 published a critical report with 30 improvement recommendations. Last, there were critical debates about excessive subsidies in the 2016 Hinkley Point C deal. These critical debates decreased public attention to climate change and weakened the climate policy consensus (Gillard, 2016).

There were also public campaigns and debates about specific technologies. In 2008–2009, activist groups such as Climate Camp campaigned against utilities' plans to expand coal-fired power stations. Demonstrations against a new plant at Kingsnorth attracted much media attention, which pressured the government to not grant licenses (Carter and Jacobs, 2014). Public opposition against nuclear expansion plans was more limited because the environmental movement was divided, with some activists (e.g., Stephen Tindale, George Monbiot, Mark Lynas) perceiving nuclear power as a necessary evil to address climate change (Verhees, 2012).

Environmental NGOs and local communities also contested the government's plans for fracking and shale gas technologies because of concerns about water and noise pollution, industries invading the countryside, and insufficient stakeholder engagement (Williams and Sovacool, 2019). Nevertheless, the government decided to move ahead in supporting fracking, with the Prime Minister personally expressing strong commitment in a letter to *The Telegraph*, dismissing protesters as uninformed NIMBY-activists (11 August 2013). Protests continued, however, and gathered wider support as increasing numbers of earthquakes in fracking areas enhanced public concerns. Combined with lower than anticipated exploration results, this led the government to halt shale gas in November 2019.

4.2.3 Policies and Governance

Governance Style

Governance styles and policy paradigms have changed substantially in the past three decades. Privatisation and liberalisation in the 1990s, which were based on neoliberal principles, introduced market-principles to the electricity system, with the specific aim of focusing utilities on low cost competition (Pearson and Watson, 2012). The government increasingly adopted a hands-off governance style, leaving decisions to the market. The Department of Energy was disbanded in 1992 and energy policy was relegated to a sub-division of the Department of Trade and Industry (DTI). To depoliticise energy governance, DTI set the regulatory framework, but left implementation to the newly created independent regulator Ofgem (Office of Gas and Electricity Markets). Ofgem's main remit was to ensure that markets were sufficiently competitive and to protect consumer interests (Kern et al., 2014a).

In the 2000s, climate change became an additional policy concern, which was layered on top of neoliberal arrangements, leading to an emphasis on market-based policies. The 2002 Renewables Obligation introduced technology-neutral trading policies, while the 2003 White Paper *Our Energy Future* emphasised carbon pricing (via European emissions trading) as the main instrument for creating a low-carbon economy.

The 2008 Climate Change Act marked a shift towards a more interventionist governance style (Kern et al., 2014a; Pearson and Watson, 2012), in which the government actively shaped markets and stimulated specific technologies. It also created new policy actors such as the Department of Energy and Climate Change (DECC) and the independent Committee on Climate Change (CCC) with responsibility for monitoring progress against climate change targets and advising the government accordingly. DECC's translation of high-level goals into more specific plans created policy delivery momentum through the UK Low Carbon Transition Plan (2009), the amended Renewables Obligation (2009), the UK Renewable Energy Strategy (2009), the Carbon Plan (2011), the Energy Bill (2012), and the Electricity Market Reform (2013), which stimulated technology implementation and deployment.

Formal Policies

In terms of specific policy instruments, the 2013 Electricity Market Reform introduced Contracts for Difference (from 2017 onwards), which aimed to attract private investors. CfDs guarantee that low-carbon electricity generators receive a stable and predictable 'strike price' for long periods.[5] While CfDs offer protection against the volatility of wholesale prices, the closed auction design makes potential suppliers compete against each other with the aim of driving prices down. The Hinkley Point C nuclear power plant, however, received a generous strike price (£92.50 per MWh for 35 years) to entice EDF as single bidder (Thomas, 2016).

The Carbon Price Floor (CPF) is another policy instrument, which was introduced in 2013 to complement the EU Emissions Trading System (ETS). Since the price of carbon in the ETS has remained too low to drive low-carbon investment, the CPF taxes fossil fuels used to generate electricity via Carbon Price Support (CPS) rates set by the government. The Carbon Price Floor thus consists of two components paid by electricity generators: ETS carbon prices and the CPS which further increases carbon prices to the carbon floor price target. The CPF was supposed to increase every year until 2020 (to a price of £30/tCO_2), but in 2014 the government decided to cap the CPS component of the floor price at a maximum of £18/tCO_2 from 2016 to 2020 to reduce energy bills for consumers. Despite this weakening, the CPF increased the relative price of coal, which contributed to its rapid decline in power generation (Figure 4.4).

Another instrument was the Levy Control Framework, which the Treasury established in 2011 to control financial spending by DECC. It did so by setting a maximum annual budget for projected costs on levy-funded schemes such as the Renewables Obligation, Feed-in-Tariff, and CfDs.

The 2015 energy reset not only reduced financial support for renewables and CCS but also signalled a desire for less interventionism, with the Secretary of State

[5] If the wholesale electricity price is below the agreed 'strike price', the generator receives a top-up payment to make up for the difference. If the wholesale price is above the strike price, the generator pays back the surplus.

for Energy and Climate Change stating: 'We want to see a competitive electricity market, with government out of the way as much as possible, by 2025.'[6] The newly elected (2015) Conservative government thus expressed its preference for restoring a neoliberal approach to electricity policy. In 2016, DECC was reorganised into BEIS (the Department for Business, Energy & Industrial Strategy), which also indicated that low-carbon energy transitions were explicitly viewed in relation to business opportunities and industrial strategy. The 2017 Clean Growth Strategy saw 'clean, smart, flexible power' as an industrial growth opportunity, focused particularly on large-scale generation options such as nuclear power, offshore wind, and large solar farms (BEIS, 2017a). The latter two options are further discussed in Sections 4.5.2 and 4.5.4.

4.3 Electricity Grid Sub-system

4.3.1 Techno-Economic Developments

Great Britain's electricity grid[7] includes a high-voltage *transmission* network, which carries electricity from power generators to sub-stations and a low-voltage *distribution* network for localised electricity delivery from sub-stations to end-users. The GB transmission network consists of 26,000 km of overhead lines, 575 sub-stations, and over 1,000 transformers that transform electricity from high to low voltage (Cotton and Devine-Wright, 2012).

Local distribution networks are organised into 14 regional area monopolies, which are managed by distribution network operators (DNOs). DNOs operate one-directional passive networks, which distribute power from electricity generators to end-users. DNOs do not measure or monitor their distribution networks, which remain relatively 'dark': DNOs cannot see technical problems or blackouts but rely on customers calling them to report problems (Lockwood, 2016).

The electricity grid took decades to build and represents major sunk investments that create material and economic path dependencies. Grid operation requires specialised technical and managerial skills to balance the supply and demand of electricity flows, which always needs to be finely tuned to prevent operational problems and blackouts. The operational model traditionally consisted of baseload generation, which operates more or less continuously, and dispatchable generation, which are more flexible sources of electricity that can be dispatched at the request of power grid operators to meet fluctuating consumer demand, including electricity

[6] www.gov.uk/government/speeches/amber-rudds-speech-on-a-new-direction-for-uk-energy-policy

[7] The electricity grid includes England, Wales, and Scotland but not Northern Ireland, which has its own grid.

peaks. To keep costs low, choices for baseload and dispatchable operation were conventionally guided by the merit order, which refers to the ranking of electricity generation sources in ascending order of short-run marginal production costs. This often resulted in the use of coal or nuclear plants for base-load generation and gas-powered plants for dispatchable generation.

Several trends are increasing pressures on the grid and the conventional operational model (Bolton and Foxon, 2015; Jenkins et al., 2015): 1) under-investment over the past few decades has led to aging grid assets (e.g., switchgear, transformers, cables) that require replacing or upgrading, 2) the creation of wind farms in remote locations (e.g., Scottish islands, Welsh coast, offshore) requires the creation of new transmission networks to connect them to the grid, 3) increasing electricity flows from Scotland and Wales (where most wind parks are situated) to England (where most electricity is used) requires upgrading, extension, and intensification of the onshore transmission grid, 4) increasing amounts of intermittent renewables (wind, solar-PV) create load management problems (matching supply and demand) and disrupt the baseload-dispatchable generation model, especially as renewables become the cheapest option and thus rank high in the merit order, 5) increasing amounts of distributed generation (e.g., solar-PV, community energy) need to be integrated into local distribution grids, which involves two-way flows instead of traditional one-directional flows, 6) possible future increases in domestic heat pumps and electric vehicle charging may create new stresses on local distribution networks that require monitoring and management.

These pressures have led to incremental changes in the high-voltage transmission grid, including: 1) extensions of Scottish onshore power cables to wind parks in remote locations; 2) the strengthening of existing transmission connections between England and Scotland and England and Wales, 3) the creation of a new west coast undersea high-voltage direct current transmission cable between Scotland and England, 4) the creation of new offshore grids to connect various wind parks, 5) the building of new interconnectors to Norway, France, Belgium, Denmark, and Iceland to increase import capacity beyond the current four interconnectors (to France, Northern Ireland, Ireland, the Netherlands). Costs for these projects between 2010 and 2020 are estimated to be around £54 billion (DECC, 2014a). Although these projects are complicated and expensive, they are incremental in the sense that they build on and extend existing technological knowledge and capabilities (Andersen, 2014). They also do not fundamentally change the transmission architecture but strengthen and extend it.

The grid pressures have also increased attention for more radical innovations in local distribution networks, including smart grids, demand-side response, and electricity storage. These radical innovations will be discussed in Sections 4.5.7 and 4.5.8.

4.3.2 Actors[8]

Policymakers: Electricity networks are highly regulated markets, in which the independent regulator Ofgem plays a central role. It provides oversight of the system operator (National Grid), Transmission Network Operators (TNOs), and Distribution Network Operators (DNOs); it implements and monitors regulations; and it approves network investment plans. Ofgem is traditionally dominated by mainstream economists (Cary, 2010; Lockwood, 2016) and focused on lowering cost (through economic competition), which is how it interpreted its original regulatory remit of 'protecting the interest of consumers'.

To minimise operational costs and improve cost-efficiency, Ofgem introduced price control regulation (called RPI-X) for electricity grids in the late 1990s. This regulation meant that the fees that TNOs and DNOs could charge electricity companies for transmission and distribution services could increase in line with an inflation index (the Retail Price Index) minus an X% reduction each year, which was intended to stimulate TNOs and DNOs to make efficiency gains (Jamasb and Pollitt, 2007).

Although mainstream economic thinking, which informed Ofgem's policies, predicted that efficiency orientations would drive innovation, this did not occur, especially not for DNOs (Bolton and Foxon, 2015). To stimulate innovation, Ofgem therefore introduced new policies for the 2005–2010 period (Innovation Funding Incentive, Registered Power Zones scheme) and the 2010–2015 period (Low Carbon Network Fund), and a new policy framework called RIIO for the post-2015 period (RIIO stands for Revenue = Incentives + Innovation + Outputs). While these new instruments stimulated DNO R&D activities (Jamasb and Pollitt, 2015), wider deployment of new technologies in distribution networks has so far remained relatively limited (Lockwood, 2016).

For grid investments, Ofgem used to have a negotiated model, in which DNOs and TNOs could make proposals that legitimated technical details and costs with regard to *demonstrated needs* (Lockwood, 2016). This approach of 'wait for proven need and then choose the optimal solution' (Shaw et al., 2010) might be efficient according to economic theory, but in the real world discouraged radical innovations that did not address well-articulated needs. In the mid-2010s, Ofgem announced that it wanted to change from the negotiated model to a competitive tendering model (the CATO regime), which was intended to commence in 2018 (see further discussion later). This change was partly inspired by experiences with the creation of new offshore transmission grids, which began using a competitive tendering model in 2009 (further discussed later).

Climate change was incorporated into Ofgem's remit in the early 2000s, but has long received far less attention than efficiency improvements and cost reduction.

[8] This section does not discuss consumers because they are not actively involved in electricity grids.

Since 2007, Ofgem has been repeatedly criticised (by the Sustainable Development Commission, the Labour Party, and Parliament's Energy and Climate Change Select Committee) for insufficiently acting on this additional goal. These criticisms have only slowly been accommodated, because Ofgem was created as an independent organisation with a substantial degree of autonomy and discretion in relation to policymakers (Lockwood, 2016).

Firms: Different kinds of companies operate different parts of the electricity grid. The transmission grid is managed by a single system operator (National Grid) and three regional Transmission Network Operators (TNOs). Most of the extension, intensification, and reinforcement of onshore transmission networks has been driven by the three TNOs and supply chain firms (e.g., ABB, Alstom), based on investment proposals approved by Ofgem. Since pressures on transmission grids were visible and imminent (e.g., connecting new wind parks, reinforcing grids connections between England and Scotland), these proposals could easily be framed in terms of demonstrated needs. But besides clear demand or regulatory pressure from Ofgem, TNOs had little incentive to innovate since they faced no competition (Jamasb and Pollitt, 2015). Most transmission grid innovations have therefore been incremental. The new CATO regime intends to change that by creating competitive tendering processes for discrete infrastructure projects. The new regime would also give the System Operator greater power in terms of overall coordination (as a 'system architect').

The local distribution system is organised into 14 regional area monopolies, run by six Distribution Network Operators (DNOs). DNOs are passive distributors, who transmit power from sub-stations to end-users. They receive a fee from the electricity companies for transmitting this power but do not have direct commercial relations with end-users. Various DNO-related lock-in mechanisms help explain the limited degree of innovation in distribution networks: 1) the RPI-X regulatory regime's focus on efficiency and short-term cost reduction stimulated TNOs and DNOs to 'sweat the assets' (by postponing network investments) and downscale R&D investments, which decreased to 0.1% of revenue by 2004 (Jamasb and Pollitt, 2008), 2) DNOs have therefore lost technical capabilities and lack the incentives for major long-term innovations (Bolton and Foxon, 2015), 3) DNOs do not have pro-active long-term innovation strategies but react to the regulatory contract or act when this is required (e.g., when distributed power generators seek connections to the distribution network), 4) DNOs also do not face much articulated need from concrete clients, which complicates the legitimation of investment proposals to Ofgem.

DNOs increased their R&D activities in response to Ofgem's Innovation Funding Incentive and Registered Power Zones scheme (Jamasb and Pollitt, 2015), but real-world implementation of new technologies has remained slow. Since the Low Carbon Network Fund and the RIIO policy framework, DNOs have become more engaged in larger, real-world demonstration projects.

A third actor, Offshore Transmission Owners (OFTOs), are consortia of large-scale investors, project developers, and construction companies that build and operate offshore transmission networks (Firestone et al., 2018). They have been attracted by Ofgem's regulations, in which attractive revenues are awarded for providing the availability of transmission infrastructure regardless of the amount of electricity that is generated and transmitted. Two dominant OFTOs are *Transmission Capital Partners* (including Transmission Capital, International Public Partnerships, and Amber Infrastructure Group) and *Blue Transmission* (including 3i Group Plc and Diamond Transmission Corporation, which is a UK subsidiary of Mitsubishi Corporation).

Wider Publics: While grid planning and decision-making is a technocratic process involving a small group of actors, on-the-ground implementation and construction affects the lives of citizens and local communities. Various infrastructure projects have encountered protests because residents and NGOs had concerns about various issues (Cotton and Devine-Wright, 2012): 1) pylons and overhead power lines that caused visual intrusion in rural and suburban landscapes and noise burdens (related to zooming sounds), 2) negative influences on property and local amenity values, 3) potential cancer risks due to electric and magnetic fields emitted by power lines, 4) distrust of large electricity companies, including National Grid, 5) limited local consultation causing feelings of being 'bulldozed over'. Examples of protests against grid infrastructure projects include the following (Cotton and Devine-Wright, 2013). In the early 2010s the John Muir Trust, a wild land charity, led fierce protests against the creation of new pylons and wires across 220 km of Scottish Highlands, which attracted much public attention. In Suffolk and Essex protesters created the Essex & Suffolk Coalition of Amenity Groups, whose protests led National Grid to decide (in 2013) to postpone its plans for new pylons until the early 2020s. There were also prolonged protests between 2011 and 2014 against new power lines from Mid-Wales through the Shropshire countryside.

The protests in Scotland and Wales led to substantial delays in consultation, approval, and construction of grid projects:

> Major delays of 2 to 4 years were announced late in 2012 for many projects in Northern Scotland and the reinforcements required in mid and north Wales remain behind schedule. Our indicators envisaged that construction would begin in 2012 (mid Wales) and this year (north Wales), but there have been continued delays in planning, largely due to local public opposition. (CCC, 2013: 92)

These social acceptance problems were one reason for constructing the west-coast undersea transmission cable between Scotland and England and considering a similar east-coast undersea cable.

4.3.3 Policies and Governance

Governance Style

The governance style in the electricity grid sub-system has characteristics of 'club governance' (Lockwood, 2016; Moran, 2003): actors meet frequently in relatively closed networks, know each other well, share mindsets and outlooks, and take each other's interest into account. The Electricity Networks Strategy Group, for instance, provides a high-level forum where the National Grid, TNOs, DNOs, Ofgem, and policymakers (e.g., DECC, BEIS) meet to discuss electricity grid challenges, policies, and plans. Ofgem has strongly shaped the grid institutions and outlooks, leading to an emphasis on efficiency and costs rather than on transformative change.

Formal Policies

The efficiency-oriented RPI-X price control regulation, introduced in the late 1990s, succeeded in decreasing costs but also hampered innovation and reduced R&D investments, especially by DNOs (Bolton and Foxon, 2015). To address this problem, Ofgem introduced two new innovation-oriented instruments for the 2005–2010 period, which were layered on top of the efficiency-oriented institutions (Lockwood, 2016). The Innovation Funding Incentive (IFI) allowed DNOs to spend up to 0.5% of their revenue on R&D and distribution system asset management. The Registered Power Zones (RPZ) scheme provided additional revenue (capped at £500,000 per DNO per year) to demonstrate innovative solutions for connecting distributed generation facilities to the network.

Although these instruments increased DNO R&D spending in subsequent years (Bolton and Foxon, 2015), they hardly influenced the broader implementation and deployment of new technologies (Lockwood, 2016). For the 2010–2015 period, Ofgem therefore created a new Low Carbon Network Fund (LCNF), which was an order of magnitude larger than IFI and allowed DNOs to bid for up to £500 million over five years for demonstration projects.

Based on an internal review, Ofgem (2013) also introduced the new RIIO-framework that aimed to stimulate innovation and promote a 'step-change' in the prominence of low-carbon futures. Coming into force in 2015, RIIO introduced several new instruments: a Network Innovation Competition, in which DNOs can bid for large-scale projects (funded from a £70m per year pot); a Network Innovation Allowance, which is an allowance each RIIO network licensee receives to fund smaller scale innovative projects that have the potential to deliver benefits to network customers; and an Innovation Roll-out Mechanism to fund the roll-out of proven low-carbon innovations for transmission owners (up to £10m).

While the RIIO-framework introduced several changes, Lockwood (2016: 120) diagnosed that it 'retained the basic structure of revenue cap regulation at its core'. More broadly, it is not guaranteed that these new instruments will be sufficient to drive actual uptake and widespread deployment of new innovations in the electricity grid:

> The understanding of innovation processes within Ofgem has evolved over time but remains incomplete. … Despite a shift away from a purist view of innovation based on Austrian economics to a more nuanced approach, it remains … to be seen if … the incentive to reduce costs in the wider regulatory framework will now be sufficient to drive network companies to take the lessons learned in LCNF trials and apply them in business-as-usual network planning, investment and operation. (Lockwood, 2016: 124–125)

Ofgem's belief in market competition also led to the introduction of competitive bidding schemes for transmission grid projects. For offshore transmission projects, DECC and Ofgem introduced the Offshore Transmission Regulatory Regime in 2009. The regulations distinguish between a 'transitional regime' (2009–2012), which forced offshore wind farm operators to sell their self-constructed transmission grids to Offshore Transmission Owners (OFTOs), and an 'enduring regime' (post-2012), in which offshore transmission grids were either built directly by OFTOs or transferred to OFTOs once construction was completed. Licenses to build and/or operate offshore grids were distributed through a competitive tendering process overseen by Ofgem. To attract private investments, the OFTO license regulations are deliberately appealing, offering investors a solid fixed 20-year return on a relatively low risk profile, underwritten by a stable regulatory framework (KPMG, 2012).

Following the 2012–2015 ITPR policy review (Integrated Transmission Planning and Regulation), Ofgem announced that it wanted to change its onshore transmission regulations from the negotiated model to a competitive tendering model, known as the Competitively Appointed Transmission Owner (CATO) regime. By providing stable, long-term (25-years), financially attractive revenues, the CATO regime aims to mobilise large amounts of private investment (e.g., from the financial community) for discrete infrastructure projects with expected capital expenditures of over £100 million. The competitive tendering process also aims to keep costs low, allowing TNOs, construction companies, and other companies to bid for the design, financing, construction, ownership, and operation of onshore transmission grid assets. Although the new CATO regime was supposed to commence in 2018, Ofgem announced in June 2017 that it had to be postponed because, in the aftermath of the 2016 Brexit decision, Parliament had been unable to make the necessary legislative changes (Ofgem, 2017).

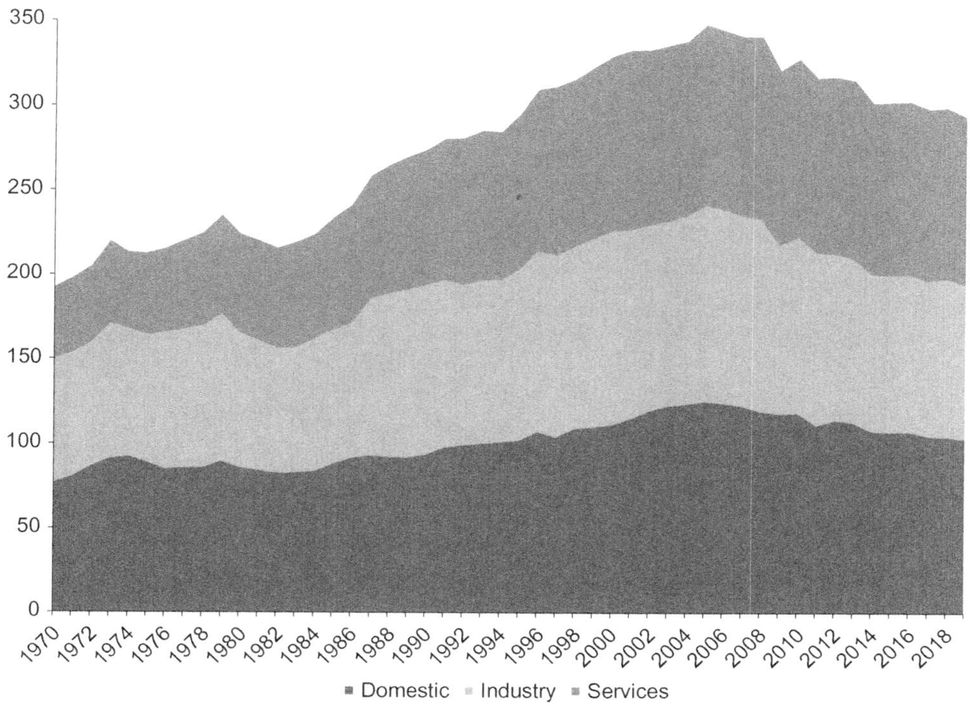

Figure 4.11 Final electricity consumption by different sectors in TWh, 1970–2019 (constructed using data from DUKES: Electricity; Electricity supply, availability, and consumption; Table 5.1.2)

4.4 Electricity Consumption Sub-system

4.4.1 Techno-Economic Developments

Electricity consumption (Figure 4.11) peaked in 2005 and then declined by 15% to 2019, owing to the combined effect of energy efficiency innovations, the financial crisis (and austerity policies), and offshoring which reduced industrial demand (Hardt et al., 2018, 2017). Although electricity consumption by households, industry, and services[9] is of roughly equal size, we focus here on domestic consumption, because of our interest in social practices and end-use functionalities.

The number of appliances in UK households has increased continuously since the 1970s (BEIS, 2019; Figure 4.12) because of the introduction of new products (e.g., juicers, microwaves, mobile phones, television flat screens), more use of existing products for different purposes (e.g., more lightbulbs to create ambiance), multiple household ownership of some appliances (e.g., fridges, TVs, computers),

[9] Services include public administration, agriculture, and commercial and miscellaneous.

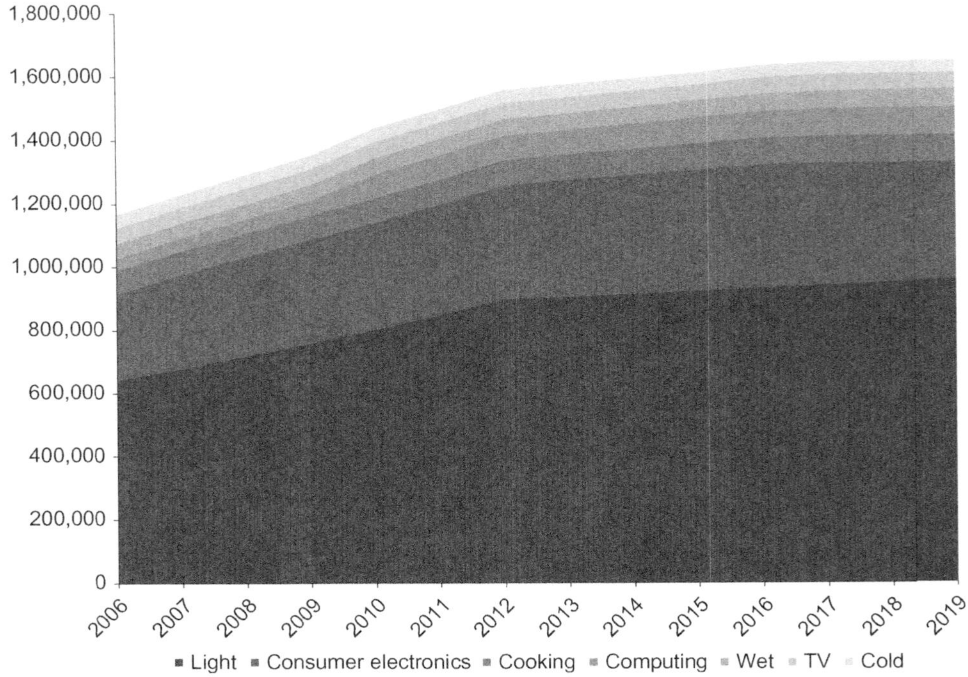

Figure 4.12 Total number of electrical appliances owned by UK households in thousands, 2006–2019 (constructed using data from Statistics at BEIS; Energy Consumption in the UK 2020; Electrical Products Data Tables; Table A2)

and increasing affordability due to manufacturing efficiencies and cost decreases (EST, 2011; McMeekin et al., 2019; Van Buskirk et al., 2014).

Household electricity consumption increased until 2005, but then declined by 17% until 2019 (Figure 4.11), despite increasing appliance use (Figure 4.12) by an increasing population. Electricity consumption by various domestic appliances has evolved in different ways in the last four decades (Figure 4.13), reflecting different diffusion patterns and innovation trajectories.

Electricity use for lighting increased until the mid-2000s, owing to expanding numbers of lightbulbs in UK households. Between 2007 and 2015, however, electricity use for lighting decreased by 38% (Figure 4.13), owing to a technological shift from ILBs (incandescent light bulbs) to CFLs (compact fluorescent lighting), halogen bulbs, and LEDs (light emitting diodes) (Figure 4.14). This unfolding transition (which is further analysed in Section 4.5.5) has reduced electricity consumption because the new technologies are more energy efficient than ILBs, which convert only 5% of electrical energy into light (Aman et al., 2013).

Electricity use by cold appliances started to decrease in the 1990s, despite continued proliferation of refrigerators, fridge-freezers, chest freezers, and upright

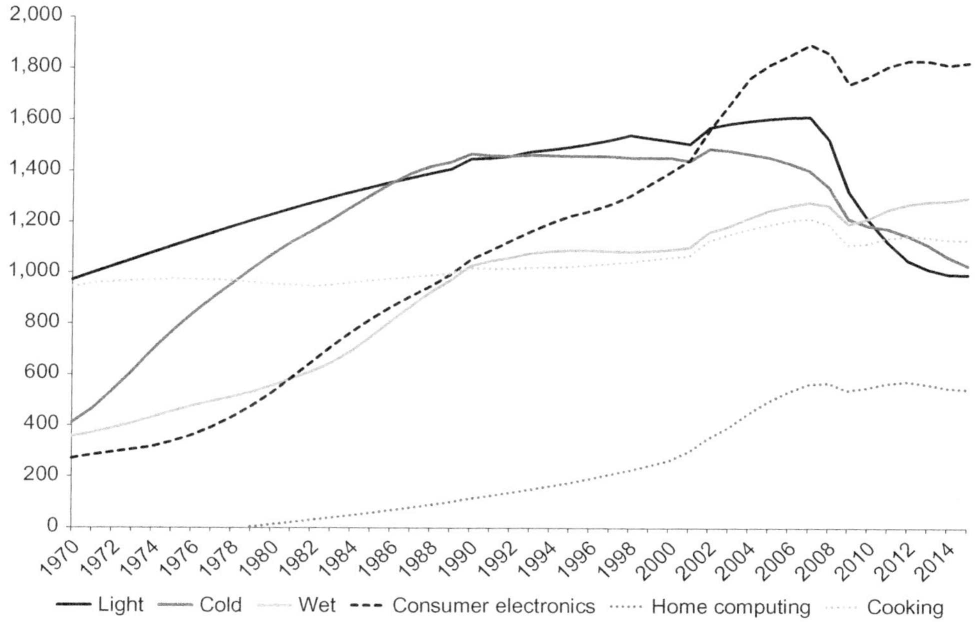

Figure 4.13 UK domestic electricity consumption by appliance category (1970–2015) in kilotons of oil equivalent (constructed using data from DUKES; Energy Consumption in the UK 2016; Electrical products tables; Table 3.08)

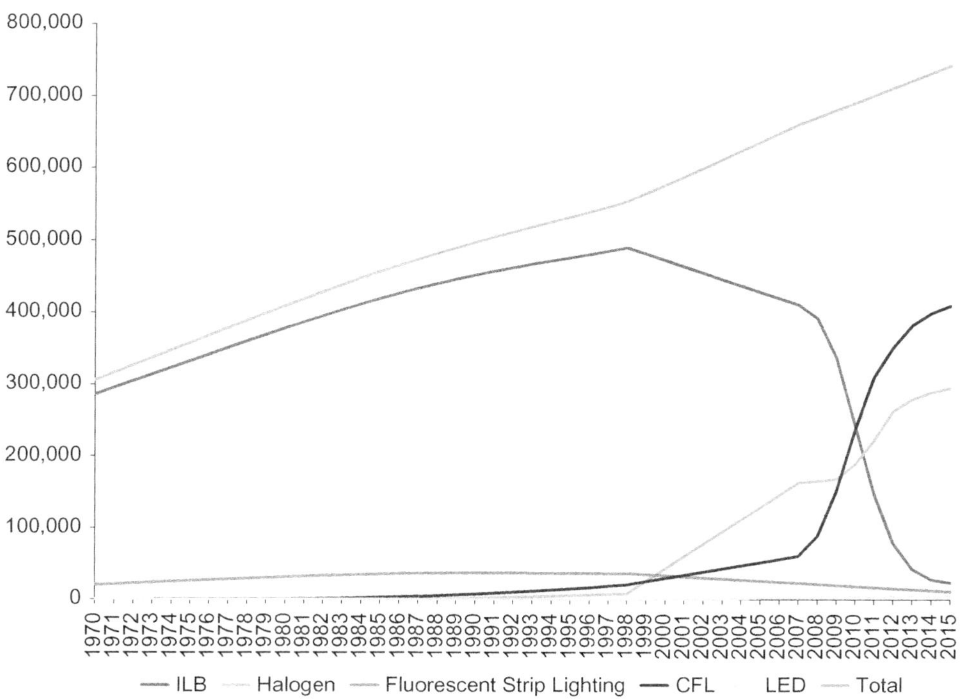

Figure 4.14 Number of light bulbs (in thousands) owned by UK households, 1970–2015 (constructed using data from DUKES; Energy Consumption in the UK 2016; Electrical products tables; Table 3.12)

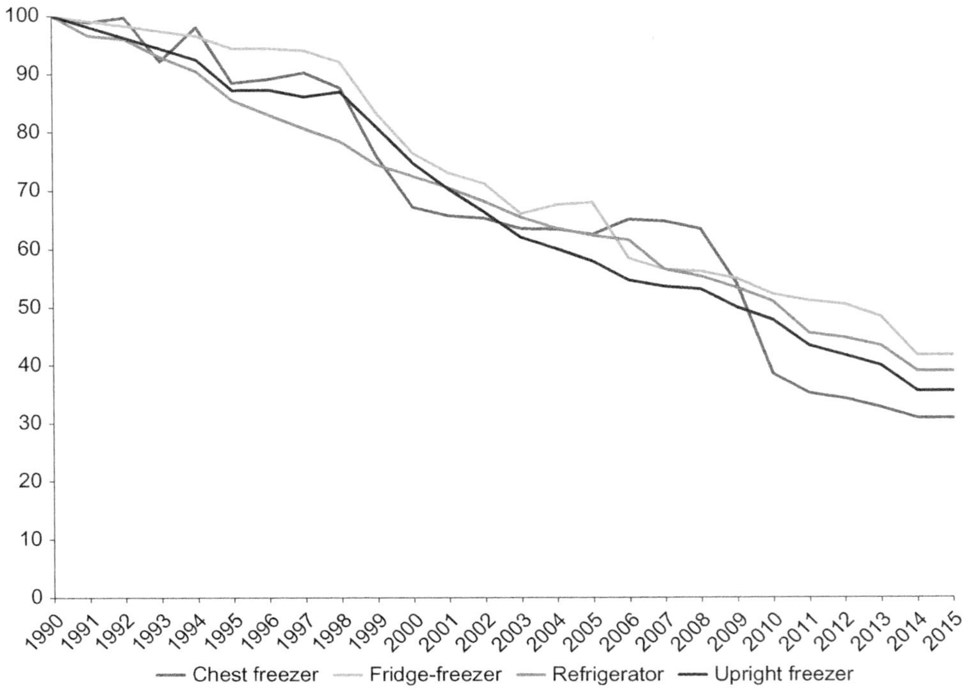

Figure 4.15 Average energy consumption of new cold appliances, 1990–2015 (index 1990=100) (constructed using data from DUKES; Energy Consumption in the UK 2016; Electrical products tables; Table 3.09)

freezers. This decrease was driven by impressive incremental energy efficiency innovations that reduced electricity consumption for new appliances by 59–69% between 1990 and 2015 (Figure 4.15).

Electricity use by wet appliances increased in the 1970s and 1980s, plateaued in the 1990s, but has increased further since the early 2000s. Electricity use by washing machines decreased between 1986 and 2015, owing to 33% energy efficiency improvements in *new* washing machines in that period (DECC, 2016). But these effects were counteracted by the diffusion of new appliances such as dishwashers, tumble dryers, and washer-dryers (Figure 4.16), which also became larger and more powerful.

Electricity use for cooking has remained largely stable since the 1970s (Figure 4.13). Electricity consumption by consumer electronics has increased substantially since the 1970s, owing to a quadrupling of TV ownership (resulting in more than two TVs per household by 2015) and the proliferation of new appliances (set-top boxes, games consoles, and DVD/VCR products). Since the mid-2000s, however, energy efficiency innovations such as increasing use of light-emitting diodes (Park et al., 2013) has helped to stabilise electricity use by consumer electronics (Figure 4.13).

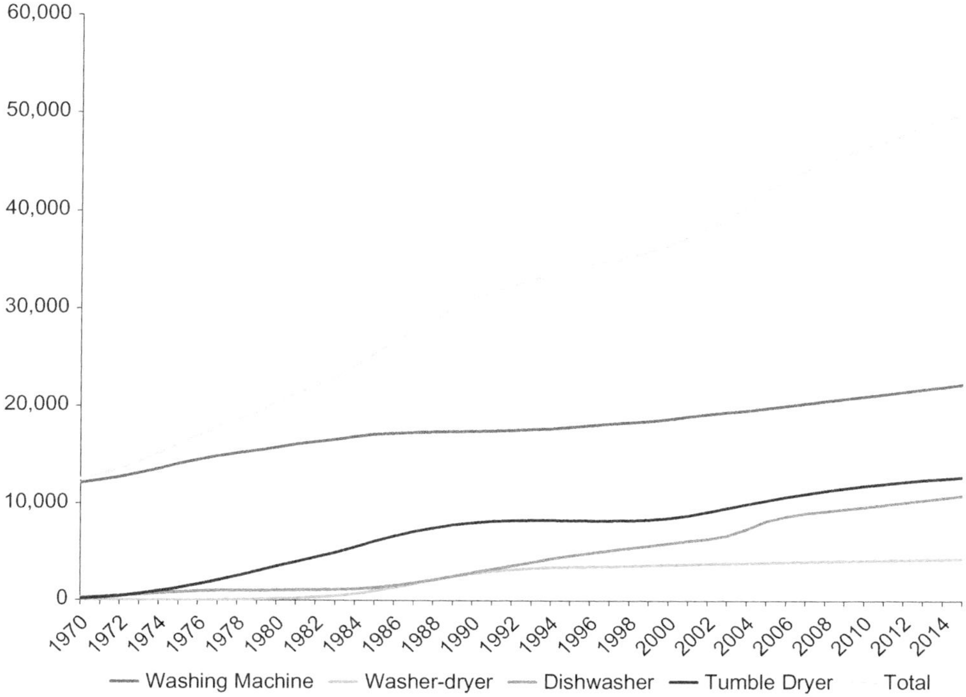

Figure 4.16 Number of wet appliances (in thousands) owned by UK households, 1970–2015 (constructed using data from DUKES; Energy Consumption in the UK 2016; Electrical products tables; Table 3.12)

Electricity use by home computing increased substantially in the 1990s and 2000s due to the emergence of an information society and the associated proliferation of computers, monitors, gadgets, and devices with higher processing power, improved connectivity, and additional functionalities (Chandler, 2001; Røpke et al., 2010). Since the late-2000s, however, their domestic electricity consumption has stabilised (Figure 4.13) because of energy efficiency innovations and the shift from desktops to laptops and from printers to multi-functional devices (Figure 4.17).

4.4.2 Actors

Firms: Most electric appliances are imported into the UK or produced by foreign-owned manufacturing facilities, leading to the decline of UK electric appliance manufacturing (Beynon et al., 2003). While appliance brands have proliferated in the UK, these are owned by a small number of multinational companies operating in a highly concentrated and oligopolistic market. Multinational appliance manufacturers (of televisions, radios, computers, refrigerators) are economically important

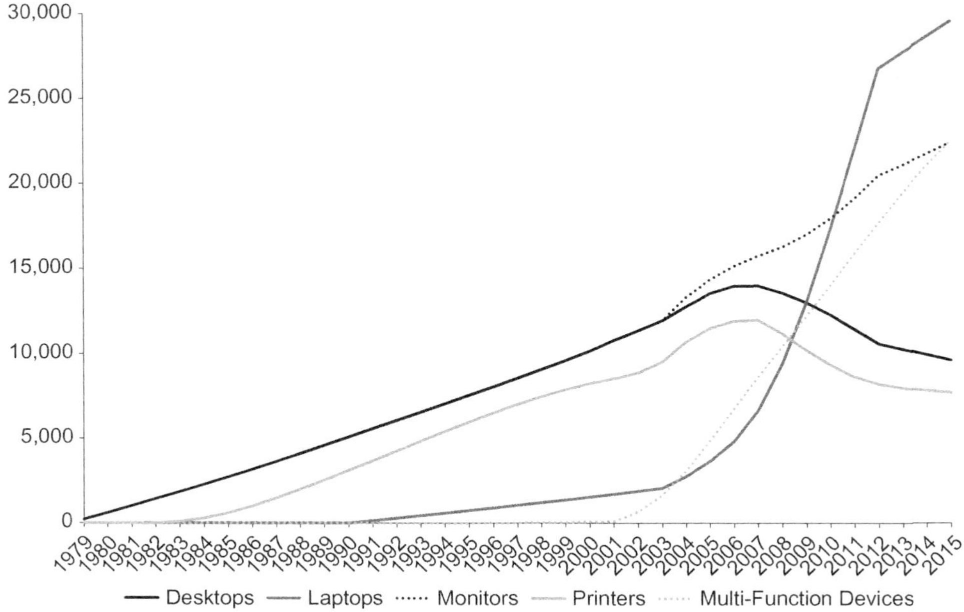

Figure 4.17 Number of home computing devices (in thousands) owned by UK households, 1970–2015 (constructed using data from DUKES; Energy Consumption in the UK 2016; Electrical products tables; Table 3.12)

and technically dynamic, generating high degrees of product innovation along many quality and performance dimensions (Chandler, 2001; Godoe, 2000). The Association of Manufacturers of Domestic Appliances (AMDEA) is the UK appliance trade association, which represents the interests of international appliance firms active in the UK.

In the early 1990s, appliance manufacturers and trade associations such as AMDEA resisted the European Union's energy efficiency initiatives (Boardman, 2004). They 'refused to supply data and cooperate in an energy efficiency study upon which to base policy, hampering the efforts of the EU to meet its obligations under the Climate Convention' (Newell and Paterson, 1998: 684). This resistance stemmed from the belief that energy efficiency was a marginal issue for consumers and that energy labels would therefore only impose costs on manufacturers (Toke, 2000).

By the mid-1990s, however, these actors changed their strategic orientation and began to engage more actively with energy efficiency as an additional consideration. By 1996, 'AMDEA had become convinced that they needed to protect their interests by acting within, rather than against the energy efficiency lobby. Their world view had changed. Peter Carver, AMDEA Director commented: "Pressure on us to improve energy efficiency is never ending. Unless we agree voluntary codes, regulations at a European level are inevitable"' (Toke, 2000: 850). Since then, AMDEA has been involved in several initiatives, further discussed later, and by the 2010s was actively lobbying for stronger energy efficiency policies: 'A change in focus is required so that policies to reduce demand for electricity receive at least as much, if not more, attention than policies to encourage low carbon electricity generation' (AMDEA, 2014: 7).

As such, the energy efficiency agenda has been actively incorporated by incumbent firms, partly as a strategy for regime protection and reproduction. Their mostly *incremental* innovation strategies substantially improved energy efficiency performance in cold appliances, wet appliances, consumer electronics, and home computing devices. In 2010, the Energy Saving Trust established a 'voluntary retailer initiative', in which eight leading retailers agreed to promote more energy efficient computers and televisions and remove the most inefficient models from their shelves by 2011 (IPPR, 2013).

Incumbent lighting companies pursued a more *radical* innovation strategy, shifting from ILBs to more efficient CFLs and LEDs (Franceschini and Alkemade, 2016). This shift, which is further analysed in Section 4.5.5, was stimulated by pressure from environmental NGOs, regulatory pressure that culminated in the 2009 European ILB ban, and competitive pressure from Chinese companies, which made incumbent companies more willing to explore the economic opportunities of new lighting technologies.

In sum, appliance and lighting companies have incorporated energy efficiency as an *additional* regime dimension, which has led to some re-orientation of industry strategies and innovation patterns. But it also left other regime rules intact, such as a focus on persistent innovation, rapid product lifecycles, increasing functionalities, differentiation, and market expansion (Wieser, 2017; Wieser and Tröger, 2018).

Wider Publics: From the main environmental UK NGOs, the Green Alliance has been the most active in pushing the consumer-focused energy efficiency agenda. Since the mid-1990s, its campaigns often aligned with AMDEA in putting pressure on the UK government (and European policymakers) to intensify policy support for energy efficiency (Boardman, 2004). At the European level, environmental NGOs such as WWF and Greenpeace started criticising the inefficiency of ILBs in the early 2000s and succeeded in establishing a new cultural framing that associated ILBs with energy waste (Franceschini and Alkemade, 2016).

Policymakers: Electricity consumption generally receives less policy attention than electricity supply. Despite its limited visibility, policymakers have been relatively effective in stimulating the development and adoption of energy efficiency innovations in various appliances, leading to a 15% reduction in electricity consumption between 2005 and 2019 (Figure 4.11), despite an increase in the number of appliances (Figure 4.12).

The push for energy efficiency innovation initially came from European regulatory frameworks, and their translation into national demand-oriented policies (Boardman, 2004; Toke, 2000). The 1992 European Directive on Energy Labelling, for instance, required manufacturers of light bulbs, white goods (e.g., refrigerators, washing machines, dishwashers), and other appliances to provide consumers with information about the energy efficiency performance of their products.

The UK government was initially slow and reluctant to implement this Directive into national policy, adopting a compliance-only approach in which it introduced labels but otherwise undertook little effort to raise public awareness (Boardman, 2004). The government did, however, establish the Energy Saving Trust in 1992 as an independent body to promote energy efficiency, publish research, and provide information to consumers. Since the mid-1990s, the UK government has begun to engage more strongly with demand reduction and efficiency improvement, which was helped by the strategic reorientation of appliance manufacturers, discussed earlier. Between 1994 and 2012, UK policymakers ran several programmes that placed energy savings obligations on energy suppliers (Rosenow, 2012), which will be further discussed later.

Regulatory pressures further increased with the 2005 European Ecodesign Directive, which introduced a new framework approach in which minimum energy efficiency standards for energy-using products would be articulated that would increase over time and thus remove the worst products from the market. These tightening standards further stimulated international appliance manufacturers to innovate and improve energy efficiency. The 2005 framework approach was elaborated by the 2009 Ecodesign Directive that specified minimum standards for more than 40 energy-using product groups (including lightbulbs, televisions, refrigerators, and boilers), which were implemented and adjusted through successive Ecodesign working plans (2009–2011, 2012–2014, 2016–2019). For incandescent lightbulbs (ILB), the increasing policy pressure even culminated in a European phase-out policy, which in 2009 banned the sale of ILBs of more than 80W, progressing to lower wattage in successive years (which is further discussed in Section 4.5.5).

The Ecodesign Directive was complemented by the 2010 EU Framework Directive on energy labelling, which updated the 1992 policy by harmonising national measures on end-user information for energy-related products via labelling and product standard information. New Ecodesign and Energy Labelling measures were agreed in 2018 and 2019, covering more product categories and raising standards further.

In the late 2000s, UK policymakers also showed some interest in behaviour change, which led the Energy Saving Trust to initiate information campaigns encouraging consumers to switch off lights when not in use, fill the kettle to the level required, and reduce the use of standby functions. These campaigns were mostly limited in their success, which led policymakers to acknowledge that: 'Currently we lack deep understanding of the complexities of what really drives energy demand and how to change it at user and provider level' (DECC, 2012a: 56). Subsequent policy efforts therefore continued to focus mostly on energy efficiency improvements in technical appliances, which have been relatively effective in reducing electricity demand, as noted earlier.

Users: Most electricity consumption is routine, taken for granted, and detached from material supply realities. Most users know little about the worlds behind the socket (how it works, where it comes from, how it is organised). They mainly interact with suppliers through meters and bills, supplier choice, and the occasional need for electrical repair. Consumer switching between suppliers was limited until 2014 but then increased rapidly (Figure 4.10), leading to more new entrants and increased competitive pressure in the electricity market.

Climate change is of less concern to consumers than electricity bills. Few UK consumers opt for 'green' electricity suppliers. Although most consumers do not actively choose renewables, they ultimately pay for the upstream investments in RETs and grid innovations through their energy bills and general taxation (which finances government subsidies to generators). This 'indirect' or 'involuntary' market demand, which has been created through regulations and billing practices, helps explain the higher speed of low-carbon transitions in electricity, compared to other domains (where consumers need to make deliberate choices to buy electric cars, insulate homes, or change food purchases).

Electricity is used by specific appliances, which relate to particular end uses and social practices (Shove and Walker, 2014). Regarding the laundry, for instance, more people have started to wash at low temperatures since the early 2000s (Mylan, 2017). But energy saving gains were counteracted by other behaviour changes such as washing clothes more frequently in smaller loads and drying them in tumble-dryers, driven by the desire for greater convenience (Mylan and Southerton, 2018).

Regarding cooling, people have also adopted more energy-efficient fridges and freezers. But here, too, energy savings have been partly eroded by increasing use of multiple cold appliances in households and shifts towards larger appliances (DEFRA, 2009). The associated behavioural trends towards storing more foods and drinks in cold appliances partly relate to the stronger preferences for chilled goods and the increasing 'cultural significance of freshness' (Evans and Mylan, 2019: 426). The increasing use of freezers, in turn, relates to the diffusion of microwaves and ready-meals and to changes in food preparation patterns towards greater convenience and flexibility, which have consequently become significantly entrenched in modern ways of living (Hand and Shove, 2007).

End-uses and social practices are thus deeply intertwined with persistent cultural conventions such as convenience (e.g., storing ready-meals in freezers, heating food in microwaves, dishwashers, drying clothes in tumble dryers), cleanliness (e.g., more frequent laundry cycles), fun and novelty (e.g., new gadgets and functionalities), and freshness (e.g., more cold drinks) as drivers of demand for domestic appliances (Hand and Shove, 2007; Mylan, 2016; Mylan and Southerton, 2018; Shove, 2003). Consumers also expect new functionalities and enhanced standards of home entertainment and digital connectivity (Crosbie, 2008). More generally, electricity has become a taken-for-granted background to modern life. Increased appliance use is associated with progress, and associated electricity consumption is rarely questioned (Shove and Walker, 2014). The increasing policy and business focus on energy efficiency does not question these cultural conventions or call for deeper behaviour change.

4.4.3 Policies and Governance

Formal Policies

Since the 1990s, European policies have been important in stimulating international appliance manufacturers to improve the energy efficiency of their products, which they actively engaged with after an initial period of resistance, as discussed previously.

National policies have been important to stimulate the adoption of energy-efficient appliances. Since the mid-1990s, UK policymakers have run several programmes that have placed energy savings obligations on energy suppliers, which incentivised them to engage with demand reduction and help diffuse energy-efficient appliances. These included the Energy Efficiency Standards of Performance (EESoP), which ran from 1994 to 2002, the Energy Efficiency Commitments (EEC) from 2002 to 2008, and the Carbon Emissions Reduction Target (CERT) from 2008 to 2012. While EESoP programmes focused exclusively on electric appliances, EEC and CERT also included gas, heating, and home insulation, which increasingly became the prime focus because of greater energy and carbon saving potential.

The various energy savings obligations set (gradually increasing) energy saving targets on energy suppliers, which could pass the costs for their actions and measures on to their customers via energy bills (although this was constrained by price control measures). This policy design fitted with neoliberal policy thinking because it was assumed that companies would compete to meet their targets at the lowest cost to consumers, who were themselves assumed to have incentives to switch to suppliers with the lowest prices (Rosenow, 2012). The successive programmes with energy savings obligations were relatively effective, leading electricity suppliers to adopt give-away programmes of energy-efficient lightbulbs and the promotion (with retailers) of energy-efficient cold appliances through discounts and targeted in-store marketing strategies (AMDEA, 2014). Between 2002 and 2005, the EEC, for instance, brought forward an estimated 4.5 million sales of fridge-freezers, compared to the existing market trend (DEFRA, 2009).

From 2013, the Energy Company Obligations (ECO) focused exclusively on heating and insulation, which thus relieved electricity companies from demand-reduction obligations.

UK policymakers implemented the 2005 and 2009 European Ecodesign Directives and subsequent upgrades through their Products Policy (DEFRA, 2009), which: a) removed the least efficient products from the market (using European minimum standards), b) encouraged the development of more efficient products (through R&D and innovation policy), and c) stimulated market uptake (through labelling, public information, producer obligations, public procurement, and voluntary initiatives). Initial implementation was relatively slow, so that by 2012 only 13 out of 25 product categories in the first tranche had regulations applied to them (Cary and Benton, 2012). In subsequent years, however, the Products Policy was further implemented, which together with tightening product standards stimulated substantial energy efficiency improvements.

Governance Style

Initial European Energy Labelling Directives were relatively weak market-based policies, which assumed that providing consumers with information (about energy efficiency ratings of appliances) would lead them to choose more energy-efficient products. When this proved limitedly effective, stronger regulatory control policies were introduced such as the European Ecodesign Directives, which articulated gradually strengthening minimum energy efficiency standards that pushed appliance manufacturers to innovate and improve their products.

The policy processes for these Directives were dominated by stakeholder consultation processes and technocratic debates about specifying the minimum level for environmental performance and the most appropriate layout of labels to communicate information to consumers (Rosenow et al., 2017). The European-level governance style thus has corporatist characteristics, with close interactions between policymakers and incumbent firms, aimed at negotiating the feasible speed of energy efficiency improvements and policies.

UK demand-focused policies also started relatively weakly but gradually strengthened over time, mostly through tightening regulations and standards.

Successive energy savings obligations (1994–2012) set gradually increasing targets on electricity companies, which stimulated them to help deploy and diffuse energy-efficient appliances. Energy saving obligations on electricity suppliers were dropped by the 2013 Energy Company Obligations, which focused exclusively on heating and insulation. But since the late 2000s, the UK's Products Policy has represented a fairly interventionist market shaping governance style, which advanced energy-efficient appliances through increasing minimum standards, labelling, public information, public procurement, and voluntary initiatives.

Energy saving policies thus focused centrally on energy-efficient products, addressing both upstream technological improvements and downstream consumer demand. This product-centred approach suits the interests of incumbent appliance manufacturers, which have therefore come to support the energy efficiency agenda.

4.5 Niche-Innovations

Radical niche-innovations have emerged and diffused in each of the three electricity sub-systems, leading to substantial reconfiguration. For the electricity generation sub-system, we will analyse four niche-innovations: 1) onshore wind, 2) offshore wind, 3) bio-power, and 4) solar PV. For the electricity consumption sub-system, we will discuss two niche-innovations: 5) energy-efficient lighting, including CFL and LEDs, and 6) smart meters. And for the grid sub-system, we will also analyse two niche-innovations: 7) smart grids, and 8) two flexibility-enhancing options: battery storage and demand-side response.

The analysis of each niche-innovation will address both techno-economic developments and actors and institutions. Several niche-innovations have experienced substantial changes in both analytical dimensions as they first emerged and then diffused. The analysis of niche-innovations in the electricity systems is therefore somewhat longer than for the heat and mobility systems.

Power-generation niche-innovations, in particular, have experienced relatively long developmental trajectories. Renewable electricity technologies (RETs) emerged in the 1990s, experienced several ups and downs in the 2000s, and diffused quite substantially in the 2010s (Figure 4.18). Onshore wind, offshore wind, and bio-power diffused fastest in recent years, because of government support and deployment by incumbent actors (utilities, project developers, foreign energy companies). Solar-PV also diffused substantially, often through deployment by new entrants such as farmers and households. Cumulatively, RETs accounted for 39% of electricity generation in 2019, which means they are taking on regime-like characteristics. Rapidly falling costs (Figure 4.19) have been an important driver of the diffusion of RETs, which are increasingly cost-competitive with coal and gas-fired power plants. Between 2010 and 2020, the global average

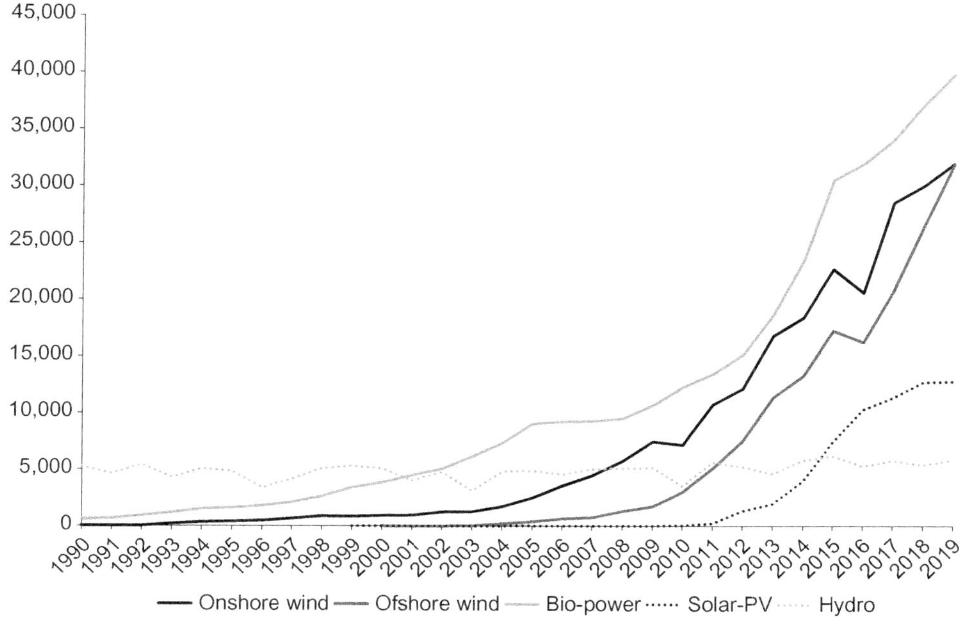

Figure 4.18 Electricity generated from renewable sources in GWh, 1990–2019 (constructed using data from the Digest of UK Energy Statistics; Electricity Statistics; Renewable sources; Table 6.6.1)

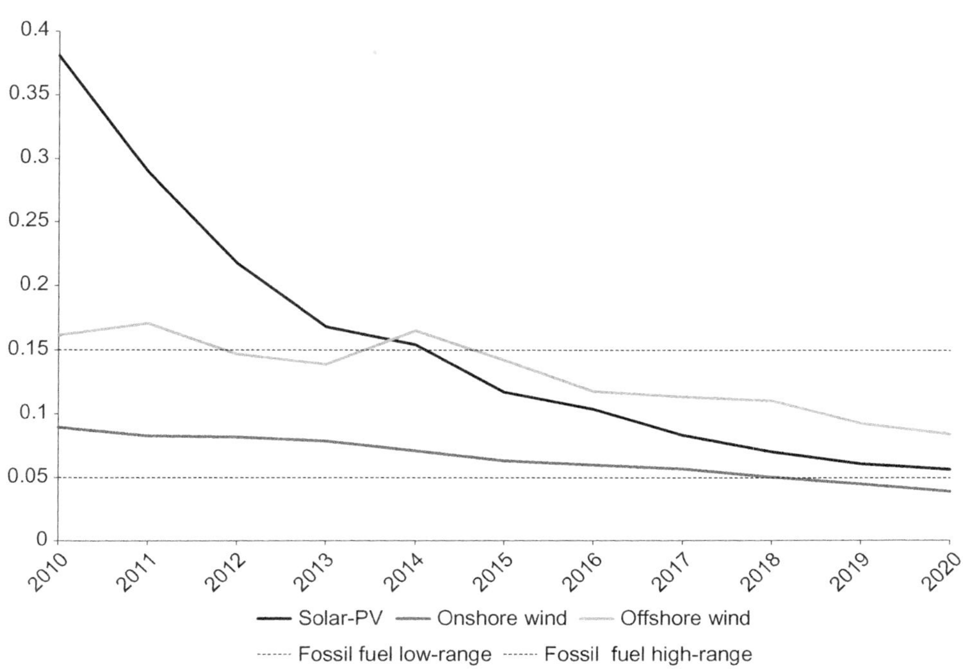

Figure 4.19 Global average levelised costs of electricity, 2010–2020 (in constant 2020 US dollars per kWh) for different technologies (constructed using data from IRENA (2021))

levelised cost of electricity[10] (LCOE) decreased by 85% for utility-scale solar-PV, 56% for onshore wind, and 48% for offshore wind (IRENA, 2021).

Other important drivers for rapid diffusion were renewable energy policies, which affected all RETs and are therefore briefly discussed here, before we analyse the individual niche-innovations. Renewable electricity policies emerged as a 'side-effect' (Toke and Lauber, 2007) of support for nuclear power via the 1990 Non-Fossil Fuels Obligation (NFFO) that required electricity companies to buy certain amounts of nuclear power. Renewables' advocates argued that renewable electricity technologies (RETs) should also qualify for non-fossil fuel subsidies, which led the government to introduce an NFFO policy for renewables in the form of a competitive auction system in five successive rounds between 1990 and 2004. Bidders could submit proposals to produce a certain amount of renewable electricity for a certain price. In each round, the government awarded contracts to the lowest bidders within particular technology bands (Mitchell and Connor, 2004).

The NFFO-bidding process was complicated and required sophisticated financial capabilities and sufficient capital to cope with economic risks and policy uncertainties. These characteristics favoured professional corporate actors and discriminated against smaller new entrants with less-developed procedural and financial capabilities and resources (Mitchell and Connor, 2004; Toke, 2005). The NFFO was limitedly effective because many accepted bids never resulted in actual RET-deployment, because many winning bidders realised their low-cost proposals were too uneconomical to be realised (Toke and Lauber, 2007). Over the whole period (1990–2004), only 30% of winning projects were actually completed (Wood and Dow, 2011).

In 2002 the Labour government introduced the Renewables Obligation (RO) as a new policy, requiring utilities to meet gradually increasing annual renewable electricity targets in one of several ways: a) generate renewable electricity themselves, b) buy Renewable Obligation Certificates (ROCs) from other generators, c) pay a 'buy-out' penalty of 3p/kWh. The RO was more market-oriented than the NFFO, because it was based on free-market trading of ROCs and abolished the NFFO's technology banding. Because all RETs received the same number of ROCs, the RO was biased towards cheaper (large-scale) technologies such as onshore wind and landfill gas (Foxon and Pearson, 2007). The RO stimulated close-to-market options, neglected innovation, and created uncertainties about longer-term policy commitment (Woodman and Mitchell, 2011). The RO

[10] LCOE refers to all the construction and operational costs to produce a certain quantity of electricity over the generation asset's lifetime. LCOE enables holistic cost comparisons between different electricity generation technologies.

also disadvantaged new entrants, because the trading of ROCs created financial uncertainties, which were easier to manage for incumbent utilities.

The increasing political salience of climate change resulted in the 2008 Climate Change Act. The translation of its high-level goals into more specific targets and plans increased policy delivery momentum, and the creation of multiple complementary instruments. For the electricity sector, the UK Low Carbon Transition Plan (2009) articulated a target of 30% renewable electricity by 2020 and an almost complete decarbonisation by 2030, which created clear directionality. Criticism of the limited effectiveness of the RO resulted in the amended Renewables Obligation (2009), which included technology bandings that allocated varying amounts of ROCs to different technologies, depending on the degree of maturity and level of risk. In 2010, the government also (reluctantly) introduced a Feed-in-Tariff (FiT) as part of a political deal with backbenchers, who wanted a stimulus for small-scale renewables in exchange for their support for nuclear and offshore wind (Smith et al., 2014). The 2013 Electricity Market Reform (discussed in Section 4.2.3) further introduced Contracts for Difference (CfD), which provided attractive long-term incentives for large-scale renewables and nuclear power from 2017 onwards (replacing the Renewables Obligation).

The 2015 energy policy reset slashed subsidies for renewables such as onshore wind, bio-power, and solar-PV. The 2016 Brexit decision resulted in 'a loss of time and policy momentum, . . . as uncertainty over Brexit continues to make it difficult to plan for the UK's energy future' (UKERC, 2019: 2). A new Energy White Paper, scheduled to be published in 2019, was delayed as a result of domestic political upheaval, which created further uncertainties. Feed-in tariffs for small-scale renewables were ended in 2019, which reinforced the UK's focus on large-scale renewables. Although the 2020 *Energy White Paper* confirmed the net-zero direction of travel across multiple systems, it did not articulate many new policies for renewable electricity technologies. It did, however, articulate the vision that a future 'low-cost, net zero consistent system is likely to be composed predominantly of wind and solar' (HM Government, 2020a: 43). This very central role of RETs is new compared to previous major policy documents in which renewables were just one among multiple generation options. Other generating options such as nuclear power and gas-with-CCS are now expected to play smaller complementary roles to RETs. The 2020 *Energy White Paper* also emphasises the future role of flexibility options such as batteries and demand-side response, which will be further discussed in Section 4.5.8.

While many of the renewable electricity policies we have discussed influenced all RETs, there were also specific policies that shaped RETs, as we will analyse next. We will first discuss four RET-niches, then two demand-oriented niche-innovations, and then two grid-oriented niche-innovations.

4.5.1 Onshore Wind

Techno-Economic Developments

Onshore wind deployment gradually increased in the 1990s (Figure 4.18), as utilities and project developers received continuous subsidy support over successive NFFO-rounds. Deployment accelerated rapidly after 2002 (Figure 4.18) because the RO and amended RO provided attractive financial support (Foxon and Pearson, 2007).

Technical developments focused on higher wind turbines with greater capacity and on improvements in rotor turbines, drive trains, and material use. Wind forecasting models also improved, leading to better turbine siting (Gross et al., 2013). Technical improvements, learning-by-doing, and scale increases decreased the cost of onshore wind by 56% between 2010 and 2020 (Figure 4.19), making it the cheapest RET.

The 2015 reductions in subsidies and the 2016 government moratorium on new onshore wind farms from 2020 onwards (further discussed later) substantially decreased annual installation rates (Figure 4.20) with a few years lag-effect, because projects in the pipeline were still being completed.

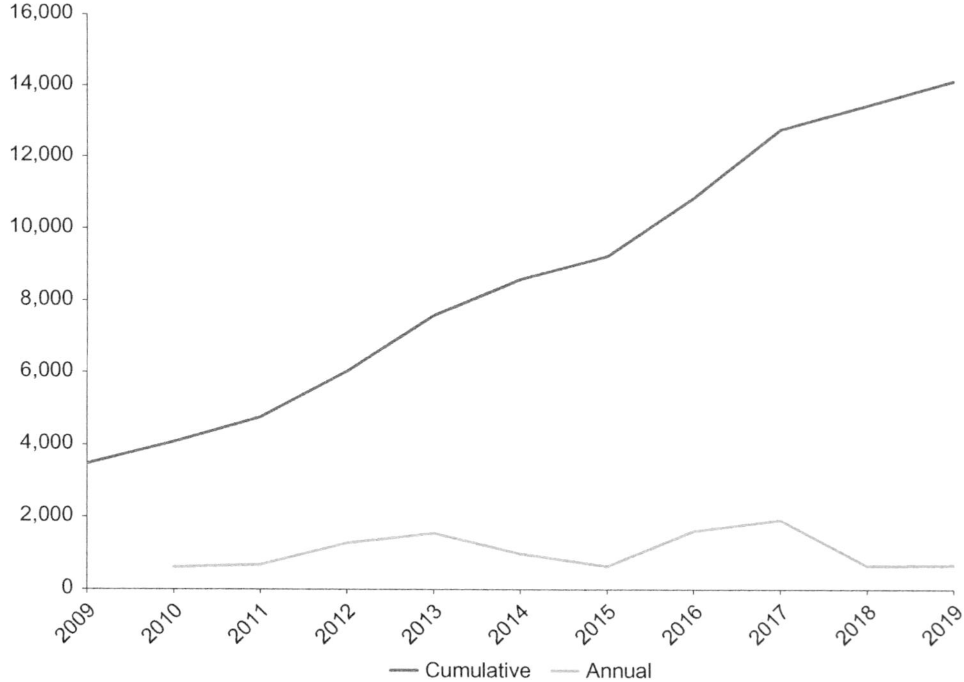

Figure 4.20 Cumulative and annual installed capacity of UK onshore wind turbines (in MW) (constructed using data from the Digest of UK Energy Statistics; Energy Trends: UK Renewables; Table 6.1 Renewable electricity capacity and generation)

Table 4.1. *Ownership of onshore wind power in 2004 by percentage capacity (Toke, 2005: 371)*

	UK	Germany	Denmark	Netherlands	Spain
Utilities, corporate independent	98	55	12	32	>99
Farmers	1	35	63	62	<0.5
Cooperatives	0.5	10	25	6	0

Actors and Policies

UK onshore wind development has mainly been a corporate economic activity, involving utilities, project developers, and independent generators who sell electricity to utilities. This corporate dominance differs from Germany, Denmark, and the Netherlands, where new entrants such as cooperatives, farmers, and communities have historically played a larger role (Table 4.1). There has been some rise of community wind energy in the UK since the mid-2000s (Walker and Devine-Wright, 2008), but this has remained comparatively small, partly because of limited organisational capacities at the community level and partly because of 'the persistence of key features of socio-technical regime for electricity provision, which continues to favour large corporations and major facilities' (Strachan et al., 2015: 106). The NFFO, RO, and CfD policies, for instance, were all designed to favour large companies, as discussed previously.

The bidding design of NFFO and RO policies also helped to create social acceptance problems, because most project developers did not start the planning permission process until after they were awarded the contract. They would then be in a hurry and often installed wind turbines without proper stakeholder consultation, which turned many stakeholders into opponents (Ellis et al., 2009). For onshore wind, this resulted in negative sentiments and perceptions of unfair distributions of costs (local stakeholders experiencing noise, visual burdens, and shadow flicker) and benefits (project developers enjoying wind resources and financial gains).

Wind farm projects therefore encountered increasing local opposition, leading to decreasing approval rates in planning procedures from 73% in 2007 to 50% in 2012 (CCC, 2013). The public wind discourse became increasingly negative, because of concerns about subsidies, visual and landscape impacts, and the perceived invasion of the countryside by corporate interests (Kern et al., 2014b). These concerns gave rise to opposition from the Campaign to Protect Rural England, which pressured local Conservative MPs, one hundred of whom wrote an open letter to the Prime Minister arguing against onshore wind subsidies (5 February 2012). These increasing socio-political problems culminated in the 2015 slashing of subsidies and the 2016 government decision to halt subsidies and not build new wind turbines after 2020.

In March 2020, however, the Johnson government overturned the moratorium, partly because nuclear power implementation problems created future capacity problems, and partly because decreasing costs enhanced the appeal of onshore wind. This U-turn enables onshore wind to participate in the 2021 CfD auction, which means that new wind farms could be up and running by the mid-2020s. The 2020 *Energy White Paper* confirmed that onshore wind (and solar) is allowed to compete in the next CfD auction in late 2021, which is likely to boost future diffusion.

4.5.2 Offshore Wind

Techno-Economic Developments

Offshore wind (OffSW) initially diffused slower than onshore wind because of greater technical difficulties and higher cost. In the 2001–2007 period, government support stimulated experimentation and learning with six successive demonstration projects (Kern et al., 2014b). Technological developments led to larger turbines with higher capacities. Operational challenges associated with salt water, stronger winds, and waves led to technical changes in materials, electronics, and gearing mechanisms. To reduce visual complaints, OffSW-farms were increasingly located further from shore in deeper waters, which required technology developments to deal with seabed foundation challenges. The installation of OffSW-farms further required specialised offshore capabilities and tools (e.g., docks, ships, platforms, cranes, drilling tools), creating a new market for UK offshore firms from the oil and gas sector.

Technical improvements, scale economies, and learning-by-doing reduced the levelised cost of electricity from offshore wind by 48% between 2010 and 2020 (Figure 4.19). In the third (2019) CfD auction round, for projects coming online in the mid-2020s, costs decreased further, reaching as low as £40 MW/h.

Offshore wind diffused rapidly after 2009 (Figure 4.18), because the amended RO provided attractive financial support (Heptonstall et al., 2012). Cumulative and annual installed OffSW capacity increased (Figure 4.21), except for 2016 when no wind farm project was completed (which relates to the lumpy nature of these mega-projects). Rapidly increasing installed capacity made the UK a world leader in OffSW. At 1.2 GW operational capacity, the newly installed (2019) Hornsea One is the largest offshore wind farm in the world.

Actors and Policies

Offshore wind is supported by a powerful network of actors with different but congruent interests (Kern et al., 2014b). Big utilities (Vattenfall, RWE, E.ON, SSE, Centrica) and energy companies (Dong, Statoil) became the dominant project developers for OffSW-farms, which enabled them to preserve their business model

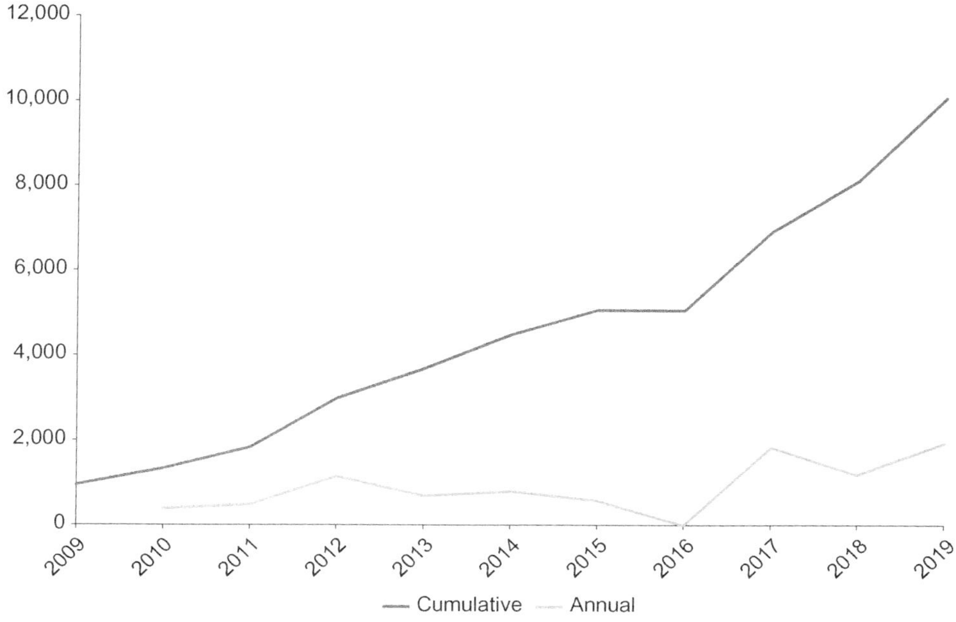

Figure 4.21 Cumulative and annual installed capacity of UK offshore wind turbines (in MW) (constructed using data from the Digest of UK Energy Statistics; Energy Trends: UK Renewables; Table 6.1 Renewable electricity capacity and generation)

based on large-scale, centralised forms of electricity generation. Research and technological innovation, aimed at OffSW cost reduction and incremental improvement, was supported by the Carbon Trust, Energy Technologies Institute, Technology Strategy Board, and the Engineering and Physical Sciences Research Council. Multiple UK ministries were involved. DECC stimulated OffSW because of climate and energy targets. The Department for Business, Innovation and Skills was interested in business opportunities and jobs related to OffSW. Ofgem was involved to arrange grid connections. And the Crown Estate, as the owner of the seabed, was involved to provide and sell wind farm concessions. Environmental NGOs also supported OffSW, which they preferred over onshore wind (Toke, 2011). And trade associations such as RenewableUK and the Offshore Wind Developers Forum provided support and coordination for the network.

Although most OffSW technologies were initially imported (e.g., turbines from Siemens and Vestas), domestic UK manufacturing capacity has gradually increased as international firms (e.g., Siemens, General Electric, Mitsubishi) set up production facilities in the UK.

Despite high technology costs, government support for OffSW has been strong since the early 2000s. Capital grants and the Renewables Obligation (RO) supported early demonstration projects. The technology banding in the amended 2009 RO, which increased the ROCs per MWh for OffSW from 1.5 to 2, created an attractive support premium estimated at £100/MWh on top of the retail price (Heptonstall et al., 2012). From 2009, the government has stimulated research and technology development projects, investing £206 million through various schemes led by DECC, Energy Technologies Institute, BIS, and the Technology Strategy Board (see Kern et al. (2014b) for details). In 2011, the government also created the Offshore Wind Cost Reduction Task Force, with the aim of reducing costs to about £100/MWh by 2020, which was markedly successful. The three CfD auction rounds (in 2015, 2017, and 2019) provided further support for OffSW but also helped to drive costs down.

In March 2019, the government and industry partners (including offshore wind developers, equipment manufacturers, and supply chain actors) agreed an Offshore Wind Sector Deal that serves as the roadmap for a 30GW installed capacity target by 2030. This would triple current capacity and enable OffSW to provide about 30% of the UK's electricity. As part of the deal, the government committed £557 million funding for bi-annual CfD auctions for the next 10 years, while industry partners committed to increase UK manufactured content to 60% by 2030, increasing exports fivefold to £2.6bn by 2030, and invest up to £250m to improve productivity and innovation in the UK supply chain. In December 2019, the newly elected Johnson government raised the 2030 target from 30GW to 40GW, making offshore wind a central plank of the future UK electricity system. The new 40GW target was confirmed in the 2020 *Energy White Paper*.

4.5.3 Bio-power

Techno-Economic Developments

Bio-power is a heterogeneous niche associated with a variety of feedstocks (such as wood pellets, wood residues, farming residues, straw, sewage sludge, waste) and conversion routes (e.g., combustion, gasification/pyrolysis, anaerobic digestion). While cumulative bio-power has grown substantially since 1990 (Figure 4.18), specific sub-categories have experienced several ups and downs (Figure 4.22):

- Landfill gas, which uses bio-technological processes to convert organic waste to methane that is burned in gas turbines, grew rapidly in the mid-1990s and early-2000s due to continuous NFFO support and the availability of many landfill sites. Deployment rates plateaued in the mid-2000s as new waste policies reduced the amount of organic landfill waste.
- Dedicated biomass plants diffused in the 1990s in the form of solid waste combustion plants (e.g., demolition wood) and other bioenergy options (e.g.,

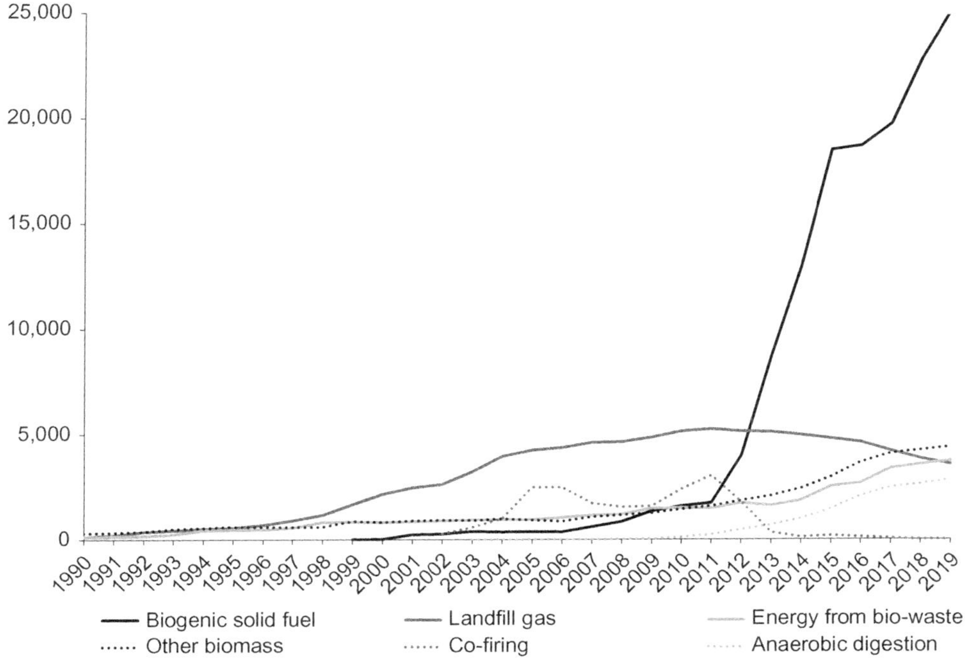

Figure 4.22 Electricity generated from bio-power sub-categories in GWh, 1990–2019 (constructed using data from the Digest of UK Energy Statistics; Electricity Statistics; Renewable sources; Table 6.6.1)

animal biomass, sewage sludge digestion, anaerobic digestion). Both categories remained relatively small because of long-term supply uncertainties and local opposition but have gained some momentum since the mid-2000s, e.g., through increased construction of small-scale dedicated biomass plants.

- Co-firing of biomass with fossil fuels in adjusted coal plants grew rapidly after the 2002 Renewables Obligation, because it was a relatively easy and cheap way for utilities to meet their renewables obligations. Co-firing decreased rapidly after 2011 and gradually disappeared, with co-firing plants converting entirely to burning biomass.
- After 2011, the conversion of several large coal plants to biomass burning (e.g., Drax, Ironbridge, Tilbury, Lynemouth) rapidly increased the use of biogenic solid fuel (such as wood chips or pellets).

Since many bio-power technologies are relatively mature, innovation efforts have focused on incremental improvements addressing problems associated with boiler injection mechanisms, ash deposition, boiler corrosion, and fuel storage and handling (Perry and Rosillo-Calle, 2008). BECCS (bioenergy with carbon capture and storage) also received attention (especially at Drax) because this would enable

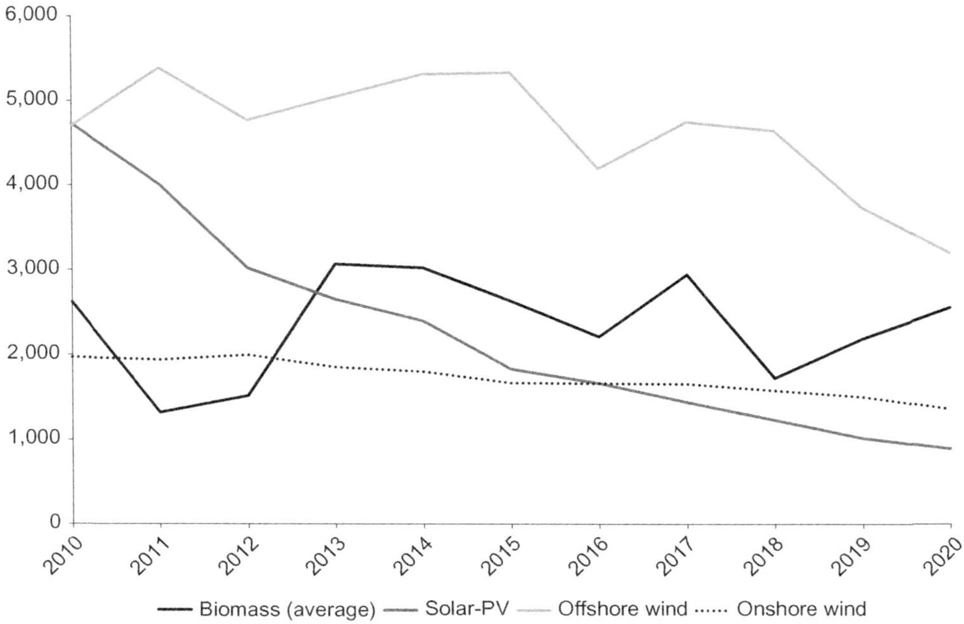

Figure 4.23 Global weighted-average total installed costs[11] of different RETs (in 2020 USD/kW), 2010–2020 (constructed using data from IRENA (2021))

negative emissions.[12] This option is hampered, however, by the lack of feasible CCS technologies and concerns about sustainable biomass supply (Fridahl and Lehtveer, 2018).

Levelised costs of different bio-power options vary substantially because of differences in technologies and supply costs of different kinds of biomass. The government's latest levelised cost estimates for 2025 put dedicated biomass at £97/MWh, biomass conversion at £87/MWh and biomass-CHP at £167/MWh (BEIS, 2016). Average installed costs/kW of electricity from biomass have not decreased over time, which contrasts markedly with other RETs (Figure 4.23).

Actors and Policies

Actor-networks in the bio-power niche are fragmented, with each sub-category having distinctive principal actors and actor constellations: *landfill gas* is mainly enacted by professional landfill site operators; *anaerobic digestion* involves a variety of actors (farmers, food and drink processors, local communities)

[11] Total installed costs refer to the final cost of designing, fabricating, and building an electricity generation asset per unit of generated electricity.

[12] Biomass absorbs CO_2 from the air when it grows. If post-combustion CO_2 could be captured and stored, BECSS could thus reduce CO_2 concentrations in the atmosphere.

associated with alternative feedstocks; *energy-from-waste incineration* projects are enacted by project developers or waste companies; *dedicated biomass* plants (which tend to be relatively small at under 50 MW) are mostly operated by new entrants (e.g., sawmills, poultry farms); *biomass conversion* involves coal plant owners converting to co-firing or 100% biomass combustion plants.

The principal actors in these sub-niches interact with networks comprising technology suppliers, local policymakers (for planning purposes), investors, energy companies, and suppliers of waste or plant biomass. Local residents, NGOs, and communities have sometimes opposed energy-from-waste incineration and anaerobic digestion projects, because of concerns about smell, unhealthy emissions, or inconvenience from waste supplying trucks. Their protests can prevent or delay project developers from acquiring local planning permissions (Upreti and Van der Horst, 2004).

NFFO and RO policies supported bio-power niches in the 1990s and 2000s, including smaller dedicated biomass plants. The *UK Bioenergy Strategy* (DECC, 2012b) represented a major policy change, which hindered the expansion of dedicated biomass and shifted the focus towards the conversion of coal plants into biomass-burning plants through stronger market shaping policies. It placed a 400MW cap on the total amount of dedicated biomass that could qualify for RO and excluded dedicated biomass from the CfD policy that replaced the RO in 2017. These restrictions did not apply to co-firing and biomass conversions, which received loan guarantees and subsidies through the RO and CfD.

Deployment of biogenic solid fuel accelerated rapidly after the 2012 policy shift. These converted biomass plants were large-scale facilities, which required some technical adjustments, but enabled coal plant operators to extend the plant's lifetime and circumvent the European Large Combustion Plant Directive.

The policy shift triggered a public controversy about the sustainability of imported biomass pellets (from forests in British Columbia and the United States). A 2012 report by the Royal Society for the Protection of Birds, Friends of the Earth, and Greenpeace (titled 'Dirtier than coal? Why government plans to burn trees are bad news for the planet') criticised DECC's assumptions for sustainability assessments, which ignored 'carbon debt' and indirect substitution emissions. The NGOs therefore campaigned against industrial-scale 'Big Biomass', including via direct protests at the 2013 opening of a converted unit of the Drax coal-fired plant and through a 2013 complaint by Friends of the Earth to the European Commission, questioning the legality of £75 million loan guarantees to Drax. In 2014, the government admitted mistakes in calculating carbon savings from large-scale biomass (DECC, 2014b) and said that biomass sustainability policies would be adjusted. In 2015, the newly elected Conservative government slashed financial support schemes for RETs, including biomass. But Drax won a CfD auction for a third biomass unit, which was again met with EU-level contestation about state aid. In December 2016, the European Commission ruled in favour of Drax, allowing its third unit to convert to wood pellets, followed by a fourth unit in 2018.

4.5.4 Solar-PV

Techno-Economic Developments

Solar photovoltaics (PV) emerged in the 1970s in space applications and then through the 1980s and 1990s in stand-alone applications without grid connection, for example, remote dwellings, boats, caravans, roadside emergency telephones. From the late 1990s, the UK saw modest deployment of solar-PV for domestic power generation. Diffusion rapidly accelerated after the 2010 Feed-in-Tariff (Figure 4.24), which provided stable long-term payments that made deployment economically feasible. Diffusion was also stimulated by positive cultural discourses and large reductions in the price of solar panels, which decreased by about 90% between 1976 and 2010 (Gross et al., 2013: 54) and another 85% between 2010 and 2020 (Figure 4.19). Price reductions in the 2010s were driven by learning-by-doing, scale economies in Chinese mass production, overproduction, and price dumping (Lauber and Jacobsson, 2016; Nemet, 2019).

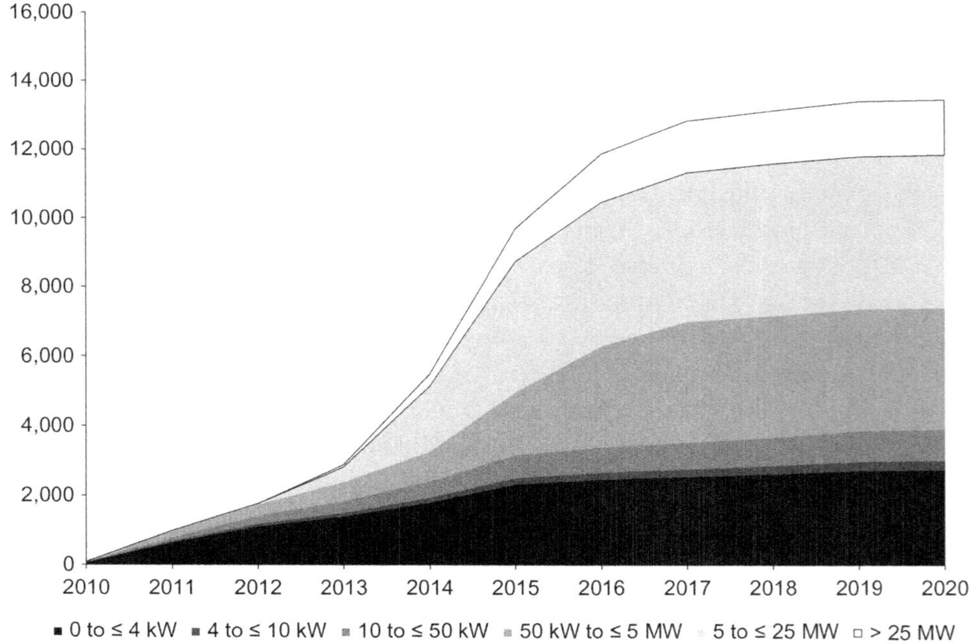

Figure 4.24 Cumulative UK installed capacity of solar-PV (in MW) by capacity size, 2010 to May 2020 (constructed using data from National Statistics; Solar photovoltaics deployment)

Solar panels are a modular technology, which can be installed in various sizes to suit the demands of different users:

- Small-scale solar-PV (<10 kW) is commonly installed on domestic rooftops.
- Mid-size building-mounted solar-PV (10–50 kW) is suitable for commercial and non-domestic properties (offices, hospitals, schools, factories, hotels, supermarkets, warehouses). Communities also installed mid-size ground-based solar farms, which delivered directly to the grid.
- Large-scale (50kW–5M) and utility-scale solar farms (>5 MW) are mostly ground-based installations in fields, which can be up to 40 hectares.

Mature first-generation technologies, such as crystalline silicon, dominate the market. Incremental innovation focuses on module design and production processes, for example, improvements in silicon wafer cutting, automation, and standardisation (Gross et al., 2013). Second-generation technologies, such as thin-film solar cells, are improving, offering the potential of lower material and manufacturing costs and easier and broader application, albeit at somewhat lower efficiencies (Nemet, 2019).

Actors and Policies

The solar-PV niche was supported by a strong network of new entrants, including environmental NGOs, roof installation companies, the solar industry, and even British Petroleum, which in 1997 pledged to increase investments in solar-PV from $100 million to $1 billion a year, as part of an energy diversification strategy (Pinkse and Van den Buuse, 2012). When the government's 2008 Energy Bill privileged nuclear energy over renewables, Friends of the Earth, Greenpeace, the construction industry, roofing contractors, and solar manufacturers mounted a visible public campaign in favour of a FiT. In this context, the government struck a political deal with pro-renewables backbenchers to support the government bill for nuclear power in return for the introduction of a FiT for small-scale renewables (Smith et al., 2014).

The 2010 FiT triggered unanticipated interest, resulting in rapid diffusion of domestic solar installations until 2016 (Figure 4.24), when more than 850,000 households had rooftop solar-PV. Increasing adoption and decreasing prices gave rise to wider cultural visions about the coming solar energy revolution and how it could transform energy systems towards decentralised production and active 'prosumers' (Barnham, 2014).

The mid-size segment (of commercial and non-domestic properties and community energy) grew only modestly after 2010. The strongest growth occurred in the large-scale and utility-scale segment, which received attractive financial support from the amended RO policy. By May 2020, this segment accounted for 70% of installed capacity (Figure 4.24). Farmers pioneered large-scale applications in the early 2010s, installing panels on barn roofs or in fields grazed by sheep.

Their success opened the way for even larger utility-scale solar farms, which were developed by commercial project developers interacting with large landowners, the financial community, energy companies, and PV system suppliers. The Shotwick Solar Park (72 MW) is presently the largest UK solar farm, but even larger ones are in the planning pipeline.

Policymakers and advisory bodies did not anticipate this boom because solar was the most expensive RET. In 2011, the Committee for Climate Change only foresaw a 'limited role for UK deployment of solar-PV' (CCC, 2011: 22), and the *UK Renewable Energy Roadmap* (DECC, 2011a) did not include a separate section on solar-PV. This changed in the next two years, when part 1 and part 2 of the *UK Solar PV Strategy* (DECC, 2014c, 2013a) saw solar-PV as an important part of the future energy mix. While these documents envisaged 10 GW solar capacity as a likely medium-range scenario for 2020, the Ministerial foreword called more bullishly for 20 GW within a decade, as part of a wider shift towards decentralised energy.

Despite these positive visions, less favourable developments in the broader political context (discussed earlier) led to the 2015 energy reset, which slashed feed-in tariffs and closed the RO for large-scale solar-PV. In 2016, the RO for small-scale solar-PV was also closed. These sudden policy adjustments sharply reduced the rate of new installations, leading to stagnated diffusion in all segments (Figure 4.24). In 2019, the government ended FiTs for small-scale renewables entirely. In March 2020, however, the government's new net-zero ambitions led it to reinstate support for large-scale solar-PV, allowing its inclusion in the 2021 CfD auction round. This inclusion was confirmed by the 2020 *Energy White Paper*.

4.5.5 Energy-Efficient Lighting

Techno-Economic Developments

The transition from incandescent lightbulbs (ILB) to compact fluorescent lamps (CFL), light emitting diodes (LEDs), and halogen (Figure 4.25) substantially reduced electricity use for lighting, as discussed in Section 4.4, despite increases in the total numbers of lamps.

Decisions to phase-out ILBs (2007 in the UK; 2009 European Commission) greatly accelerated this transition, but techno-economic developments were also important.

- ILBs and halogen have long been the cheapest per bulb, but their lighting efficacy (in lumen per Watt) improved little over time (Aman et al., 2013).[13]
- Fluorescent lighting uses three to five times less energy per lumen than ILBs (Aman et al., 2013), but its linear (strip-lighting) form was long confined to

[13] Halogen (tungsten) lamps are incandescent lightbulbs that operate somewhat more efficiently at higher temperatures, because small amounts of halogen have been added.

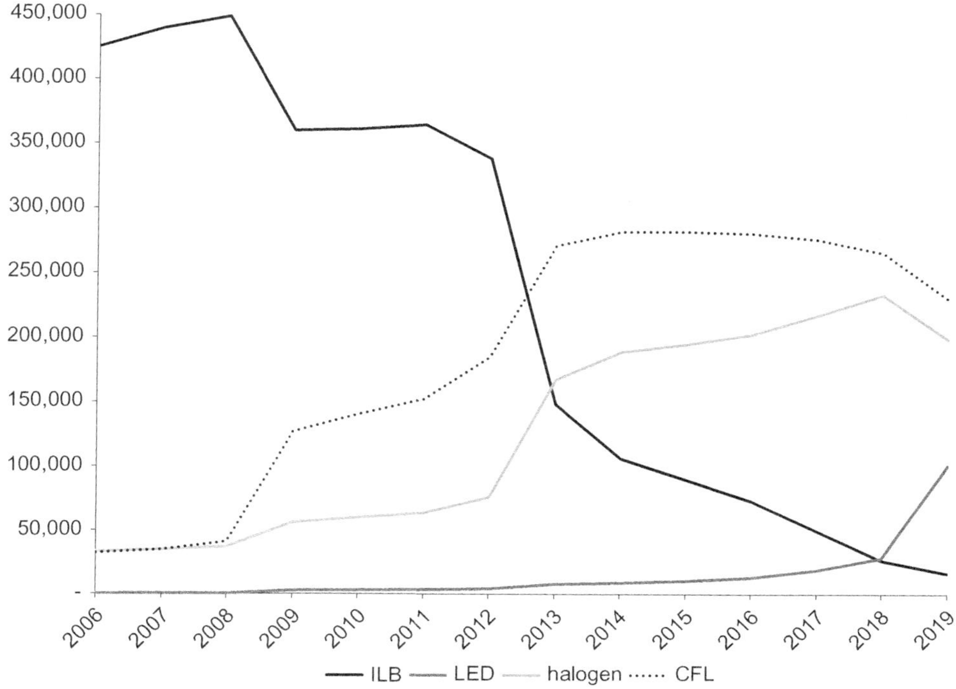

Figure 4.25 Number of different kinds of bulbs (in thousands) for non-directional lighting, owned by UK households, 2006–2019 (constructed using data from DUKES; Energy Consumption in the UK 2019; Electrical products tables; Table A2)

large-space lighting niches (e.g., schools, offices) where bright light was required. General Electric (GE) developed a more compact, spiral-shaped fluorescent lamp in the 1970s (Franceschini and Alkemade, 2016), but the purchase price was initially much higher (up to 20 times) than ILBs (Menanteau and Lefebvre, 2000), which hindered CFL diffusion into households. Their lifetime costs were lower, however, because they lasted considerably longer and used less energy.

- LEDs, which are based on solid state electronics, use seven to 10 times less energy than ILBs (Aman et al., 2013) and efficiencies are still improving (Sanderson and Simons, 2014). LEDs long remained much more expensive per bulb, which limited domestic use to small market niches such as directional, spot lighting, and decorative (e.g., Christmas trees) purposes. Broader use was stimulated by rapidly falling LED prices, which decreased 96% between 2008 and 2015 (Figure 4.26) due to scale economies, standardisation and commoditisation of LED chip technology, and improved manufacturing techniques (Sanderson and Simons, 2014). LEDs also last long, which by 2011 had already brought

Table 4.2. *Cost comparison in 2011 of different lamps (Aman et al., 2013: 488)*

Lamp	Cost ($/kilo-lumen)	Rated life (h)	Cost ($/kilo-lumen-h)
Halogen (750 lumen)	2.5	5,000	0.0050
CFL (800 lumen)	2	12,000	0.00017
Linear fluorescent lamp	4	25,000	0.00016
LED (800 lumen)	30	25,000	0.00120
		50,000	0.00060

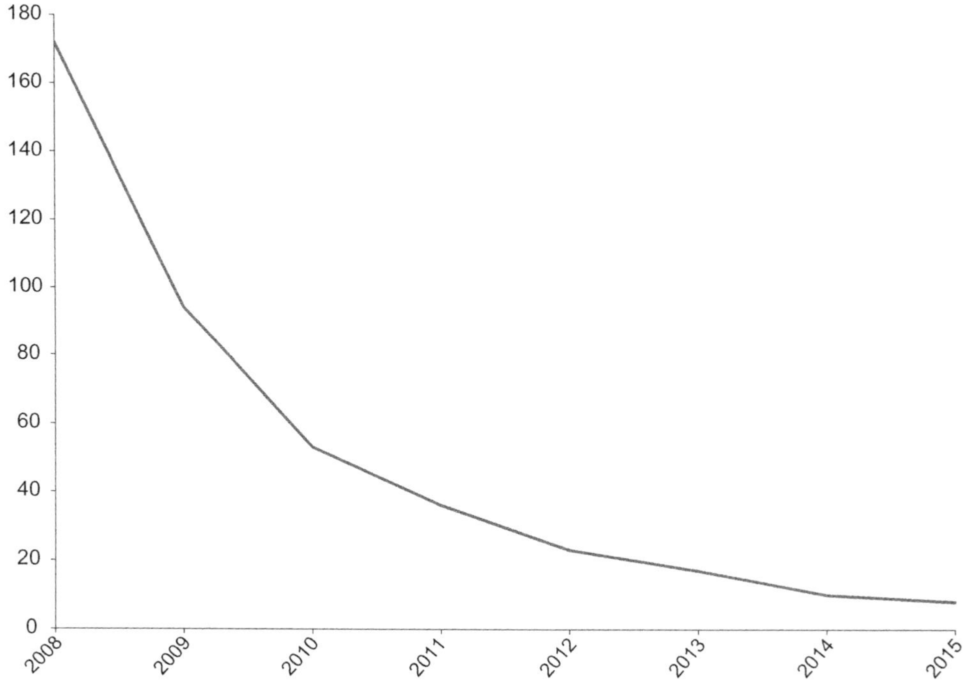

Figure 4.26 Decreasing prices of light emitting diodes, in USD/kilolumen (constructed using data from Thielemans et al. (2017))

their whole life costs ($/kilolumen-hour) close to CFL and halogen (Table 4.2), which stimulated wider uptake. LEDs are also versatile and can generate different types of lighting, which opens up the possibility for consumers to control colour and light intensity by connecting LEDs to in-house ICT systems, leading to 'smart lighting systems'.

Actors and Policies

Incumbent lighting firms (e.g., Westinghouse, General Electric, Philips) developed and marketed CFLs in the 1980s and 1990s, but cheaper Chinese firms overtook them in the 2000s (Khan and Abas, 2011), which lowered CFL prices. LEDs were initially developed by electronics firms (e.g., Bell Labs Hewlett-Packard, National Semiconductor, Fairchild, Samsung), but established lighting companies (e.g., Philips, Osram, GE) also acquired semi-conductor capabilities in the 1990s to position themselves for the anticipated LED market (Franceschini and Alkemade, 2016). UK firms play only a minor role in lighting innovation.

As attention on energy efficiency increased in the 1990s, policymakers attempted to stimulate energy-saving lamp diffusion through voluntary measures. The 1992 European Directive on Energy Labelling, for instance, required lighting companies to provide consumers with information about the energy efficiency performance of their products. UK policymakers also introduced subsidised give-away programmes and rebates to encourage retailers and utility firms to promote energy-efficient lighting, especially CFLs (Howarth and Rosenow, 2014; Martinot and Borg, 1998).[14] These policies had limited effects, however, because consumers perceived CFLs as giving 'cold' light; being unattractively shaped; taking too long to achieve full brightness; and being unsuitable for many fittings (Wall and Crosbie, 2009).

These problems galvanised further CFL technology development, aimed at reducing flickering, re-engineering shapes, and improving durability. Philips led these CFL developments, capturing 50% of the European market by 2000 (Menanteau and Lefebvre, 2000). LED-innovation in the 2000s was also driven by western multinationals, with Philips again acting as leading firm.

Increasing political attention for climate change and criticisms from environmental NGOs (e.g., WWF, Greenpeace) strengthened debates about the inefficiency of ILBs, which were increasingly framed in terms of energy waste (Franceschini and Alkemade, 2016). The 2005 European Eco-design Directive increased regulatory pressure by stipulating minimum energy efficiency standards for light bulbs and other energy-using products. UK policymakers also tightened Energy Efficiency Commitments (from 2002 to 2008) on utilities, mandating them to promote CFLs (Rosenow, 2012).

Incumbent lighting firms began to abandon ILBs, which was a low-profit margin market with tough international competition. In 2006, Philips announced support for a possible ILB ban, with the European Lamp Companies Federation (an industry association including Philips, Osram, and GE) following suit in 2007. The UK government announced in 2007 that it would phase out ILBs by 2011, while other European member states expressed support for the idea (Howarth and Rosenow, 2014). In 2009, combined pressures from the lighting industry, NGOs,

[14] Between 1994 and 1997, the Energy Savings Trust coordinated a give-away programme, involving 800,000 household CFLs (Martinot and Borg, 1998).

and member states led the European Commission to introduce a ban for ILBs of more than 80W, progressing to lower wattage in successive years.

While the ILB ban initially mostly boosted halogen and CFLs, it also stimulated LED-uptake (Figure 4.25). Further LED diffusion was driven both by declining costs and by new functionalities that allowed consumers to control colour and light intensity to create ambience (Monreal et al., 2016). Because of these rapid developments, LEDs came to be seen as the future of domestic lighting (Franceschini et al., 2018). The 2016 and 2018 European bans on directional and non-directional halogen bulbs explicitly intended to further accelerate the LED-transition and reduce energy consumption for residential lighting.

4.5.6 Smart Meters

Techno-Economic Developments

First discussed in the 1970s, discussions about smart meters gained momentum in the 1990s, when ICT-devices became more sophisticated and available. Smart meters can measure exact gas and electricity usage by customers and send the information to the energy supplier and the customer through in-home displays (IHDs) or other ICT-devices. A 2006 UK study suggested that such information feedback could lead to behaviour change and energy demand reductions of between 5 and 15% (Darby, 2006). Based on a positive cost-benefit analysis (BERR, 2007), the UK government decided in 2008 to roll out 53 million smart gas and electricity meters for all households and small businesses.

Depending on their technical configuration, smart meters vary in costs, from £50–60 for simple clip-on customer display units to £240 for more complex, multi-functional meters (BERR, 2007). After years of negotiations, the UK chose for a smart meter configuration towards the more complex end of the scale, costing around £215 (DECC, 2014d). Rising cost estimates were accompanied by increasingly positive estimates of benefits (Table 4.3), which in 2014 were

Table 4.3. *Successive cost-benefit estimates of smart meter introduction programme (data collected from successive government impact assessments)*

Year	**Costs (£billion)**	**Benefits (£billion)**	**Net Present Value (£billion)**
2009	8,110	11,700	3,590
2010	9,119	14,154	5,035
2011	10,575	15,827	5,070
2012	10,850	15,689	4,839
2013	12,114	18,774	6,660
2014	10,927	17,141	6,214

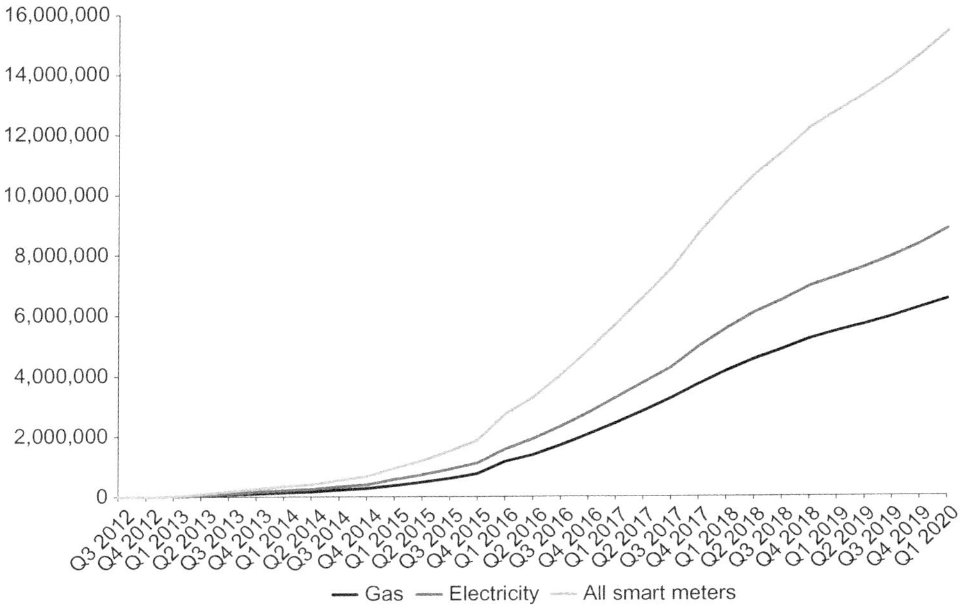

Figure 4.27 Domestic smart meters operated by large suppliers, 2012–2020 (constructed using data from Statistics at BEIS; Smart Meter Statistics; Table 1)

estimated at £17,141 billion: 48% in the form of supplier cost savings (by removing the need for meter reading visits, simplifying consumer switching, improving theft detection and debt management), 33% consumer energy savings, 8% UK-wide benefits (carbon), 5% peak load shifting, and 6% network benefits (DECC, 2014d: 15).

Despite the positive ex-ante estimates, smart meter diffusion has been disappointing (Figure 4.27), with only 15.5 million smart meters being installed by 2020, which is 29% of the original target.

Actors and Policies

Wanting to act on climate change, the Labour government's interest in smart meters was triggered by the aforementioned 2006 study. From 2007 to 2010, the government sponsored a large-scale smart meter trial in the Energy Demand Research Project. But instead of waiting for the trial outcomes, it decided in 2008 for a 100% roll-out of smart meters by 2020. The government also decided to make energy suppliers responsible for the roll-out and ruled that they would bear the up-front costs, which they could subsequently recover through consumer energy bills (Sovacool et al., 2017). This deviates from most other countries, which allocated this responsibility to Distribution Network Operators, who could install meters on a street-by-street basis in their regional monopolies (Geels et al., 2021). The government's 2008 decision also stipulated that smart meters should include IHD-units to provide consumers with information about their energy use, which was intended to trigger behaviour change. Observing Dutch social acceptance problems with a mandatory roll-out model, UK policymakers announced in 2012 that their smart meter roll-out would be voluntary, based on an 'opt-in' model, in which consumers had to ask energy companies for a smart meter (Geels et al., 2021).

Early implementations showed that smart meters did not work well in high-rise flats, basements, and rural areas (Sovacool et al., 2017). Evaluations also suggested that energy savings from information feedback might be more limited than was initially assumed, leading to reductions in estimated energy savings from 5–15% to 1–3% (Shipworth et al., 2019). A further problem was that first-generation ('SMETS-1') meters were incompatible with some suppliers, which complicated consumer switching.

These technical problems were widely covered in the media. Public debates also expressed concern about rising costs and the 'business case' for consumers. Additionally, libertarian and conservative media warned that the government could use smart meters to increase its control and surveillance over citizens, sowing doubts among the public about privacy and security consequences (Sovacool et al., 2019). These negative debates generated substantial scepticism around smart meters, which hampered the roll-out (Hielscher and Sovacool, 2018).

Energy companies also questioned the need for IHDs, arguing instead for cheaper apps that would allow phones, tablets, or personal computers to capture meter readings with no additional hardware cost (IoD, 2015). But government officials were committed to a standardised roll-out of IHDs and reluctant to consider changes. The logistics of the supplier-led roll-out and the fragmented nature of energy markets also created extra costs and delays because installation had to be done through individual site visits (Sovacool et al., 2017). Limited public trust in energy suppliers further undermined the roll-out, with growing concerns since 2016 about the health impacts of smart meters (Hielscher and Sovacool, 2018).

Large-scale roll-out commenced in 2016, but installation failures remained commonplace, with more than 10% of homes requiring multiple visits to complete the installation (Utility Week, 2017). After two relatively successful years, quarterly installation rates decreased markedly after 2017 (Figure 4.28). The problems with technical functionalities and SMETS1 standards, combined with years of negative debates, tarnished the public reputation of smart meters. The (un) reliability of energy companies and home installation problems also became issues of public debate (Connor et al., 2018). Although the first SMETS2 meters began to be installed in 2017, public trust remained problematic (Meadows, 2018), resulting in lukewarm consumer interest.

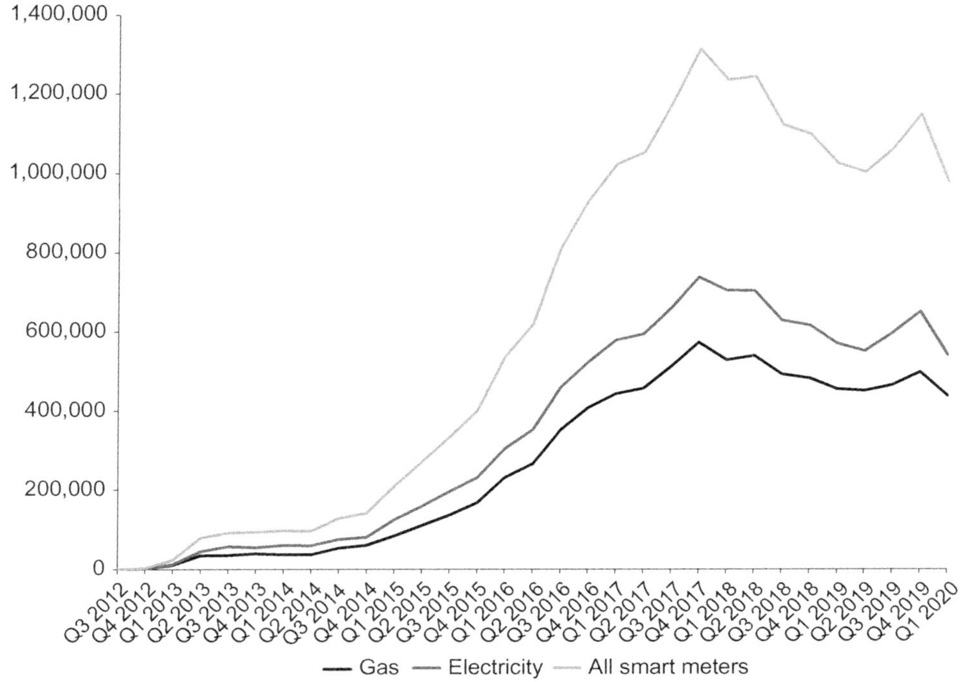

Figure 4.28 Domestic smart meters quarterly installation by large suppliers (constructed using data from Statistics at BEIS; Smart Meter Statistics; Table 2)

Acknowledging the delays and problems, the government announced in 2019 that it would postpone the roll-out deadline from 2020 to 2024, which would increase costs but allow for a greater focus on consumers (BBC, 2019). Although demand-reduction expectations have been downscaled, proponents made new promises about the future role of smart meters, smart grids, peak shifting, and demand-side response (DRS), in which information feedback from smart meters and new kinds of tariffs may modulate demand to accommodate fluctuations in electricity supply (Hielscher and Sovacool, 2018).

4.5.7 Smart Grids

'Smart grids' is a catch-all concept that can refer more narrowly to the inclusion of information and communication technologies (e.g., sensors, automatic switches, power electronics, digital controls) into the grid to enhance the visibility, control, and management of electricity flows or, more broadly, to wider functionalities that the ICT-enriched grid may enable and support, including demand-side response, storage, and flexible capacity management. We discuss the former in this section and the latter in Section 4.5.8.

Techno-Economic Developments

Interest in smart grids has increased since the 2000s, first, because of the increasing proliferation of information and communication technologies (ICT) and the shift to an information age, and, second, because the use of ICTs could potentially help address some of the increasing pressures on electricity grids, discussed in Section 4.3.1, particularly load management problems (due to increasing amounts of intermittent renewable power generation) and bi-directional electricity flows in distribution networks (due to increasing amounts of distributed power generation). Both problems require increased monitoring, control, and management of electricity flows (especially in distribution networks), which is something that ICT-devices could help with.

Smart grids are still a protected technological niche, constituted by a collection of projects. Both the number of UK projects and the proportion of real-world demonstration projects have increased since the early 2000s (Figure 4.29). Between 2002 and 2015, there were 95 R&D projects and 103 demonstration projects, covering a wide variety of functions and technologies such as sensors,

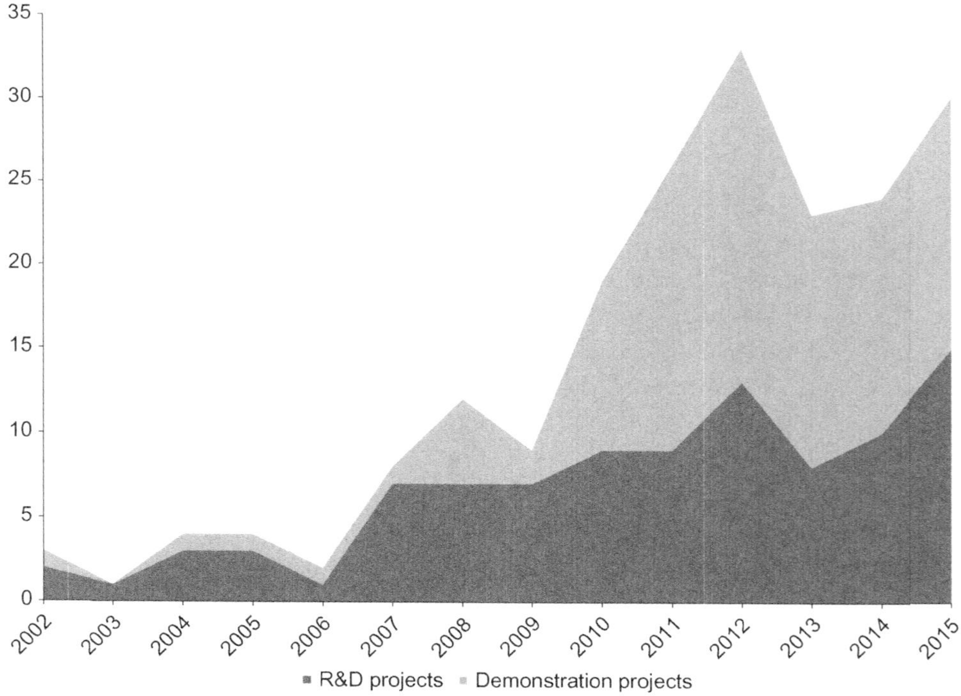

Figure 4.29 Annual number of new R&D and demonstration projects with smart grid technologies in the UK (constructed using information from the European smart grid projects outlook database JRC (2017))

automatic switches, power electronics devices, network management, and voltage and thermal constraints (Jenkins et al., 2015). Illustrative examples from the 2010s include Orkney Islands active network management project (exploring integration of high loads of wind, wave, and tidal energy into local grids), E.On MK2000 (exploring relations between smart homes and smart grids), and the Smart Hooky project (which connected monitoring nodes in 40 households in an Oxfordshire village to a smart hub at the distribution sub-station).

While these subsidised projects enabled learning processes and the build-up of new technical capabilities, they have not yet been followed by widespread diffusion and deployment of smart grid technologies (Lockwood, 2016).

Actors and Policies

DNOs were initially reluctant to engage with radical innovations such as smart grids, because of limited technical capabilities, limited future planning skills, and business models focused on efficiency and cost reduction (discussed in Section 4.3.2). To stimulate network innovation, Ofgem introduced the Innovation Funding Incentive (2005–2010), which increased DNO engagement in smart grid R&D projects. To also stimulate larger demonstration projects, Ofgem introduced the Low Carbon Network Fund (2010–2015), which allowed DNOs to bid for up to £500 million over five years. This instrument aimed to engage DNOs in real-world learning-by-doing processes and also pushed them to be 'more outward-looking by requiring bids for LCNF funding to include partnerships with suppliers, ICT firms, local communities and universities' (Lockwood, 2016: 120). The LCNF clearly accelerated the number of demonstration projects (Figure 4.29), which involve different configurations of actors. The Orkney Island project, for instance, involved Scottish and Southern Energy Power Distribution, University of Strathclyde, Smarter Grid Solutions, local renewable generators, and community energy groups. The E.On MK2000 project involved E.On, households, and Greenwave. And the Smart Hooky project involved Western Power Distribution, Hook Norton Local Authority, local community groups, Ranesas Electronics, and Tech Research (Owaineh et al., 2015).

Incumbent energy companies have been prominent in articulating future visions for UK smart grids. The Electricity Networks Strategy Group,[15] for instance, produced a high-level plan and a cost-benefit analysis for smart grids (ENSG, 2009), followed by a 'route map' (ENSG, 2010) and scenarios (ENSG, 2012). In 2011, DECC and Ofgem established the Smart Grid Forum (SGF) to address technical and economic drivers and barriers. Although the Forum accommodated multiple kinds of stakeholders, 'membership is oriented towards the electricity

[15] The ENSG includes Ofgem, National Grid, TNOs, DNOs, and policymakers (DECC, BEIS).

sector, and includes lower representation from the ICT sector' (Hiteva and Watson, 2019: 147). The SGF's 2014 *Smart Grid Vision and Routemap* is broad in terms of technologies and functionalities (including demand management and storage) and does 'not attempt to outline a precise path' (SGF, 2014: 7). It does, however, focus mostly on distribution networks, which smart technologies aim to make more 'flexible and efficient', leading to lower costs, empowerment of consumers, economic growth, and jobs (p. 6). A key aspect of this vision concerned significant organisational change for DNOs to become Distribution System Operators (DSOs), with much wider responsibilities as *active* managers of a smart distribution system.

The focus in these smart grid visions on efficient, low-cost improvements of existing grids is unsurprising given the dominance of incumbent energy companies. It does, however, mean that visions of local semi-autonomous micro-grids are hardly discussed in the UK (Lockwood, 2016). This contrasts with other countries (e.g., United States, Germany, Denmark, Netherlands), where visions of micro-grids in relation to decentralised energy systems and socio-cultural changes (Meadowcroft et al., 2018; Stephens et al., 2013; Van Mierlo, 2019) are more common.

It is also striking that policymakers have, so far, been relatively absent in the articulation of visions or policy plans for smart grid development. Some analysts therefore diagnosed that: 'Regulatory incentives like LCNF encourage piecemeal solutions without a clear UK strategy' (Connor et al., 2018: 6).

In sum, there are increasing numbers of smart grid projects, which enable learning processes, network building, and development of new technical capabilities. But future visions are very broad and oriented towards large-scale applications suiting incumbent interests. Given DNO resistance to major change, it remains unclear if current impulses and incentives are sufficient to overcome the DNO lock-in mechanisms and lead to widespread deployment of smart grid technologies in distribution networks (Connor et al., 2018; Hiteva and Watson, 2019; Lockwood, 2016).

4.5.8 Flexibility-Enhancing Options: Battery Storage and Demand-Side Response

To alleviate increasing load management problems, grid managers have become more interested in flexibility-enhancing options that help match electricity supply and demand. Conventional power generation (e.g., gas turbines) or energy storage technologies can be used to provide back-up capacity, which can be activated when demand peaks. Demand-side response (DSR) provides flexibility by reducing electricity demand at certain periods to either 'shift the peak' or follow fluctuating supply, which would entail a *reversal* of the current operational principle (in which supply follows demand). While some of these flexibility-enhancing options have been used for some time (e.g., energy storage through pumped hydro facilities or paying large firms to disconnect from grid supply during peak demand), it is the emergence of newer options (e.g., battery storage and household DSR, mediated by smart meters) that has generated new enthusiasm and will be discussed next.

Techno-Economic Developments

Battery Storage: While a range of energy storage technologies (e.g., flywheels, compressed air, liquid air, hydrogen) is under development (Schmidt et al., 2017), battery storage is already in the deployment stage. The use of home batteries (in the 4–20 kWh range) in combination with rooftop solar-PV has remained relatively small in the UK, compared to countries such as Germany, Spain, and Italy (Gardiner et al., 2020). Larger-scale battery storage has experienced more growth, especially in relation to providing grid-oriented flexibility services. Since 2013, the number of UK battery storage projects larger than 150 kW and annual installed capacity has increased substantially (Figures 4.30 and 4.31). Despite a drop in 2019 (which will be discussed next), future growth prospects are high, with 13 battery storage projects (with 343 MW capacity) under construction and 192 projects (with 13,152 GW capacity) being granted planning permission (information from the Renewable Energy Planning Database at Statistics at BEIS). While battery storage can be operated as stand-alone facilities, they can also be co-located with fossil fuel plants or large-scale renewable energy (RE) generators (which means that excess capacity can be stored and sold to the grid at a later stage). Most residential and utility storage projects use lithium-ion batteries, which have reduced substantially in price in recent years, owing to a learning rate of about 12% (Schmidt et al., 2017).[16] Residential battery storage is relatively more expensive than utility storage. Gardiner et al. (2020) estimate that a 4kWh home-storage system (including battery plus inverter and other electronics) cost around £3,497 in 2017, which has restricted its use to a small market niche.

Demand-Side Response: DSR already exists for UK major commercial energy users, who have bespoke contractual arrangements about the payments and conditions under which they will reduce electricity demand in response to a signal or incentive from the grid operator (Grünewald and Torriti, 2013). In the early 2010s, UK smart-meter roll-out plans stimulated visions about applying DSR to household demand (Parrish et al., 2019). Various combinations of smart meters, new tariffs, and smart appliances resulted in different residential DSR-visions

[16] Learning rates refer to relative price decreases for a doubling of installed capacity.

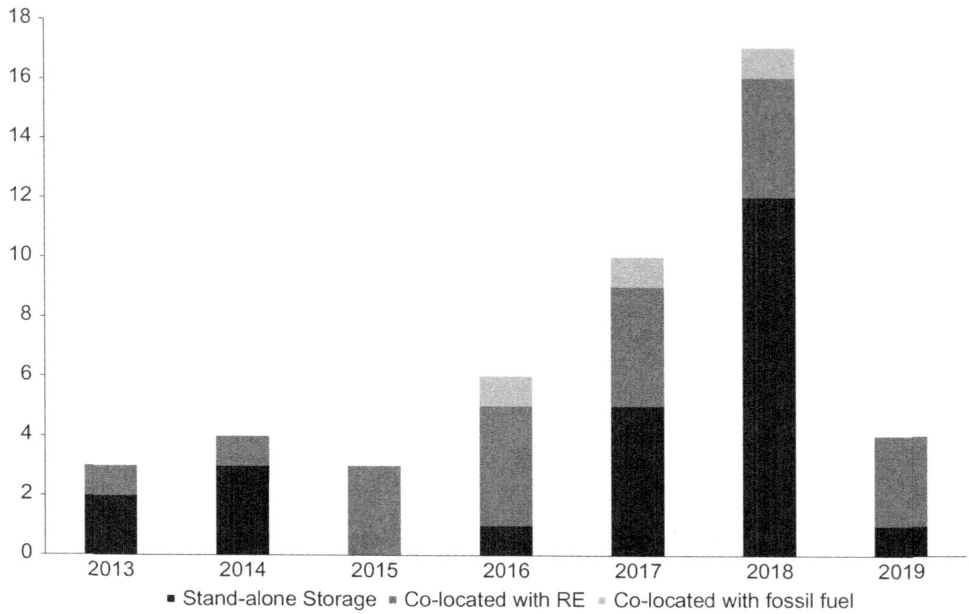

Figure 4.30 Number and type of UK battery storage projects larger than 150 kW, 2013–2019 (constructed using information from the Renewable Energy Planning Database at Statistics at BEIS) (RE refers to Renewable Energy)

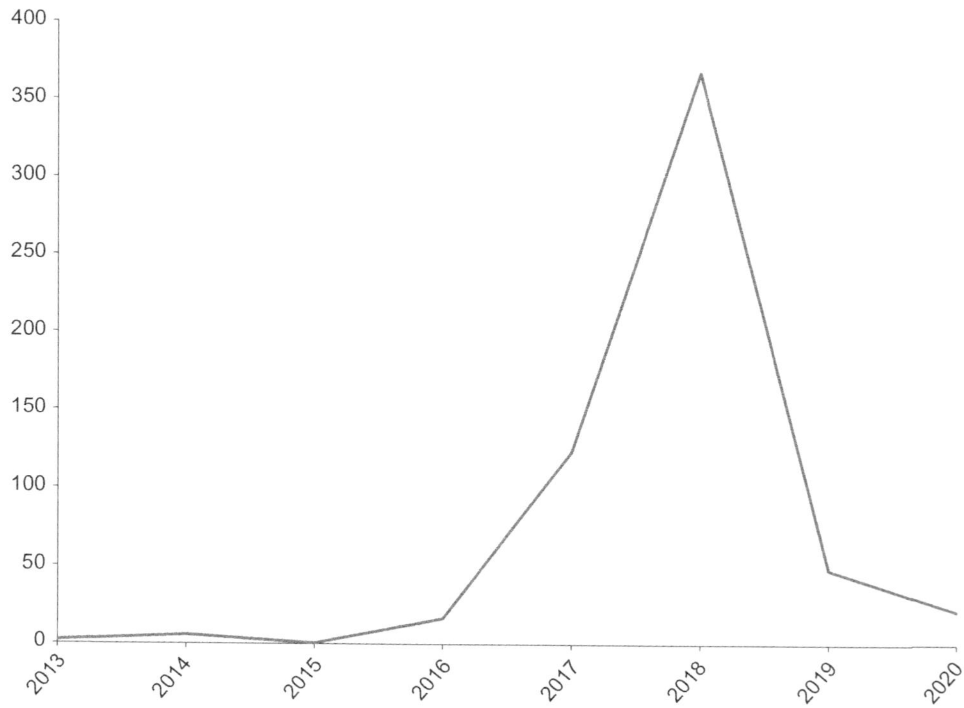

Figure 4.31 Annual installed capacity (in MW) of UK battery storage projects larger than 150 kW, 2013–2019 (constructed using information from the Renewable Energy Planning Database at Statistics at BEIS)

(Darby and McKenna, 2012), which to varying degrees propose economic principles and technological controls to bring about desired behaviour change:

- *Static peak shifting*: Smart meters enable the introduction of time-of-use tariffs that incentivise consumers to change the timing of certain activities (e.g., doing the laundry at night). Timers could be added to automatically shift loads away from peak times.
- *Dynamic load shifting*: Smart meters enable the introduction of real-time ('dynamic') tariffs that would incentivise households to change electricity demand on an hourly basis in response to price fluctuations driven by supply and demand.
- *Continuous balancing* (or 'direct load control'): Smart meters, smart appliances, and new tariff structures allow grid managers to remotely control appliances (e.g., freezers, refrigerators, washing machines) to adjust electricity demand.

While there have been some trials, implementation of household DSR has, so far, been limited (Langendahl et al., 2019; Parrish et al., 2019).

Actors and Policies

Battery Storage: Home batteries are provided by new companies such as Tesla, Powervault, Moixa, and Sonnen, but, given high prices and lack of dedicated policy, the residential battery storage market has remained small. Without policy intervention, residential batteries are not currently financially viable in the UK (Gardiner et al., 2020: 1).

The number of commercial storage companies, who purchase batteries from international suppliers and provide flexible energy services to grid companies, has increased quickly in recent years. New entrants such as KiWi Power, Renewable Energy Systems, and UK Battery Storage were followed by incumbent utilities (such as EDF, Vattenfall, and E.ON) and big firms from other sectors (e.g., Jaguar Land Rover, Statoil). This rapid growth was stimulated by Capacity Market (CM) contracts and Enhanced Frequency Response (EFR) contracts from National Grid.[17] The drop in 2019 was caused by CM rule changes for the 2017/18 auctions (which de-rated batteries depending on their duration, halving CM revenues for most batteries) and National Grid halting EFR in 2018. Since 2019, however, battery storage was able to access the National Grid's broadened Balancing Mechanism[18] and other forms of trading, which opened up new revenue streams

[17] EFR is an energy provisioning service that achieves 100% active power output at 1 second (or less) of registering a frequency deviation. This new high-speed service helps National Grid fine-tune grid frequencies closer to 50Hz.

[18] The Balancing Mechanism is a trading instrument used by the National Grid to balance electricity supply and demand with less than an hours' notice. In 2019, National Grid lowered the minimum threshold to take part from 100 MW to 1 MW, which broadened access to smaller participants, including battery storage providers.

and attracted more companies and projects. With the aim of further boosting the large-scale storage market (particularly in relation to utility-scale wind and solar farms), the government in July 2020 relaxed project size limitations from 50 MW to 250 MW in England and 350 MW in Wales.[19] Decentralised energy storage and visions of energy independence (in which households mostly consume self-generated power) or virtual power plants (in which local communities combine and aggregate self-generated and stored power) do not receive much policy attention in the UK (Ambrosio-Albalá et al., 2019).

Demand-Side Response: Increasing amounts of intermittent renewables and anticipated smart meter rollouts led economists, engineers, and policymakers to consider possibilities of residential DSR. Because demand-side flexibility could support load management and reduce the need for network upgrades and peaking plants, policymakers saw DSR, smart meters, and new tariffs as interesting possibilities: 'Smart meters make time-varying and other sophisticated type of tariffs possible by recording the time when electricity is used, and by allowing two-way communications. Such tariffs can incentivise demand-side response (DSR) or load-shifting, which can potentially bring significant benefits to the electricity system' (DECC, 2014d: 58). DSR was therefore allowed to participate in the Capacity Market (CM) auctions, which pay power station owners or providers of DSR or storage for the *availability* of electricity generation or reduced demand in agreed periods (often related to peak demand). Between 2014 and 2020, however, DSR won only 1.5% of contracted resources in these auctions (Lockwood et al., 2020).

There are various reasons that DSR has so far remained a 'niche activity' (Langendahl et al., 2019: 3). First, incumbent actors remained hesitant about DSR, which 'represents something very different from more established forms of the electricity industry's asset management approaches' (Langendahl et al., 2019: 7), which are based on technological solutions and the building of new capacity. Contractual DSR relationships and incentives through new tariffs also created uncertainties about the reliability of delivery of demand reduction. Second, lobbying from incumbent power generators (who had invested substantially in flexible gas-powered plants) shaped CM auction rules so that these marginalised DSR and favoured conventional back-up capacity (Lockwood et al., 2020).[20] 'The protection of the existing system thus appears to have been a higher priority than the development of a new, more flexible and demand-side focused system'

19 Before that date, projects larger than 50 MW had to go through national infrastructure planning regulations.

20 'It is clear that some Big Six and second-tier generating companies lobbied at certain points to close down what they framed as special treatment for DSR, for example in bid bond levels and contract length, and to change the proposed triad option for cost pass through. In this lobbying, incumbents deployed a set of ideas about DSR, especially that it was inferior to generation capacity in both duration and reliability of delivery' (Lockwood et al., 2020: 8).

(Lockwood, 2017: 17). Third, households may be less interested in DSR than economic modelling studies assume. Based on a systematic review of international evidence on trials, surveys, and programmes of residential DSR, Parrish et al. (2019: 107) conclude that 'the high levels of demand response modelled in some future energy system scenarios may be more than a little optimistic'. Specifically, households tend to have limited knowledge of DSR, may not accept high degrees of external appliance control, and may not respond predictably to price signals because most electricity consumption relates to routinised practices (Verbong et al., 2013).

Although DSR holds potential, we suggest that its momentum has weakened, especially compared to battery storage, which is expected to provide large amounts of flexible capacity in the coming years.

4.6 Low-Carbon Transition through Whole System Reconfiguration

Pulling together information from the sub-system and niche-innovation analyses, this section first assesses low-carbon whole system reconfiguration through the three lenses (techno-economic, actors, institutions) and then addresses speed, scope, and depth of change.

4.6.1 Low-Carbon Innovations Driving GHG Emission Reductions

Based on the investigations in this chapter, we conclude that the substantial reduction in greenhouse gas emissions between 1990 and 2019 resulted from four main changes:

- the diffusion of renewable electricity technologies (e.g., onshore wind, offshore wind, bio-power, and solar-PV), which displaced coal in the power generation sub-system; this change represents niche-innovations replacing existing system technologies.
- the diffusion of CFLs and LEDs, which replaced incandescent light bulbs in the consumption sub-system; this change also represents niche-innovations replacing existing system technologies.
- the switch from coal to gas in the power generation sub-system, which represents a substitution between existing system technologies;
- efficiency innovation in appliances, which helped to reduce electricity demand, despite appliance proliferation in households; this change represents existing incremental improvements in existing system components.

4.6.2 Techno-Economic Reconfiguration

These low-carbon innovations represent either 'modular incrementalism' or 'modular substitutions', which improved or replaced parts of the existing sub-systems without challenging traditional boundaries between them. Summary

Table 4.4. *Mapping the winds of whole system reconfiguration in the UK electricity system (adapted from McMeekin et al. (2019: 1226))*

<table>
<tr><th colspan="6">Core elements</th></tr>
<tr><th rowspan="4">Linkages (coupling) between system components</th><th></th><th>Reinforced</th><th colspan="3">Substituted</th></tr>
<tr><th rowspan="2">Unchanged</th><td rowspan="2">Modular incrementalism
- efficiency innovation in appliances</td><td colspan="3">Modular substitution</td></tr>
<tr><td>System-system switching
From coal to gas</td><td>Niche-system hybridisation
coal–biomass conversion</td><td>Niche-system replacement
RETs for fossil fuels; CFLs/LEDs for ILBs</td></tr>
<tr><th>Changed</th><td>Architectural stretching
- grid expansion for offshore renewables and imports</td><td colspan="3">Architectural reshaping
- smart grid, smart meters
- battery storage
- demand-side response (DSR)</td></tr>
</table>

Table 4.4 positions these and the other innovations we have discussed in the techno-economic mapping framework that we developed in Section 2.2.1. RET diffusion was accompanied and enabled by incremental innovations in transmission grids (extensions, strengthening, new offshore grids, new interconnectors), which represent 'architectural stretching' to incorporate remote RETs such as onshore and offshore wind farms.

For a few years, however, increasing amounts of intermittent RETs in the generation sub-system have been having knock-on effects in the consumption and infrastructure sub-systems, because they create problems (such as intermittency and load balancing) that have increased interest in radical niche-innovations (such as smart meters, battery storage, smart grids, demand-side response), which improve grid management by enhancing flexibility and bi-directional electricity flows. These knock-on effects and niche-innovations create new linkages between the three sub-systems. The sub-system boundaries are thus becoming more porous, suggesting the possible emergence of a new phase of whole system reconfiguration (McMeekin et al., 2019), including changes in the system architecture, for example, intelligent and flexible load management, peak shifting, decentralised power generation (including 'prosumption', which is the use of self-generated electricity by consumers), and different operational principles (e.g., demand-follows-supply). This 'architectural reshaping' is still in the making and uncertain because most associated niche-innovations are still in early developmental stages and because of ongoing institutional struggles (which are further discussed next).

Modular changes within generation and demand sub-systems have, so far, accounted for most of the low-carbon progress. This modular reconfiguration pattern was possible because electricity generation and demand sub-systems are separated from each other by the grid. These loosely coupled sub-systems can thus have independent operational and innovation patterns that do not interfere with each other. It is only in recent years that modular changes in the electricity generation system have been having knock-on effects on the grid and demand sub-systems, which create the prospect of a new phase of architectural reshaping, as indicated earlier.

4.6.3 Actor Reconfiguration

Focusing on actors and social networks, we conclude that the unfolding low-carbon electricity system transition was enabled by substantial changes in goals, agendas, interests, strategies, and capabilities. This means that constitutive dimensions of some actors changed during the transitions, as they learned, struggled, debated, and interpreted changing contexts, barriers, and opportunities. This does not mean, however, that all actor dimensions changed in low-carbon directions. Some routines, conventions, capabilities, or interests remained relatively 'locked-in' and unchanged, which hindered the speed and depth of change. Some actors also became worried about new concerns that hindered or detracted from low-carbon transitions.

Regarding the three sub-systems, Tables 4.5 to 4.7 provide interpretive assessments of the main actor changes, lock-ins, and new concerns. For each actor category, we have added evaluative qualifiers in capital letters to indicate the relative importance of actor changes or lock-ins for the unfolding low-carbon transition.

In the electricity *generation sub-system*, the main actor changes in support of low-carbon transitions were: a) gradual reorientation of utilities through adjustments in investment strategies, and technical and operational capabilities, b) relatively interventionist government policies that shaped markets and supported utilities' reorientation through attractive financial incentives for large-scale low-carbon technologies, c) relatively high, but fluctuating, public attention on climate change, negative discourses about coal, and positive discourses about renewables.

The main lock-ins and new issues that hampered low-carbon transitions were: a) sunk investments in existing coal- and gas-fired power plants, and large-scale business models, which utilities aimed to protect by shaping the pace of change, b) stable and closed networks between policymakers and utilities (including official platforms and informal consultation channels), which enabled frequent discussions and coordination, c) increasing concerns (reinforced by conservative politicians and utilities) about rising electricity consumer bills.

Table 4.5. *Changes and lock-ins for actors in the electricity generation sub-system*

	Actor changes supporting low-carbon transition	Actor lock-ins and competing issues constraining low-carbon transition
Firms	MEDIUM-LARGE - Utilities accommodated climate change mitigation as important goal - Gradual low-carbon reorientation through changing innovation strategies: a) retiring coal-fired power plants, b) expanding gas-fired power plants, c) attempted nuclear fleet expansion, d) moving into large-scale RETs (biomass conversion, onshore and offshore wind parks).	MEDIUM - Protect sunk investments in existing power plants by shaping the pace of change. - Maintain large-scale operations and business model - Competitive pressure from new entrants is new issue that strengthens low-cost focus.
Policymakers	LARGE - Climate change rose on policy agendas, leading to new Ministries (first DECC, then BEIS), goals, plans, and policies that provided direction for low-carbon transitions. - New policy instruments provided (attractive) financial support for nuclear, gas, and large-scale RETs, which underpinned reorientation by utilities. - Less and more fluctuating policy support for small-scale RETs, community energy, and households.	MEDIUM - Relatively stable networks between policymakers and utilities, enabling deliberations about direction and pace of change. - Competing issues such as affordability and energy security also rose on policy agendas. - Concerns about rising energy prices contributed to a negative discourse around low-carbon innovations ('green crap') and led to down-scaled support ('energy reset').
Users	SMALL - Limited direct involvement of consumers in power generation	MEDIUM - Consumers indirectly pay for low-carbon electricity generation (through energy bills or taxation) and care about rising electricity prices.
Civil society organisations, public debate	LARGE - Fluctuating but relatively high public attention to climate change kept the issue on policy agendas - NGOs campaigns supported low-carbon transition and stronger policies. - NGO protests delegitimated (proposals for) coal-fired power plants and hampered shale gas.	MEDIUM - Public debates about rising energy costs constrains low-carbon transition - NGO protests and negative debates about some RETs (Big Biomass, onshore wind farms) - Increasing distrust of utilities and complaints about dysfunctional markets.

Table 4.6. *Changes and lock-ins for actors in the electricity consumption sub-system*

	Actor changes supporting low-carbon transition	Actor lock-ins and competing issues constraining low-carbon transition
Firms	LARGE - Appliance manufacturers have accepted the energy efficiency agenda and reoriented their innovation strategies (to protect their core business model).	LARGE - Strong lock-in to business model of selling more appliances that are continuously improved and differentiated along multiple quality and performance dimensions, including the addition of new functionalities.
Users	SMALL-MEDIUM - Consumers relatively disengaged, but showed some willingness to buy energy-efficient appliances such as light bulbs or refrigerators (although less than in other European countries).	LARGE - Electricity use results from routine practices that are hardly questioned as such. - Cultural conventions regarding convenience, cleanliness, freshness, and rising expectations for connectivity and entertainment are more important considerations than climate change and underpin the dynamics of domestic practices, including the purchase of more and larger appliances (e.g., TVs, fridges). - Smart meters triggered less demand reduction or DSR than hoped.
Policymakers	LARGE - Strengthening European and UK energy efficiency regulations for appliances. - New visions of flexible demand (DSR, new tariffs, smart meters) to support load management in grids.	MEDIUM - Electricity consumption less salient than electricity supply issues. - No desire for policies to stimulate deeper low-carbon behaviour changes beyond efficient appliance purchase.
Civil society organisations, public debate	SMALL - Only a few NGOs campaign on electricity use. - These are largely supportive of efficiency agenda, calling for deeper and quicker implementation.	SMALL - Muted public debate on energy efficiency (because of limited disagreement). - Limited public debate about electricity consumption levels or behaviour change - Debates around digital transitions and smart homes legitimate further spread of ICT-devices.

Table 4.7. *Changes and lock-ins for actors in the electricity grid sub-system*

	Actor changes supporting low-carbon transition	Actor lock-ins and competing issues constraining low-carbon transition
Independent regulator (Ofgem)	SMALL/MODERATE - Ofgem reluctantly incorporated climate change into its remit but has been criticised for insufficiently acting on this goal. - Since the mid-2000s, Ofgem introduced new policy instruments to stimulate network innovation, but these focused mainly on R&D and demonstration projects rather than wider deployment.	LARGE - Ofgem's focus is on low cost and efficiency, which hampered network innovation and subsequently led to add-on policies. - Climate change layered on top of traditional goals but remained less important than low-cost focus.
Transmission Network Operators	MODERATE TNOs and National Grid gradually reoriented through incremental grid changes (e.g., reinforcements, onshore and offshore extensions).	LARGE Transmission network operators have deep sunk investments, and are oriented towards stability, rent seeking, and incremental change.
Distribution Network Operators (DNOs)	SMALL/MODERATE DNOs engage in demonstration projects (e.g., smart grids, battery storage, DSR), but are not yet committed to wider deployment of radical innovations.	LARGE DNOs hesitant about radical change because of risk-averse orientation, traditional business model (around passive distribution), atrophied technical capabilities, and limited long-term planning skills.
Civil society organisations, public debate	SMALL Public debate about low-carbon electricity infrastructure upgrades relatively muted.	MODERATE Amenity and landscape concerns resulted in some protests from local communities and NGOs against new cables and pylons, leading to delays.

In the electricity *consumption sub-system*, the main actor changes in support of low-carbon transitions were: a) tightening European and UK energy efficiency regulations, including strong policies such as a ban on incandescent light bulbs, b) gradual reorientation of appliance manufacturers, who embraced the energy efficiency agenda and adjusted their innovation strategies.

The main lock-ins that hampered the depth of low-carbon reorientation were: a) adherence to the business model of selling more electrical appliances, which are continuously improved and differentiated along multiple dimensions offering better and more functionalities to consumers, b) cultural conventions and assumptions with regard to consumption (e.g., electricity use as an unquestioned background assumption of modern societies; desires for convenience, cleanliness, freshness, entertainment, and connectivity that underpin domestic practices and increased appliance use).

In the electricity *grid sub-system*, the actor changes were less substantial than for the generation and consumption sub-systems: a) the National Grid and Transmission Network Operators (TNOs) gradually reoriented in response to grid pressures, making incremental changes that build on existing capabilities, b) the regulator Ofgem reluctantly accepted climate mitigation as an additional goal and introduced some add-on instruments to stimulate infrastructure innovation (mostly in the form of R&D and demonstration projects).

The main actors are reluctant to commit to radical change because of strong lock-in mechanisms: a) TNOs and Distribution Network Operators (DNOs) have deep sunk investments in electricity infrastructures, b) DNOs are also locked into their traditional business model (around passive distribution) and have lost technical capabilities and long-term planning skills in recent decades, owing to a low-cost 'sweat the assets' orientation, c) Ofgem mainly focused on low cost and efficiency, and shaped grid regulations accordingly, d) National Grid, DNOs, TNOs, and Ofgem formed stable closed-knit networks and shared mindset and orientations (operating a form of 'club governance').

This analysis shows that that the unfolding low-carbon transition in the UK electricity system has evolved into what currently is a negotiated and controlled transformation process, driven by the reorientation of incumbent actors (e.g., utilities, grid actors, appliance manufacturers, policymakers), who gradually adjust their goals, capabilities, strategies, and instruments. Civil society organisations and public debates played important roles in raising the profile of climate change and in shaping perceptions of electricity generation technologies (e.g., delegitimating coal, supporting RETs). The role of consumers and households in the low-carbon transition has, so far, remained more limited (e.g., stagnated rooftop solar-PV adoption and limited 'prosumption', limited DSR, limited behaviour change), although there has been some change towards buying more energy-efficient appliances. The most important consumer role, so far, is that they have ultimately paid for low-carbon power generation and grids through their electricity bills and general taxation. But this was not a deliberate choice since electricity bills are opaque, and consumers did not explicitly consent to utilities passing on extra costs through their bills.

One actor-related risk to the transition concerns frequent policy changes and U-turns, and the recently weakening support policies, which may erode investor confidence. Another risk is the presence of social acceptance problems of particular innovations, which are partly caused by a top-down technocratic policy style. A third risk is that the planned retirement of coal and nuclear plants by the mid-2020s may create capacity problems, especially if RET expansion and new nuclear power plants and CCS proceed more slowly than anticipated. A fourth risk is that the slow and reluctant reorientation of grid actors (especially DNOs) may create reverse salients that hamper deeper reconfiguration of the entire electricity system. A fifth risk relates to the challenge of mobilising £200–300 billion investments that the low-carbon electricity transition is estimated to require between 2010 and 2030 (Watson et al., 2014). While investments have increased substantially in the 2010s, the further roll-out of low-carbon options and infrastructural reconfiguration will require much greater expenditure in the next 10 years, which may be challenging in a post-COVID climate of economic recession and high debt.

4.6.4 Policy Reconfiguration

Formal Policies and Regulations

We conclude that the unfolding low-carbon electricity system transition was supported by strengthening policies, especially in the generation and consumption sub-systems.

The electricity generation sub-system was institutionally reconfigured through many policy changes. A target of 30% renewable electricity by 2020 was introduced by the UK Low Carbon Transition Plan (2009), creating a general sense of direction and speed. The direction of travel was further elaborated by multiple policy plans and strategies that altered regulatory frameworks, for example, the UK Low Carbon Transition Plan (2009), the amended Renewables Obligation (2009), the UK Renewable Energy Strategy (2009), the Carbon Plan (2011), the Energy Bill (2012), the Electricity Market Reform (2013), and the energy reset (2015). Several generic financial instruments shaped economic frame conditions, for example, Feed-in-Tariffs, Renewables Obligation, Contracts for Difference, and Carbon Floor Price. And various technology-specific plans and strategies addressed more specific implementation issues, for example, White Paper on Nuclear Energy (2008), UK Bioenergy Strategy (2012), UK Solar PV Strategy (2013; 2014), and Offshore Wind Sector Deal (2020).

In the electricity consumption sub-system, policy changes were, more limitedly, focused on the energy efficiency of appliances. But since regulatory policies strengthened over time, they increasingly shaped markets and innovation strategies. The European Directive on Energy Labelling focused on information provision to consumers, while successive European Ecodesign Directives (2005, 2009) articulated minimum efficiency standards for appliances, which increased over time. The UK Products Policy (from 2009 onwards) translated these European Directives to UK contexts. Between 1994 and 2012, UK policymakers also imposed several energy savings obligations on energy suppliers to help disseminate more efficient appliances. The incandescent light bulb ban (2007 UK,

2009 EU) was a strong market shaping policy that accelerated the transition towards energy-efficient CFLs and LEDs. The UK's smart meter roll-out decision (2009) was also a strong (technology-forcing) policy, but this encountered implementation problems.

Policy changes in the grid sub-system have remained more limited and weaker. While there are some broad visions and ad-hoc decisions (e.g., recent relaxing of project size limitations for battery storage), there are few policy strategies or dedicated instruments for smart grids, DSR, or battery storage. And the Capacity Market policy, which potentially could have supported grid transformation, was designed in a way that favoured conventional back-up capacity and marginalised alternatives such as DSR and storage. So, while the technologies for architectural reshaping are becoming available (as discussed in Section 4.6.1), socio-institutional changes in the grid sub-system are lagging behind. There thus seems to be increasing tension between technological and institutional reconfiguration. The independent regulator Ofgem, which itself initially resisted inclusion of climate change in its remit, layered a few innovation-oriented instruments on top of its primary, efficiency-oriented regulations, but these have, so far, done little to bring about wider system change. Deeper reconfiguration would require a shift in the grid sub-system from being a buffer to an active integrator of the whole system. But existing grid-actor roles and operational routines have, so far, not changed much since the late 1990s.

While these strengthening policies in the generation and consumption sub-systems supported low-carbon transitions, they did not threaten vested interests. Instead, these policies enabled incumbent actors (e.g., utilities, appliance manufacturers, TNOs, National Grid) to gradually reorient. In fact, policymakers made many *political* choices that favoured large-scale options, which suited incumbents, over smaller-scale options linked to new entrants:

- large-scale biomass (e.g., co-firing and biomass conversion of coal-fired plants) was favoured over smaller-scale dedicated biomass plants (by sawmills or poultry farms), even though the latter had higher carbon performance (but was somewhat more expensive);
- large-scale solar-PV farms (operated by landowners, investors, project developers) were favoured over small-scale roof-top solar-PV (by households);
- large-scale wind farms (by energy companies, investors, project developers) were favoured over smaller-scale community wind projects;
- large-scale battery storage (by new and incumbent companies) was favoured over decentralised batteries by households with rooftop solar-PV;
- smart grids in relation to the flexibility agenda were favoured over micro-grids in relation to decentralisation and energy independence;
- the Capacity Markets policies were designed to favour conventional back-up capacity (by utilities) rather than domestic DSR;

- large-scale offshore wind parks and nuclear power (by utilities) received generous financial (and political) support.

Another characteristic of UK electricity policy, particularly in the generation sub-system, is the high degree of flux and churn over time: RET policies such as the RO (2002), amended RO (2009), and CfD (2013) succeeded each other in the space of a few years; CCS support policies were introduced and removed several times (2007–2010; 2012–2016); a moratorium on onshore wind was introduced in 2016 and removed again in 2020; feed-in-tariffs for small-scale renewables were introduced in 2010, reduced in 2016, and scrapped in 2019. The CCC (2020: 99) characterised this policy flux and churn as 'shortcomings', noting that 'frequent changing of policy should be avoided' because it 'can damage faith in Government policy and reduce business willingness to invest'.

Governance Style

Substantial policy changes in the electricity generation and consumption systems were complemented by substantial changes in governance style. In the generation sub-system, the governance style shifted in the late 2000s from a hands-off, technology-neutral approach towards a more interventionist and technology-specific style (Carter and Jacobs, 2014; Kern et al., 2014a). In the consumption sub-system, the governance style also became increasingly interventionist as stronger energy efficiency standards and phase-out policies increasingly shaped markets and innovation strategies. These changes in governance style happened in tandem with changes in policy goals. Although climate change was initially layered on top of existing goals (e.g., low cost, energy security), they became increasingly integrated, leading to concepts such as the 'energy trilemma' that recognised all three goals as important, while acknowledging potential trade-offs.

Some other governance style dimensions remained unchanged, however, which helps explain the persistent political privileging of incumbent interests and large-scale options. The centralised style of policymaking and close-knit policy networks provided more access to incumbent utilities and grid actors than to new entrants. The technocratic, top-down policy style with limited interest in stakeholder engagement contributed to social acceptance problems for onshore wind, Big Biomass, and shale gas, which were pushed through with limited consultation of citizens and societal actors. And neoliberal political ideology explains the preference for low-cost, market-based policy instruments (e.g., auctions), which often favour incumbents over new entrants.[21] It thus remains to be seen if the post-

[21] The government dropped this low-cost preference to push large-scale offshore wind and nuclear power.

2009 interventionist 'market shaping' approach will last or if the government will return to a hands-off approach.

The changes in policy instruments and governance styles did not affect deeper arrangements such as privatised market organisation, large-scale operational models of utilities, the business model of appliance manufacturers (selling more and better appliances and stimulating consumer upgrading to the latest ICT-models), consumer sovereignty, or cultural conventions (e.g., desires for convenience, entertainment, novelty, connectivity). New goals (such as climate mitigation) and policies have thus been layered on top of existing arrangements (McMeekin et al., 2019), which suits incumbent interests (e.g., utilities, appliance manufacturers, grid-actors) because it enables them to survive the unfolding energy transition.

Although these deeper arrangements have not been disrupted or overhauled, significant decarbonisation has been achieved. This finding contradicts some of the neo-institutional literature, which tends to see policy change as relatively 'superficial' (Scott, 2008) and suggests that deeper arrangements need to change to bring about transitions. In contrast to this theoretical claim, our empirical analysis of the UK electricity system shows that major low-carbon improvements can be achieved without radical change in deeper arrangements, provided that policies are strong enough to stimulate the reorientation of incumbent actors.

4.6.5 Scope, Depth, and Speed of Reconfiguration

The scope of techno-economic reconfiguration has been substantial for the generation and consumption sub-systems, where substantial changes have occurred in many power generation technologies and appliances. The scope of techno-economic change remained more limited in the infrastructure sub-systems, where most niche-innovations have remained small. Reconfiguration scope is increasing because the diffusion of intermittent renewables is having knock-on effects in the other two sub-systems, leading to the emerging deployment of battery storage, smart meters, smart grids, and demand-side response.

The scope of actor reconfiguration in the generation sub-system has been substantial, as utilities, policymakers, and wider publics experienced substantial change. It remained medium to low in the consumption sub-system, because only appliance manufacturers and policymakers enacted substantial change, while users simply adopted more efficient appliances, which involved limited change. The scope of actor reconfiguration remained limited in the grid subsystem, because few actors enacted substantial change.

The scope of policy reconfiguration has been substantial for the generation sub-system because multiple strategies, plans, and instruments have been used since the early 2010s to drive change. It was somewhat less for the consumption sub-system, where the policy instrument mix was less comprehensive, relying mostly on regulations (e.g., efficiency standards, energy supplier obligations). It remained limited in the grid sub-system, where Ofgem only introduced some ad-hoc instruments that have not supported broad deployment of new technologies.

Techno-economic reconfiguration has been moderate-depth in the generation sub-system because modular substitution mostly occurred between large-scale options (e.g., from coal to gas and large-scale RETs), while small-scale RETs, which represent a deeper change towards decentralised generation, have remained relatively marginal. Techno-economic reconfiguration remained limited in depth in the consumption and grid sub-systems, where incremental change dominated and niche-innovations remained marginal.

The depth of actor reconfiguration has been substantial for policymakers in the generation and consumption sub-systems, who changed both policies and governance styles, including goals. The depth of actor reconfiguration has been moderate for utilities and appliance manufacturers, which changed their investment strategies and technical capabilities but mostly retained their core business models, and for wider publics, who increasingly acknowledged climate change as an important issue but also maintained traditional concerns (e.g., affordability, energy security). Actor reconfiguration has been less substantial for consumers, who adopted energy-efficient appliances but did not alter their social practices or cultural conventions, and for grid-actors, who have not deeply changed their roles and operational routines, although they are extending their technical capabilities. The depth of actor reconfiguration has gradually increased over time, as the diffusion of renewables and the knock-on effects on grid and consumption sub-systems have required various actors to make deeper adjustments.

The depth of policy reconfiguration was substantial in the generation and consumption sub-systems because of changes in policy goals and governance style, and the creation of a new ministry (DECC, later BEIS) with budgets and policy responsibilities. It was limited in the grid sub-system, where climate mitigation remained an add-on policy goal and economic efficiency thinking and incentive tinkering prevailed.

The speed of reconfiguration has been substantial in the generation sub-system, especially in the last 10 years when RETs diffused rapidly (Figure 4.18) and coal-fired power generation almost disappeared (Figure 4.4). Substantial speed was driven by: a) strong policy interventions (including large financial incentives), b) strategic reorientation of incumbent actors towards low-carbon innovations, in response to both policy incentives and perceived economic opportunities, resulting in high investments (in the order of tens of billions of pounds), c) rapid cost decreases in RETs, particularly for onshore and offshore wind, and solar-PV

(Figure 4.19), d) high public attention for climate change and positive discourses about RETs, which incentivised politicians; e) ('indirect' or 'involuntary') market demand because consumers (ultimately) paid for low-carbon innovations through energy bills and taxation, even though they do not make active purchase decisions. Rapid change in power generation was also enabled by the electricity system's architecture, in which the grid infrastructure separates the generation and consumption sub-systems. This structural characteristic means that consumers are not directly involved in upstream changes in power generation (even though they do pay for them).

The speed of change has been gradual but sustained in the consumption sub-system, where the energy-efficiency of many appliances incrementally improved due to tightening standards, labelling, public information campaigns, producer obligations, public procurement, and voluntary initiatives. The speed of change has been moderate for transmission grids (e.g., extensions and new offshore grids) and relatively low for distribution grid transformation (e.g., smart grids, bi-directional flows) due to various DNO lock-in mechanisms.

4.6.6 *Future Outlook*

The decarbonisation of the UK electricity system has progressed well in the past decade, and this is likely to continue in the coming years, potentially resulting in a decarbonised system in the 2030s. Increasing amounts of intermittent renewables increasingly have knock-on effects on the electricity grid and consumption sub-systems, potentially leading to further substantial reconfiguration of the entire system. This low-carbon transition is thus beginning to have characteristics of a Great Reconfiguration with deep and broad changes across the entire system.

The depth of this reconfiguration could be even more substantial if multiple smaller-scale options would diffuse, but this is being prevented by incumbent firms and policymakers, who systematically privilege large-scale options. Reconfiguration of local distribution networks has not yet progressed much because of techno-economic lock-in effects, hesitation by Distribution Network Operators, and limited policy push. This may become a bottleneck in the coming decade when the likely diffusion of electric vehicles and potential diffusion of heat pumps will increase both electricity demand and local network electricity flows. If demand-side response and domestic rooftop solar-PV, which has stalled in recent years, also diffuse more widely in the coming decade, local electricity networks will certainly need to be deeply reconfigured to accommodate more flows, bi-directional flows, and flexible management. It remains to be seen if these grid reconfigurations will be implemented at sufficient speed and scale.

There are also ongoing uncertainties about the composition of the future electricity generation mix. While future onshore and offshore wind power generation will likely be substantial, it is uncertain how large the contributions of solar power (probably at utility scale) and bio-power will be. On the one hand, biomass could potentially be reoriented to other sectors with fewer decarbonisation options (e.g., biofuels in aviation, heavy goods vehicles, or shipping). On the other hand, bio-power with CCS offers the potential of negative emissions, although there are various feasibility uncertainties. Beyond renewables, there are also uncertainties about the future role of nuclear power and CCS. Despite ambitious goals and plans in the past decade, the actual realisation of nuclear power expansion has lagged far behind stated intentions (due to cost concerns and financing problems), while several attempts to kickstart CCS for power generation have failed. Continued cost reductions of renewable electricity technologies (and battery storage) may further undermine the business case for these options.

Despite these uncertainties and possible bottlenecks, low-carbon reconfiguration of the UK electricity system is likely to continue and deepen in the coming decade, providing a concrete example of the feasibility and effectiveness of Great Reconfigurations.

5
Passenger Mobility Systems

5.1 Introduction[1]

Land-based passenger mobility, which is the focus of this chapter, has several special characteristics. First, it is not one system, but multiple systems, which can be distinguished by transport modes and technologies (such as trains, trams, cars, bicycles, and buses) and ownership, including public transport (bus, train, trams) and private transport (cars, bicycles, motorcycles). The various transport modes are associated with specific socio-technical systems that have their own technologies, industries, markets, and user practices. Automobility, bus, and bicycle-systems partly overlap because of shared road infrastructure use (Figure 5.1). Railways and trams have their own dedicated infrastructures.

Second, although multiple systems co-exist, the automobility system is, by far, the largest in England (and most other Western countries), both in terms of passenger kilometres and number of trips (Table 5.1). Rail, bus, and cycling systems are much smaller but have a high degree of stability in underlying actor coalitions, institutionalised rules, and social practices. They should therefore be seen as subaltern systems rather than as niche-innovations (Geels, 2012).

Third, the post-war mobility explosion (Figure 5.3) fundamentally transformed society by enabling increasing geographical and spatial dispersion, which deeply embedded cars in social practices. Whereas many families had traditionally lived in the same city or neighbourhood, the post-war mobility explosion enabled friends and families to live further apart, often in different cities. People's homes also became more spatially separated from where they worked, went to school, shopped, enjoyed leisure activities, and went on holidays. The increasing spatial separation was not only enabled by cars but also made people more dependent on cars to sustain daily life practices, thus creating deep social, spatial, and cultural lock-ins (Urry, 2004).

[1] Parts of this chapter draw on Geels (2018), but update, elaborate, and refocus the analysis.

Table 5.1. *Mode share of trips in passenger kilometres and number of trips in England in 2019 (DfT, 2020a: 2)*

Transport mode	% of passenger kilometres	% of number of trips
Car/van (driver + passenger)	77	61
Rail	10	2
Bus	4	5
Bicycle	1	2
Walk	3	26
Other (e.g., tram, subway)	5	4

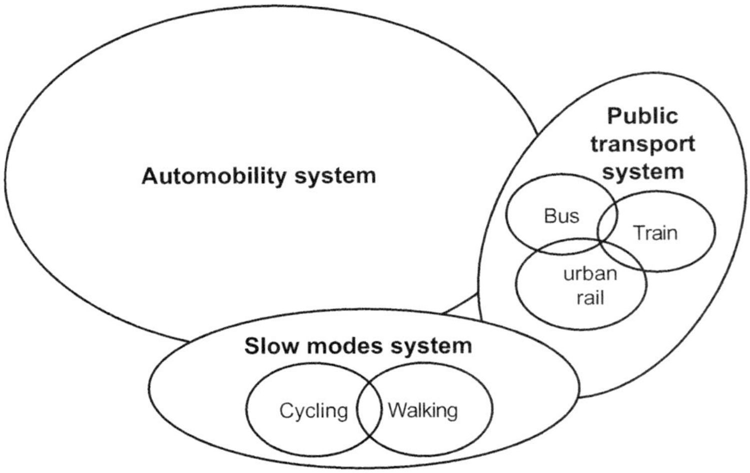

Figure 5.1 Schematic representation of different land-based passenger mobility systems

Fourth, land-based passenger mobility systems are infrastructure heavy. Although mobility systems are organised around artefacts (such as cars, trains, buses, bicycles), their use depends on the presence of roads, railways, tunnels, and bridges as well as fuel infrastructures. Infrastructures are thus deeply entwined with the *use* of artefacts, which gives mobility systems a different architecture than in the electricity system, where grid infrastructures are located *between* production and use.

Figure 5.2, which schematically portrays the automobility system, illustrates this inter-penetration of road (and fuel) infrastructures and use.

Fifth, the spatial spread and density of infrastructures and mobility systems varies substantially. Road infrastructures are spatially very extensive, ranging from a backbone of high-speed motorways to very dispersed minor roads. This enables car drivers to go almost everywhere and reach any human settlement. Although buses (mostly) use the same roads as cars, bus transport services are less evenly

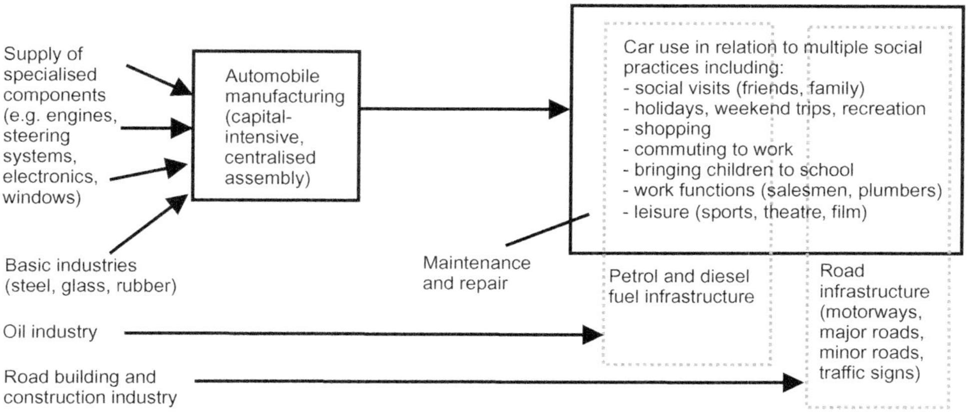

Figure 5.2 Schematic representation of the material elements and flows in the automobility system

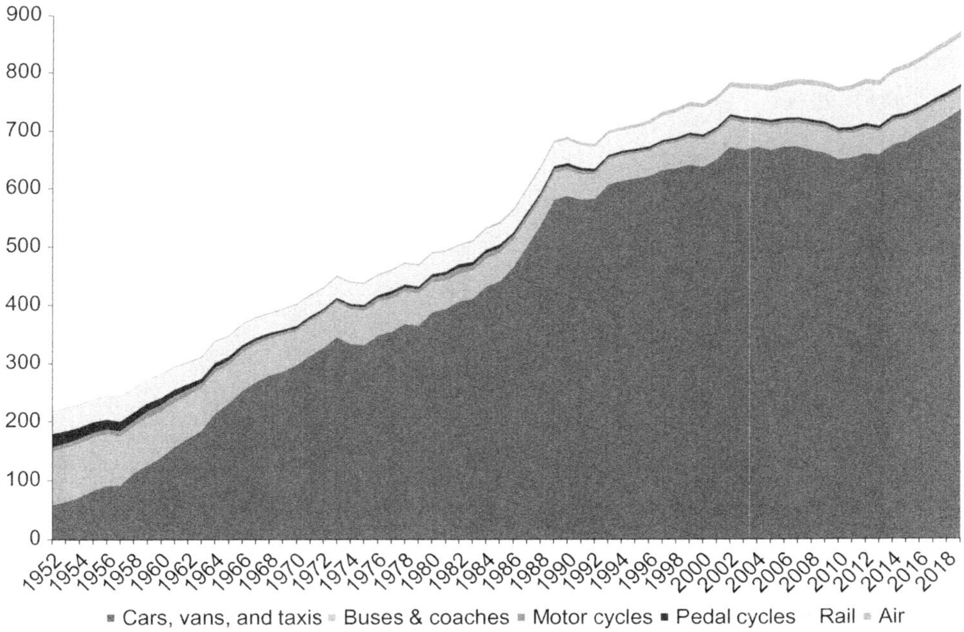

Figure 5.3 Domestic passenger mobility (in billion passenger-kilometres) by transport mode, 1952–2019 in Great Britain (constructed using data from Department for Transport Statistics; modal comparisons; Table TSGB0101)

spread, having higher frequency in high-density areas for commercial reasons. Rural and sparsely settled regions are therefore not well covered by bus systems, which thus increases car dependency of people living there. Rail systems and services also focus on mobility between high-density locations such as cities, which thus creates gaps in spatial coverage. As a slow transport mode, cycling is mostly used for local, especially urban, transport.

Sixth, automobility is an energy-intensive system, generating greenhouse gas (GHG) emissions directly (through driving cars) and indirectly (because manufacturing cars involves large amounts of energy and metals, while road building uses large amounts of concrete and asphalt). Bus, rail, and cycling systems generate fewer GHG emissions per passenger-kilometre, which is why modal shifts from cars to other transport modes represent one climate mitigation option (although some of the considerations above imply that such modal shifts are not equally feasible for all people everywhere in the country).

The automobility system expanded very rapidly after the Second World War, when people increasingly bought private cars which they used to cover larger distances (Figure 5.3). Overall passenger mobility quadrupled from 218 billion passenger-kilometres in 1952 to 873 billion passenger-kilometres in 2019, while Great Britain's population increased by 32% in the same period (from 49.05 million to 64.90 million).

In 2020 and 2021, the COVID-19 pandemic strongly affected the use of transport modes (Figure 5.4), because three national lockdowns (March–June

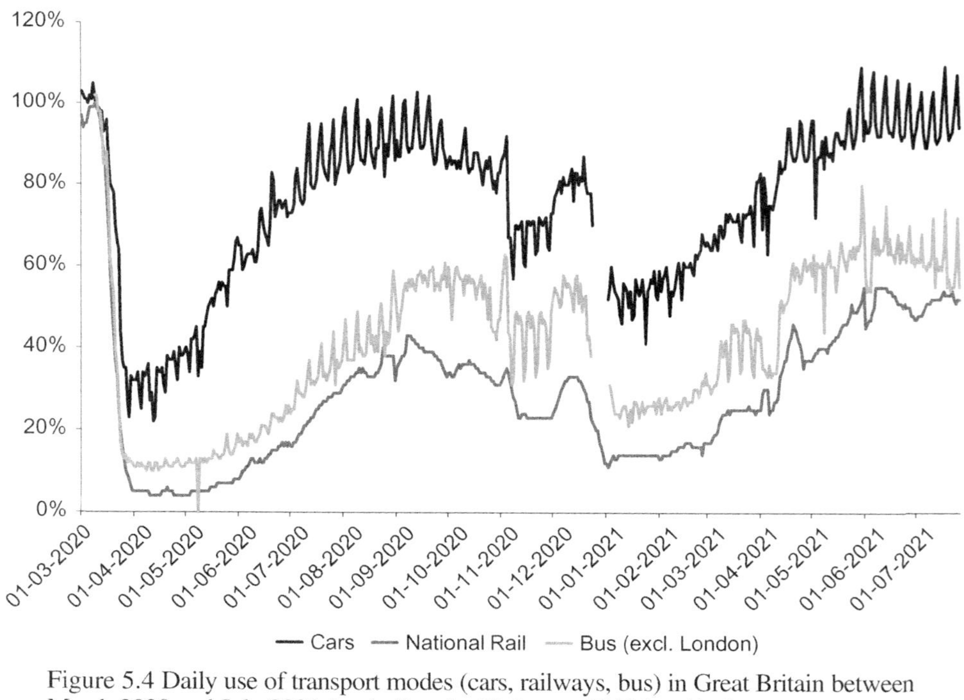

Figure 5.4 Daily use of transport modes (cars, railways, bus) in Great Britain between March 2020 and July 2021 (excluding the Christmas 2020 break); figures are percentages of an equivalent day or week (constructed using data from Department for Transport statistics; Transport use during the coronavirus COVID-19 pandemic)

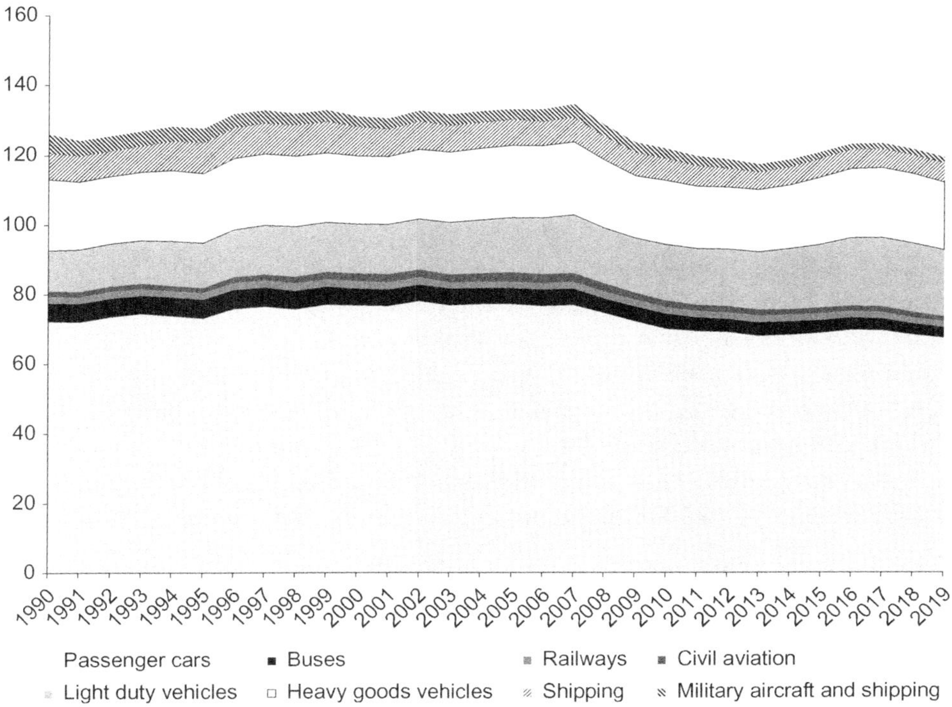

Figure 5.5 UK domestic transport-related greenhouse gas emissions 1990–2019, in $MtCO_2e$ (constructed using data from BEIS, 2020 Final UK greenhouse gas emissions national statistics 1990–2019)

2020, November–December 2020, and January 2021–March 2021) restricted people's mobility. The shocks were mostly temporary for passenger car use, which rebounded to pre-pandemic levels when restrictions were lifted. Railway and bus travel experienced deeper declines during the lockdowns and rebounded less strongly after re-openings. Bicycle travel, which is discussed in Section 5.5, increased strongly during the first lockdown in 2020 but has since declined.

Because of our interest in climate mitigation, this chapter focuses on land-based passenger transport modes (passenger cars, buses, railways, cycling), which in 2019 accounted for the majority (59%) of direct, domestic transport-related GHG emissions in the UK (Figure 5.5). Heavy goods and light duty vehicles both accounted for 16% of GHG emissions in 2019. Domestic shipping generated 4.5% of emissions and domestic civil aviation for 1%. The focus on *domestic* GHG emissions means that international emissions (from aviation and shipping) are not included in our analysis. By 2017, however, these had become quite significant: international aviation generated 35 $MtCO_2$ (roughly 29% of domestic transport-related GHG emissions) and international shipping 7.8 $MtCO_2$ (about 6% of domestic GHG emissions).

Figure 5.5 shows longitudinal trends in *total* domestic transport-related GHG emissions. Between 1990 and 2007, these emissions gradually increased from 128.1 to 137.5 million tons CO_2-equivalent ($MtCO_2e$). Following the financial-economic crisis, emissions decreased by 12.7% to 120 $MtCO_2e$ between 2007 and 2013. As economic activity picked up again, emissions increased between 2013 and 2017 to 126.1 $MtCO_2e$. Emissions then decreased somewhat to 122.2 $MtCO_2e$ in 2019, and declined by 29% in 2020 because of the COVID-related lockdowns (CCC, 2021). Because this unprecedented decline is likely to be temporary, 'we can expect a significant rebound in transport emissions' in 2021 (CCC, 2021: 19). For that reason, our analysis of structural emission trends and underlying drivers goes up to 2019 and excludes 2020. For actors, policies, and some techno-economic developments such as sales, we do, however, include COVID-19 in our analysis.

From their 2007 peak, total domestic transport-related GHG emissions decreased by 11% to 2019. Developments varied for different transport modes.

- Emissions from passenger cars, which is the single largest category, decreased by 12% between 2007 and 2019 (from 77.1 $MtCO_2e$ to 67.7 $MtCO_2e$). This is a significant decrease considering that passenger car mobility increased by 10% in the same period (Figure 5.3).
- Railway emissions remained unchanged between 2007 and 2019 (at 2 $MtCO_2e$), while emissions from domestic aviation decreased (from 2.4 to 1.4 $MtCO_2e$) in the same period.
- Emissions from bus/coaches also decreased in the 2007–2019 period (from 4.8 to 3.1 $MtCO_2e$), which is partly due to a 20% decrease in bus passenger mobility in this period (Figure 5.3).
- GHG emissions from light duty road freight transport *increased* by 15% between 1990 and 2019 (from 16.8 MtCO2e to 19.2 $MtCO_2e$), at least partly due to an increase in online shopping and home delivery. Emissions from heavy freight decreased by about 7% in the same period. Because of this chapter's focus on passenger mobility, these freight-related developments are not further discussed.

Passenger mobility from cars, railways, and buses was 11% higher in 2019 (854 billion passenger-kilometres) than in 2007 (771 billion passenger-kilometres). Nevertheless, combined GHG-emissions from these modes decreased by 14%, from 83.3 $MtCO_2e$ in 2007 to 71.8 $MtCO_2e$ in 2019 (Figure 5.5). This chapter aims to assess the underlying change processes that caused this reduction, while also analysing other relevant socio-technical developments in passenger mobility systems.

To that end, Sections 5.2, 5.3, 5.4, and 5.5 respectively investigate the main developments in automobility, rail, bus, and cycling systems. For each system, we first analyse techno-economic developments and then actors and institutions.

Section 5.6 analyses six niche-innovations, which are not only seen as having considerable carbon reduction potential but also represent different transition pathways, aimed at changing different parts of mobility systems. Electric vehicles and biofuels are two technological niche-innovations that aim to reduce GHG emissions from cars and buses. Tele-working is a niche-innovation that aims to reduce mobility by removing the need to commute to work. Car sharing and intermodal transport (including smart cards and Mobility-as-a-Service) are two niche-innovations that aim to reduce car ownership and increase intermodal travel. And self-driving personal cars promise to radically alter mobility, reduce accidents, enhance traffic flow efficiency, and thus reduce GHG emissions. The six niche-analyses are also divided into techno-economic developments and actors and institutions, which are, however, more fluid and less articulated than for regimes.

Section 5.7 provides interpretive assessments of low-carbon transition and degrees and kinds of whole system reconfiguration.

5.2 The Auto-Mobility System

5.2.1 Techno-Economic Developments

The automobility system is a large and stable system, centred on a primary artefact: the internal combustion engine (ICE) car. The system's functioning also relies on a wide-ranging configuration of material components and infrastructures such as a complex roads network (which includes both roads and traffic management and signalling), petrol and diesel distribution, car manufacturing plants, and maintenance and repair facilities (Figure 5.2).This large material configuration is deeply embedded in the physical environment, particularly in urban settings where the car has co-evolved with the building and lay-out of cities and conurbations. These interactions between the automobility system and the built environment are a source of material lock-in but also sites where changes can be negotiated (e.g., traffic regulation, repurposing of roads).

ICE cars are complicated artefacts, requiring the assembly of many components, including vehicles frames (chassis), engines, steering wheels, brakes, glazing, and increasingly elaborate interiors and electronics. The car industry has therefore developed significant logistical innovations, oriented towards the assembly of heterogeneous components in highly automated facilities, relying on extensive networks of specialised suppliers, and the optimisation of component shipping ('just in time') to minimise idle stocks.

The automobility system expanded rapidly after the Second World War, as people bought more automobiles and passenger car mobility exploded (Figure 5.3). Growth slowed in the 1990s. Car mobility decreased by 2% between 2007 and

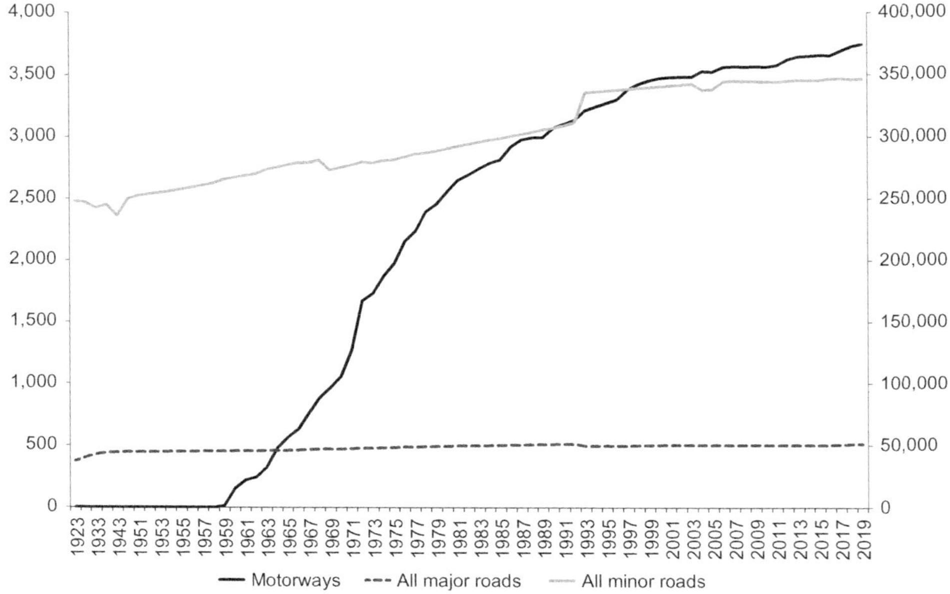

Figure 5.6 Length of different road types (in kilometres) in Great Britain, 1923–2019: Motorways on left-hand axis, all major and minor roads on right-hand axis (constructed using data from Department for Transport Statistics; Road lengths statistics; Table RDL0203)

2013 because of high oil prices and the financial-economic crisis and subsequent recession. But between 2014 and 2019, passenger kilometres by cars, vans, and taxis bounced back and increased to a level that was 10% higher than in 2007.

The increase of passenger car mobility has been enabled (and driven) by expanding road infrastructures. The network of *minor roads*, which distributes traffic to urban and regional localities at relatively low speeds, gradually increased from 250,001 km in 1947 to 346,404 km in 2019 (Figure 5.6). To facilitate traffic flows at very high speeds over long distances, *motorways* were constructed as a new type of road in the late 1950s, reaching 3,742 km in 2019. The network of *major roads*, which includes motorways and trunk roads (which are both maintained by national highway agencies) and principal roads (which are maintained by local authorities), increased from 44,591 km in 1947 to 51,191 km in 2019. The different types of roads thus have different functions (local, medium- or long-distance traffic) and are maintained by different kinds of actors. The majority of total motor vehicle traffic (62% in 2019) is accommodated by major roads, including motorways, where cars travel at high speed and high density (Figure 5.7).

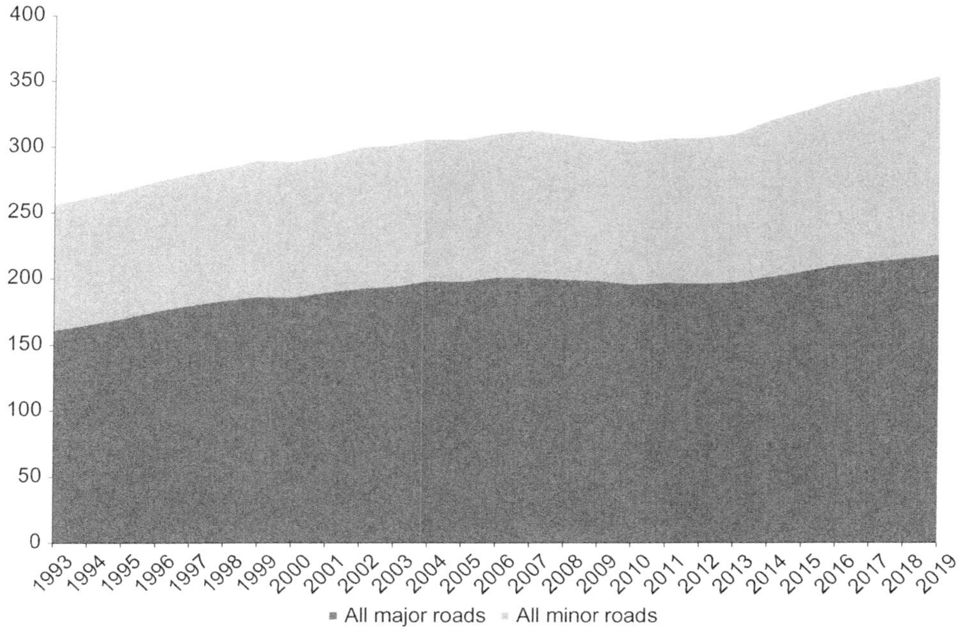

Figure 5.7 Motor vehicle traffic (vehicle miles) by road class in Great Britain, 1993–2019 (constructed using data from the Department for Transport statistics; Road traffic statistical tables; Table TRA0102)

Annual car sales have increased rapidly since the 1950s (Figure 5.8), but experienced recurring fluctuations due to macro-economic developments such as recessions and oil price changes (Figure 5.9). The 2007/8 financial crisis (and subsequent recession and austerity politics) and high oil prices depressed car sales until 2012. Sales increased again until 2016, when economic uncertainties following the Brexit referendum caused a new decline since 2017. Car sales plummeted by 28% in 2020 due to COVID-lockdowns.

The reverberations of the 2015 'Diesel-gate' scandal (in which automakers were found to have cheated emission tests for many years) led to a particularly strong decline in diesel car sales (Figure 5.10). The sales of 'other' cars (which mostly include electric vehicles), which will be further discussed in Section 5.6.1, has gradually increased over the past decade, reaching 21% of all sales in 2020 (Figure 5.10).

Real car purchase prices have decreased over time, while running costs (maintenance, fuel, insurance, taxes) have substantially increased (Figure 5.11). Total motoring costs have increased more slowly than the cost of living, however, which means that car travel has become relatively more attractive in the past two decades. Total motoring costs have also relatively increased less than bus and train fares, which rose faster than RPI (Figure 5.11).

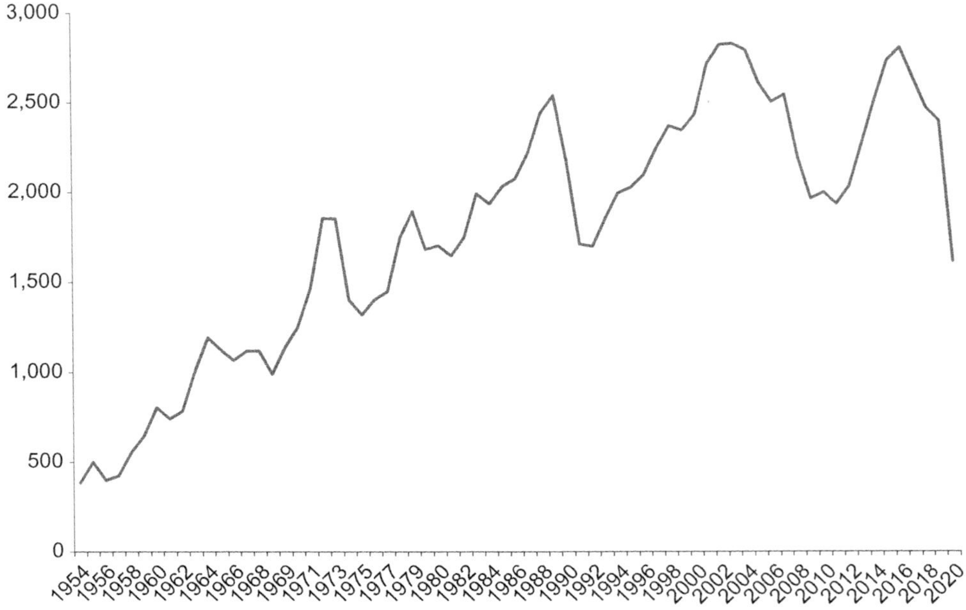

Figure 5.8 Annual car sales (new registrations) of private and light goods vehicles in Great Britain, 1954–2020, in thousands (constructed using data from Statistics at DfT; Vehicle Licensing Statistics; Table VEH0153)

Figure 5.9 Crude oil price from 1978 to 2020, in 2020 constant dollars per barrel (constructed using data from Table 11.05 from the U.S. Department of Energy's Transportation Energy Data Book)

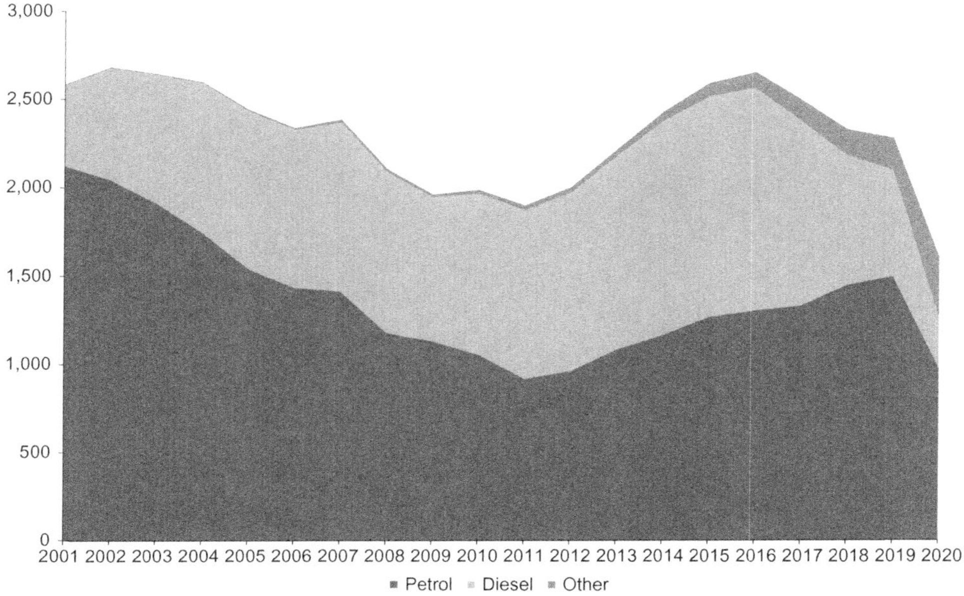

Figure 5.10 Annual car sales (new registrations) of petrol, diesel, and 'other' cars in Great Britain, in thousands, 2001–2020 (constructed using data from Department for Transport Statistics; Vehicle Statistics dataset; Table VEH0253)

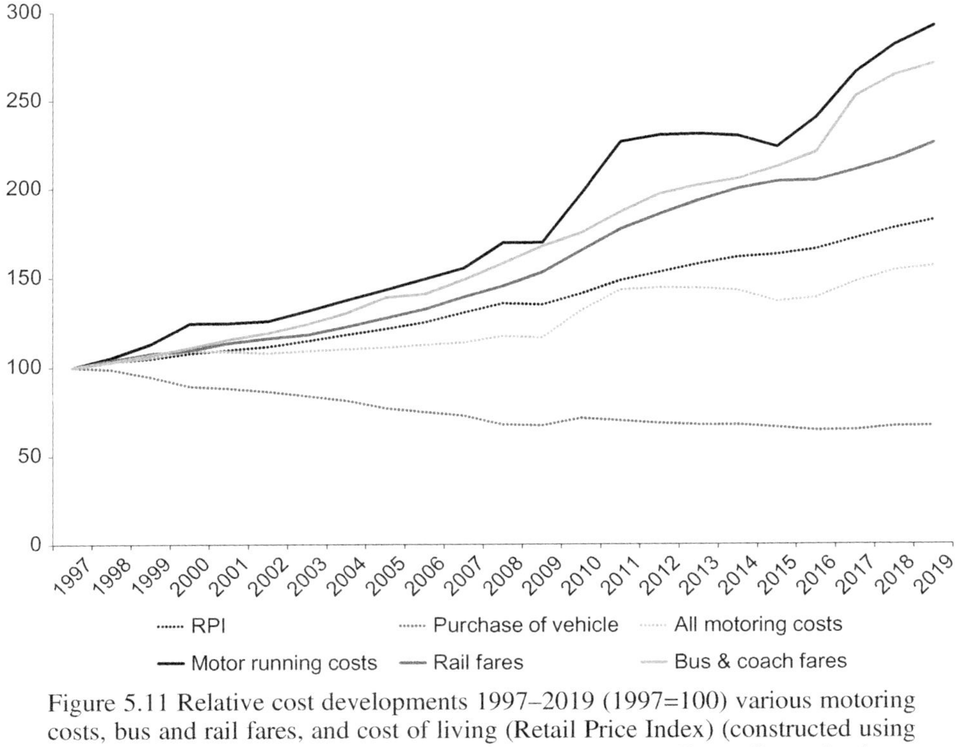

Figure 5.11 Relative cost developments 1997–2019 (1997=100) various motoring costs, bus and rail fares, and cost of living (Retail Price Index) (constructed using data from Department for Transport Statistics; Transport Expenditure database; Table TSGB1308)

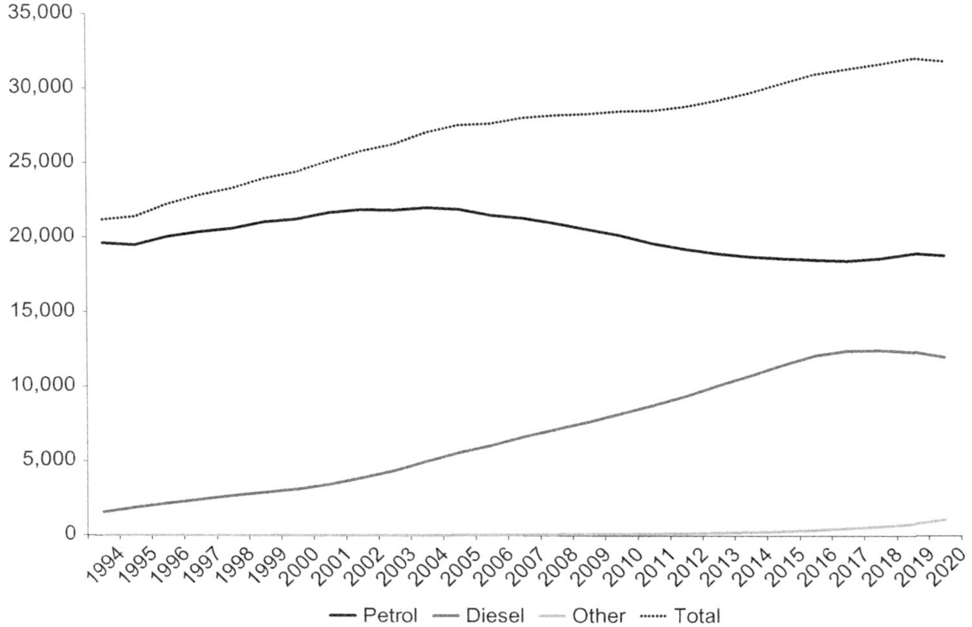

Figure 5.12 Passenger car fleet (total number of licensed vehicles) in Great Britain by fuel type 1994–2020 in thousands (constructed using data from Department for Transport Statistics; Vehicle Statistics dataset; table VEH0203)

The total passenger car fleet has grown steadily since the 1950s, and in the last two decades experienced a relative shift from petrol to diesel cars (Figure 5.12) until Diesel-gate reversed this trend. 'Other' cars (mostly electric vehicles) represented 3.4% of the fleet in 2020.

Two gradual developments contributed to CO_2 emission reductions in the car fleet. First, the relative consumer demand shift from petrol to diesel cars (Figure 5.10) helped to reduce emissions, because diesel cars are more fuel-efficient (but more polluting). Second, fuel efficiency performance of *new* diesel and petrol cars has improved substantially in the last 20 years due to many incremental innovations (Figure 5.13). An important caveat with these fuel efficiency numbers is that automakers have increasingly 'gamed' laboratory tests, leading to discrepancies of 30–40% with real-world driving conditions (CCC, 2015; Fontaras et al., 2017).

Two other developments first blunted and then partly reversed these fuel efficiency gains. First, people increasingly bought heavier cars with lower fuel efficiency. The percentage of heavy SUVs (Sports Utility Vehicles) in passenger car sales increased from 6.6% in 2009 to 13.5% in 2015 and then jumped to 21.2% in 2018 (UKERC, 2019). Second, in response to the 2015 Diesel-gate revelations, people shifted from diesel to petrol cars, which are less fuel-efficient.

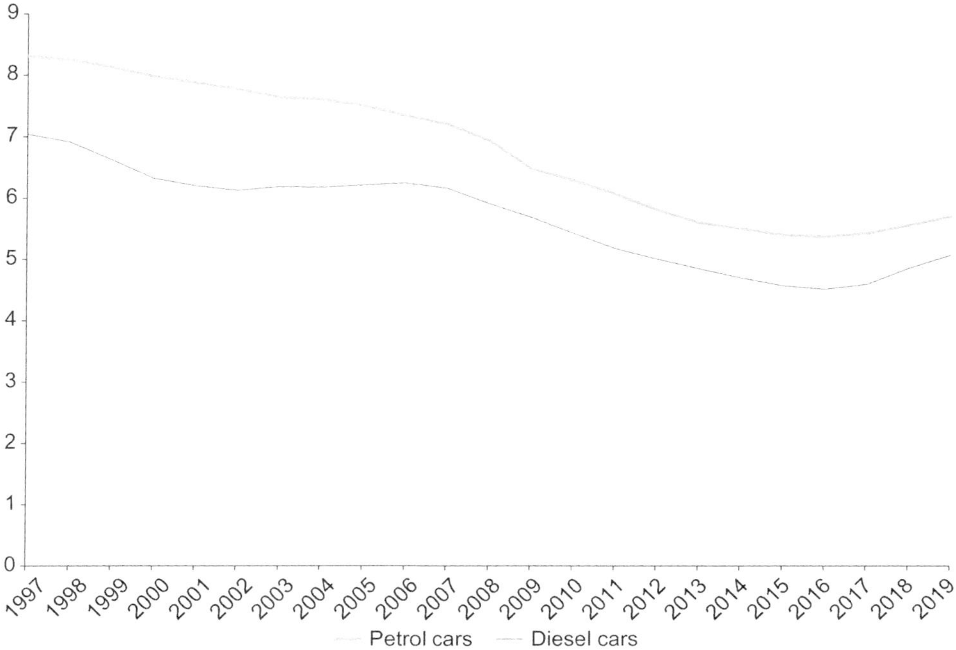

Figure 5.13 Sales-weighted[2] average new car fuel consumption in Great Britain, 1997–2019, in litres per 100 km (constructed using data from Statistics at DfT; Energy and Environment data; Table ENV0103)

5.2.2 *Actors*

Firms: The car industry operates at a global scale, with multinational companies managing manufacturing plants in many countries. Six foreign-owned volume car manufacturers (Jaguar Land Rover, Nissan, Mini, Toyota, Honda, Vauxhall) have manufacturing plants in the UK, which produced 1.3 million cars in 2019 for both domestic and export markets, generating £79 billion turnover and £15.3 billion gross value added. Exported cars were worth $42 billion, accounting for 13% of the UK's total export goods (SMMT, 2020). Although the UK ranks only 16th in global car manufacturing (SMMT, 2020), its car industry is still relatively important for the country. Around 180,000 people are directly employed in car manufacturing, while 860,000 employees work in the wider automotive industry, which also includes component suppliers, car dealerships, and petrol stations (SMMT, 2020). The industry's economic clout gives it substantial lobbying power with the UK government (Shaw and Docherty, 2014).

[2] The numbers in Figure 5.13 are weighted to account for the relative sales of different models of vehicles.

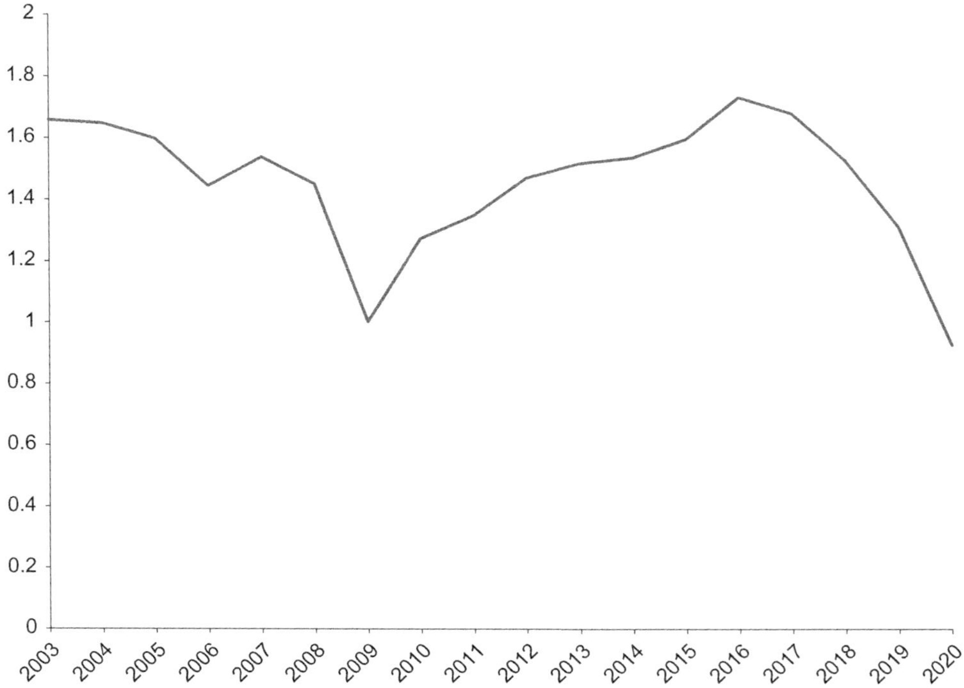

Figure 5.14 Passenger cars produced in the UK in millions, 2003–2020 (constructed using data from SMMT, 2020)

UK car manufacturing steadily increased after the 2007/8 financial crisis but decreased markedly after the 2016 Brexit referendum (Figure 5.14), which created deep uncertainties for UK-based automakers, who are deeply entwined with Europe for imports and just-in-time deliveries of components as well as for exports (since the EU accounts for more than half of UK car exports). Automakers therefore worried about the risk of a 'no deal' Brexit and the imposition of trade tariffs of 10% or more, which would have threatened the long-term competitiveness of UK-based manufacturing plants. Investments in new projects and UK plant upgrades, which are normally around £2.5 billion per year, decreased to around £590 million in 2017 and 2018, as global car companies delayed or diverted spending (Campbell and Inagaki, 2021). Ford and Honda decided to close UK manufacturing plants as part of a wider global restructuring move. Although the 2020 Brexit deal removed the tariff risk, Toyota and Vauxhall's new owner Stellantis, formed by a merger in 2021 of Fiat Chrysler and PSA, are still reviewing their options for future investments and closures. Both companies operate vehicle plants in mainland Europe, which they may decide to expand. Car manufacturing plummeted by 29% in 2020 due to the COVID-19 pandemic

(Figure 5.14), which led to plant shutdowns and reduced production in response to shrinking demand.

Operating in a global environment, automakers face several problems that are more important to them than climate change (Geels, 2012; SMMT, 2020): 1) survival in cut-throat competition, 2) under-utilisation of factories and cost pressures, 3) market saturation in developed countries, 4) declining passenger car sales and manufacturing output since 2017 in the UK (Figures 5.8 and 5.12) and globally, which erodes profitability (Miller, 2019); the COVID-19 pandemic further exacerbated these sales problems in 2020.

In response to these pressures, automakers have focused on cost-savings, mergers and collaborations, factory efficiency improvements, sales in emerging economies, and continuous incremental product innovation (in engines, safety devices, air-conditioning, occupant comfort, and entertainment). The combination of earlier ICT devices (such as on-board electronics, anti-lock braking, real-time information technologies, and navigation technologies) with improved sensing devices and faster computers has given rise to high expectations about driverless cars, which are discussed in Section 5.6.6.

Automakers also face climate change pressures and have therefore incrementally improved internal combustion engines (with variable valve timing, direct fuel injection), leading to fuel efficiency improvements (Figure 5.13). Automakers have also dedicated efforts towards developing alternative fuel cars (e.g., electric, hybrid, and plug-in hybrid vehicles). Automakers initially pursued these new technologies reluctantly because of uncertainties and to protect their sunk investments (Penna and Geels, 2015). Since 2015, however, most automakers have seriously committed to strategic reorientation activities (Bohnsack et al., 2020).

The large investments in electric vehicles, driverless cars, and ventures into car sharing (further discussed in Section 5.6.4) are financially challenging for automakers, whose profits have been squeezed by declining passenger car sales since 2017. An article in *The Economist* (19 January 2019), titled 'The big freeze: Carmakers scramble to prepare for a chilly future', summarised the challenge as follows: 'Even if coping with these [economic] problems were not enough, carmakers also need to make big investments in electric cars, autonomous vehicles and "mobility" services, such as car-sharing and ride-hailing.'

Users: Cars are deeply embedded in everyday life and used for many different purposes, for example, commuting to work, shopping, leisure, social visits, escort of other people (e.g., children), business, and personal business (Sheller, 2012).[3]

[3] Leisure includes visiting friends, entertainment, sport, holidays, and day trips. Personal business includes visits to services (e.g., hairdressers, laundrettes, dry-cleaners, betting shops, solicitors, banks, estate agents, libraries, and churches); or for medical consultations or treatment; or for eating and drinking, unless the main purpose was entertainment or social. Business refers to personal trips in the course of work.

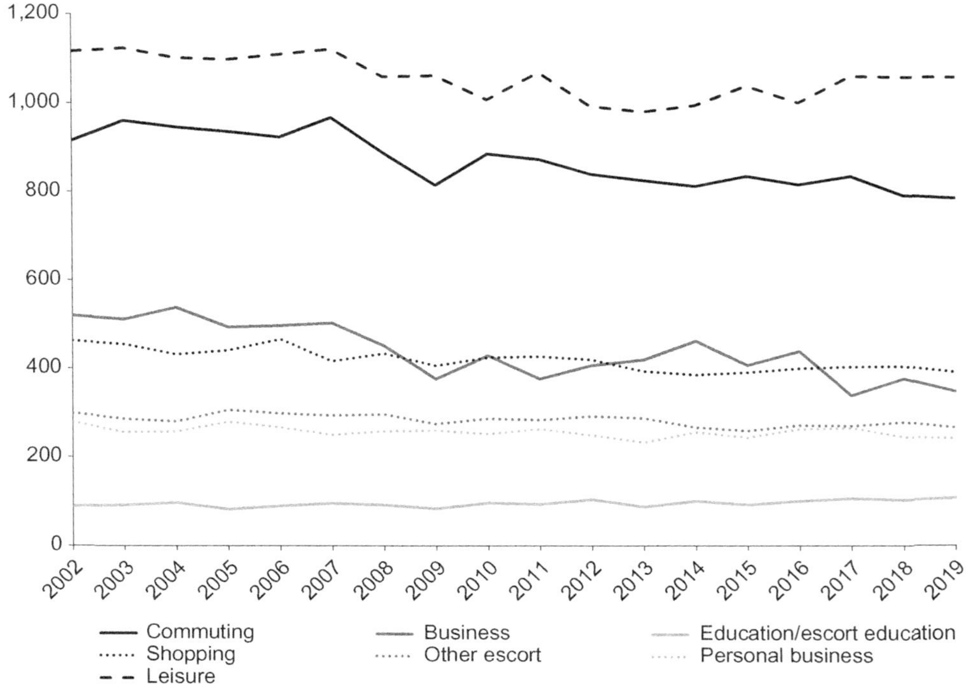

Figure 5.15 Average distance travelled by car/van for different purposes (miles per person per year, England), 2002–2019 (constructed using data from Department for Transport Statistics; National Travel Survey; Table NTS0409b)

Commuting and leisure account for most passenger-kilometres, although average travel distance for these purposes has decreased somewhat since 2002 (Figure 5.15). Car use decreased sharply in response to COVID-lockdowns in 2020 and 2021 (Figure 5.4) but also rebounded quickly when restrictions were lifted, because many people depend on cars to support many daily life practices.

Since the mid-1980s, the percentage of households without cars has steadily decreased, while double car ownership has increased (Figure 5.16). Nevertheless, average per capita passenger car mobility *decreased* by about 9% between 2002 and 2011, but then increased somewhat (Figure 5.17). In 2019, however, per capita passenger car mobility was still 3% below 2002, providing some support for the ‘peak car’ hypothesis (Metz, 2010; Millard-Ball and Schipper, 2011).

There is ongoing debate in the literature about the underlying causes for this phenomenon, with some scholars (e.g., Goodwin and van Dender, 2013; Metz, 2013; Newman and Kenworthy, 2011; Wadud and Baierl, 2017) emphasising social, cultural, and demographic factors such as changes in lifestyle and cultural attitudes, aging of the population, and younger generations showing less interest in cars. Other scholars, however, emphasise economic factors such as rising fuel

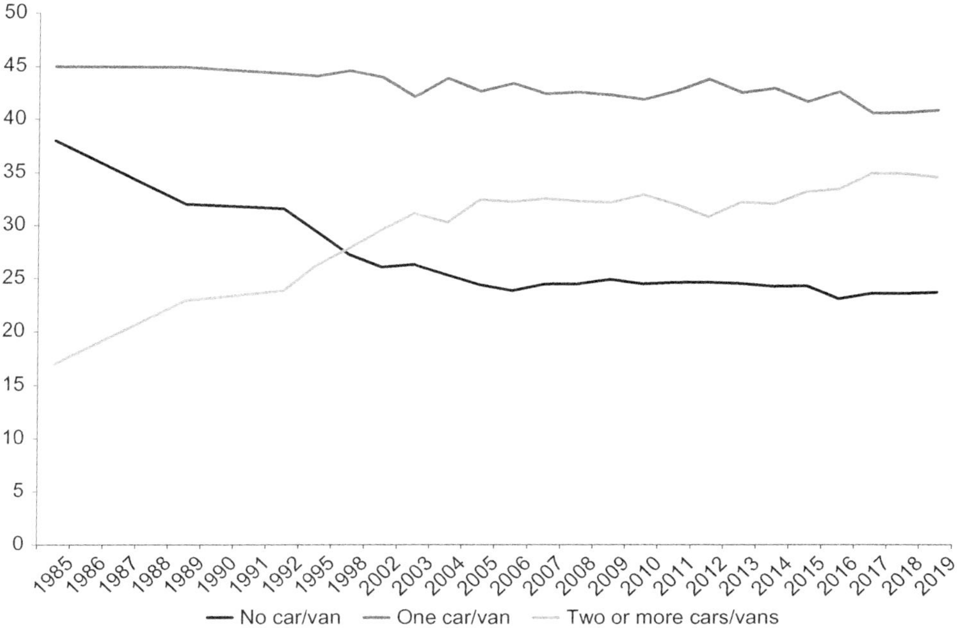

Figure 5.16 Household car availability as percentage of the population in England, 1985–2019 (constructed using data from Statistics at DfT; National Travel Survey; Table NTS0205)

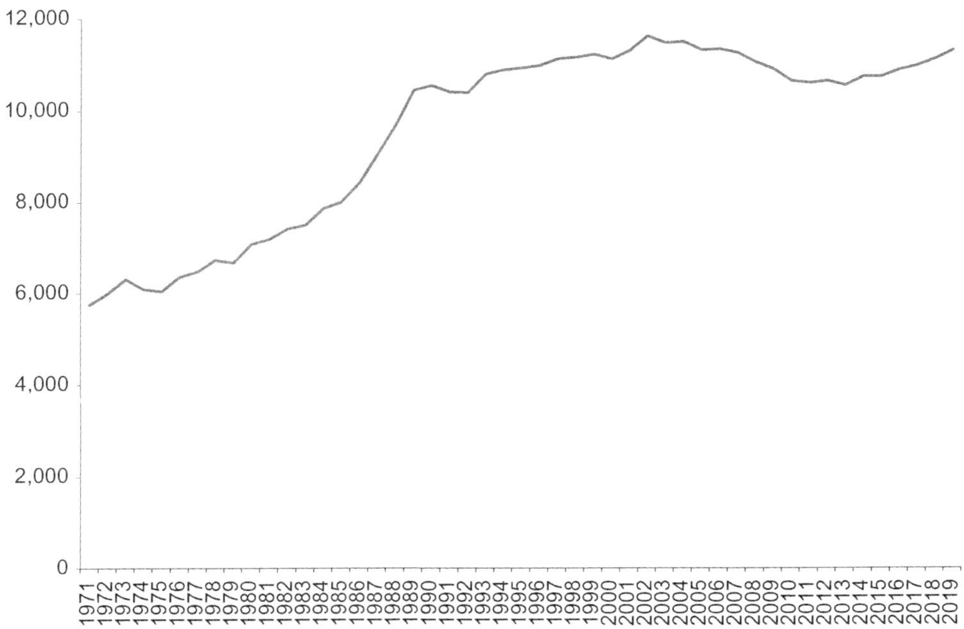

Figure 5.17 Average per capita passenger travel by cars and light vans in Great Britain, 1971–2019 (kilometres per capita per year) (calculated using data on total car passenger kilometres from Table TSGB0101, divided by population)

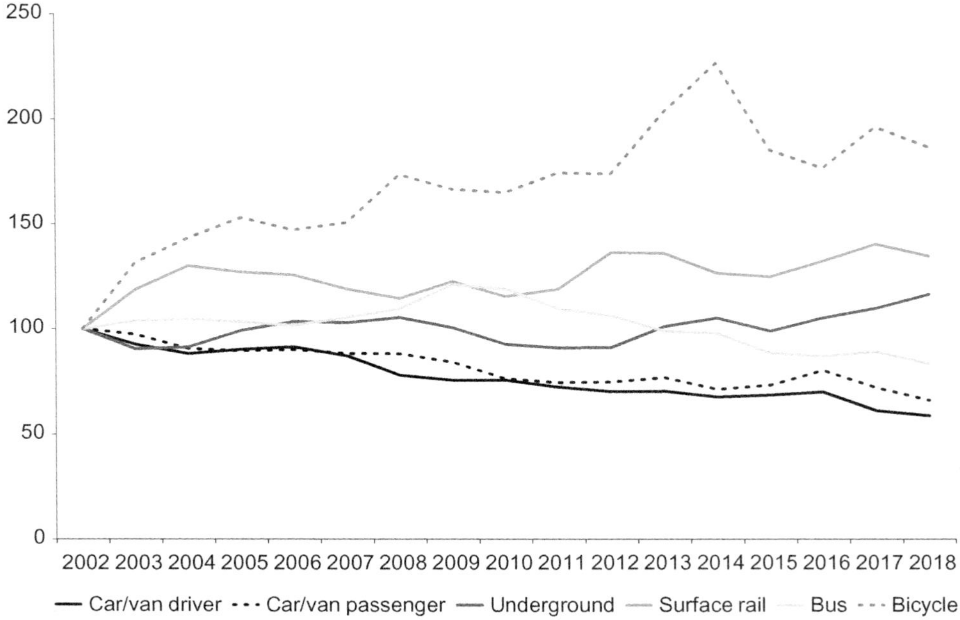

Figure 5.18 Indexed transport modes in London, 2002–2018 (in terms of distance) (constructed using data from Department for Transport Statistics; National Travel Survey; Table NTS9904)

costs, lower incomes due to the financial-economic crisis, and changing tax rules for company car use (especially increased taxation on fuel provided for private use) (Bastian et al., 2017, 2016; RAC, 2012).

Although people frequently complain about congestion and rising motor running costs, most of them like the act of driving (Sheller, 2004) and the associated 'feelings of liberation and empowerment' (Shaw and Docherty, 2014: 75). Many consumers also see cars as the most practical transport mode (in terms of speed, convenience, carrying capacity), except in large cities such as London, where car passenger mobility has been declining since the late 1990s (RAC, 2012), because of increasing congestion, parking problems, and costs. The London congestion charge, introduced in 2003, made automobile use in the city centre increasingly expensive as the charge gradually increased from £5 in 2003 to £8 in 2005, to £10 in 2011, to £11.50 in 2014, and £15 in 2020. Between 2002 and 2018, passenger car and van use in London declined by 35–40%, while underground, surface rail, and cycling increased over the same period, suggesting a substantial modal shift is under way (Figure 5.18).

Table 5.2 further reinforces the specificities of London's passenger mobility system, where public transport (bus, tube, rail) is used much more for commuting

Table 5.2. *Commuting trips in different areas by mode or multimode in percentage (short walks excluded): England, 2012–2016 (data from DfT (2018))*

	All areas	London	Urban areas	Rural areas
Car/van	69	36	75	90
Bus	7.5	13	6.7	1
Tube/London Rail	2.5	11	0.2	0.1
Rail	2.3	7.1	1.4	0.3
Walk	5.4	4.7	6.2	1.2
Bicycle	4.3	4	4.5	2.5
Taxi	0.7	0.5	0.9	0.2
Other	1.3	1.7	1.3	2
Multi-mode	7	22	3.8	2.5

purposes than in the rest of the country. Strikingly, 22% of London's commuting trips involve multiple transport modes, which suggests the presence of an effective intermodal mobility system.

When buying a new car, most consumers find conventional criteria (e.g., price, car size, reliability, comfort, safety, running costs, appearance, engine performance, image) more important than climate change mitigation (Geels, 2012). Increased adoption of SUVs, for instance, is driven by appreciation of safety, comfort, size, and reliability (UKERC, 2019).

Among young people, however, the prevalence of cars and driver's licenses has decreased since the 1990s (DfT, 2017). These trends led to speculations about changing cultural attitudes about cars among young people (McDonald, 2015). Recent studies, however, suggest these trends may be more related to higher fuel costs, youth unemployment, and a tendency for millennials to delay adult life stage decisions such as having children or buying a house (Delbosc, 2016; Garikapati et al., 2016). Quantitative analyses of several large European datasets between 2014 and 2018 found that young people ('Millennials') do own and use cars less often than post-war Baby Boomers, but that the difference is diminishing as improving economic conditions have led Millennials to buy and use more cars (Colli, 2020).

Policymakers: A prime consideration of UK transport policymakers has long been to stimulate economic growth by facilitating the smooth flow of goods and people. From the 1960s to the 1990s, the so-called predict and provide policy paradigm (Goulden et al., 2014) therefore aimed to provide enough road capacity to match forecast increases in car and freight traffic. This resulted in the construction of an extensive road network (Figure 5.6). The promotion of UK-based car manufacturing has been another policy priority in recent decades, aimed at supporting jobs and economic growth (Wolmar, 2016).

UK transport policy has also aimed to address several negative externalities of car traffic such as congestion, traffic accidents, air pollution, and climate change (Geels, 2012). In the 1990s, the predict-and-provide paradigm ran out of steam, as policymakers realised that the construction of more roads stimulated the increase of road traffic, which exacerbated congestion and other problems (Owens, 1995). The Labour government's (1997–2010) *New Deal for Transport: Better for Everyone* White Paper (1998) explicitly abandoned 'predict and provide' as unsustainable and instead introduced new principles. One new principle was traffic management, which aimed to improve the flow of vehicles on *existing* roads using speed limitations, priority lanes, roundabouts, and ICT devices (e.g., video cameras, communication technologies, electronic signalling devices, computer networks) to monitor and influence traffic (Geels, 2007b). Another new principle was demand management, which aimed to reduce the amount of road traffic by changing travel habits and promoting alternative modes. Plans were made to stimulate and better align public transport modes, captured by new concepts such as 'integrated transport' and 'sustainable transport'.

Although these plans suggest that transport policy became less 'pro-car', road building did not stop. In fact, between 2000 and 2019, the motorway infrastructure was lengthened by 275 km or 8% (Figure 5.6). Policy plans also encountered opposition. The Labour government's plans to increase fuel duties more rapidly, for instance, encountered fuel protests and road blockades in 2000, which threatened to paralyse the country. This, in turn, led to an institutionalised fear not to go against what the public wants and to the postponement of the ambitious policy plans in the 1998 *New Deal for Transport*. Explicit efforts to shape travel habits and stimulate modal shifts (from cars to other transport modes) were gradually dropped and increasingly replaced by a focus on technological improvements (in cars, roads, railways, busses), leaving it to consumers to choose the options they preferred. Although policymakers did stimulate public transport (to some extent), 'successive administrations have backed away from doing anything that they thought would be construed as overtly "anti-car"' (Shaw and Docherty, 2014: 175).

The continuing importance of cars was also visible in responses to the financial-economic crisis, which led policymakers to not only support financial institutions but also the car industry. To prop up demand, policymakers introduced scrappage schemes that provided £2,000 rebates for replacing old vehicles with new ones. They also invested in new road building, signalling a return of 'predict and provide' (Goulden et al., 2014). In 2014, the government announced a £15 billion roadbuilding programme. The 2020 Spending Review further invested £27 billion through a five-year Road Investment Strategy aimed at building 4,000 miles of new strategic roads and motorways.

Austerity policies after the 2007/8 financial crisis also led to major reductions in government funding for local authorities, which are responsible for parking policy, local traffic plans, and for maintaining minor and principal roads, which make up the bulk (87%) of the total road infrastructure (Figure 5.6). Reduced local infrastructure spending resulted in deteriorating conditions of local roads (Wolmar, 2016).

Climate change has climbed rapidly on transport policy agendas since the late 2000s. Although public transport, cycling, biofuels, and other options received some (fluctuating) support, the policy emphasis overwhelmingly came to focus on electric vehicles (further discussed in this chapter). The focus on the 'greening of cars' came from both European and UK policymakers. European policymakers introduced CO_2 emission performance standards for new cars in 2009, which were subsequently tightened in 2014 and 2019. And UK transport policymakers introduced a raft of policies to stimulate electric vehicles (further discussed later), aiming to both address climate change and stimulate domestic vehicle manufacturing (Mazur et al., 2015; Skeete, 2019).

In the context of Diesel-gate and climate change debates, the government announced plans in 2017 to phase-out petrol and diesel cars by 2040. In February 2020, the phase-out date was brought forward to 2035 and in November 2020 to 2030. This phase-out policy will create mass markets for electric vehicles. Although the stimulus and phase-out policies thus became increasingly interventionist and disruptive in some ways, they also signalled that cars would remain central in future low-carbon transitions and UK transport policy.

This continued commitment to cars was reinforced by the government's *Road to Zero* report (DfT, 2018b) that almost exclusively focused on road transport and electric vehicles, which were explicitly linked to industrial strategy (Brand et al., 2020). Partly in response to criticisms of this narrow focus, the Department for Transport launched a consultation paper in March 2020 (DfT, 2020a), which announced intentions to develop the 'first comprehensive action plan' (p. 3) for decarbonising the whole transport system. In July 2021, the government published the resulting new strategy, the Transport Decarbonisation Plan (DfT, 2021a), which indeed addresses all transport modes and freight. The strategic vision also strikes a new tone because two of its six strategic priorities emphasise modal shift towards public and active transport (in line with the new bus, rail, and cycling strategies announced in 2020 and 2021, which are discussed later) and place-based solutions focused on local transport systems, which have remained marginal in UK transport policy for the past two decades. The other priorities, however, continue to focus on low-carbon technologies for passenger vehicles, freight transport, aviation, and shipping, and aim to position the UK as a 'hub for green transport technology and innovation' (p. 40). Electric vehicles remain a core plank of the strategy, with the government planning to consult on the introduction of a Zero

Emissions Vehicle mandate (which would impose sales targets on automakers) and phase-out plans for small and heavy diesel trucks by 2035 and 2040, respectively.

Although the new discursive emphasis on modal shift and place-based solutions is interesting and welcome, the plan did not announce new money for these options compared to earlier statements in February 2020, which are further discussed later. The plan's ambivalence is also clear in the Ministerial foreword, which on the one hand says that 'we must make public transport, cycling and walking the natural first choice' (p. 6), but on the other hand states that the plan is 'not about stopping people doing things: it's about doing the same things differently. . . . We will still drive on improved roads, but increasingly in zero emission cars' (p. 4). So, while the strategic vision is full of good intentions, it remains unclear if there will be sufficient policies to achieve the stated goals.

Wider Publics: Concerns about air pollution, the countryside, and the quality of urban space led to strong anti-car and anti-road narratives and protests in the 1980s and 1990s (Roberts and Geels, 2018), which succeeded in halting the road-building programme announced in the 1989 White Paper *Roads for Prosperity*. But in wider public debates, cars remained culturally associated with positive values such as freedom, individuality, and success (Sheller, 2012), although there appear to be generational differences, with young adults showing more varied attitudes (Colli, 2020; McDonald, 2015).

Public debates also remain concerned with congestion and fuel prices, which create pressures on policymakers to address these issues. Climate change has become an important issue in public debates about transport, but discourses focus more on electric vehicles than on reduced car use or ownership (Bergman et al., 2017). Local air pollution, which is responsible for between 28,000 and 36,000 premature deaths in the UK each year (PHE, 2019), has also risen high on public agendas, because many UK cities breached air pollution standards for many years. The 2015 Diesel-gate scandal caused public outrage and anger, because cheating automakers clearly privileged car sales over air pollution and public health considerations. Negative public debates not only contributed to declining diesel car sales but also prepared the ground for later phase-out policies of diesel and petrol cars.

5.2.3 Policies and Governance

Policies

Many formal rules and regulations shape automobility and road transport, including traffic rules, drivers' licenses, road taxes, excise duties, road infrastructure design and construction rules, parking rules, and vehicle standards (on emissions, safety, noise, recycling, materials). Regulations and policies with regard to climate change have gradually strengthened over time, initially stimulating incremental changes (such as engine efficiency improvements), but increasingly also stimulating more radical technical changes (e.g., electric vehicles).

The UK government introduced CO_2-banding in its Vehicle Excise Duty in 2001, which reduced vehicle sales taxes for fuel-efficient cars. This stimulated the adoption of diesel cars (Figure 5.10). In 2015, however, the new Conservative government replaced this CO_2-banding with a flat-rate annual fee, starting in 2017. The unintended consequence of diesel car diffusion was an increase in local air pollution (NOx, particulate matter), which has remained a topic of heated public health debates since 2016, leading to decreasing diesel car sales (Figure 5.10) and phase-out regulations.

In 2009, the European Commission introduced new car CO_2-regulations of 130 g/km for 2015 fleet average company sales. These regulations accelerated fuel efficiency improvements (Figure 5.11). In 2009, the European Commission also introduced the Renewable Energy Directive, which stipulated that 10% of transport energy should come from renewable sources by 2020, which stimulated bio-fuel deployment (further discussed in Section 5.6.2). In 2014, the CO_2-emission target was tightened to 95 g/km in 2020/21. In 2019, the European Commission not only confirmed that automakers had to meet this target for the average of all their sales but also added stiff fines for companies that would miss the target, namely €95 (£83) for every gram they are over the limit, multiplied by the number of cars sold that year. As a transitional arrangement, the highest-polluting 5% of new cars registered in 2020 are excluded for the 2021 calculation of fines over the previous year. For 2021, however, all major carmakers are expected to miss their emissions targets, which may lead to €20 billion fines in 2022 (which are likely to be especially large for Volkswagen and PSA). To lower their fleet average emissions, automakers are therefore rushing to market new electric vehicles, or even pool emission numbers with companies such as Tesla (Jolly, 2020a).

UK policymakers also introduced policies to stimulate electric vehicles (further discussed in Section 5.6.1). In 2009, they created the Office for Low Emission Vehicles (OLEV), which disbursed £400 million of government funding between 2011 and 2015 on R&D, consumer subsidies, and recharging infrastructure. Between 2015 and 2020, a further £500 million was spent on supporting ultra-low emission vehicles (ULEVs), which emit less than 75 gCO_2/km, that is, electric vehicles and plug-in hybrids. The 2030 phase-out policy of diesel and petrol cars is another policy to support ULEV markets.

London's policymakers introduced a congestion charge in 2003, which aimed to reduce traffic flows, air pollution, and noise pollution in the central London area. In 2017, London also introduced a toxicity charge in response to increasing air pollution concerns in central London. The £10 charge applies to older and more polluting cars and vans that do not meet Euro 4 standards. In 2019 this charge was replaced by the Ultra-Low Emission Zone (ULEZ), which in Central London charges fees to pre-2006 petrol cars and vans and pre-2015 diesel cars and vans.

Governance Style

Beyond specific regulations and policy instruments, the UK transport governance style has several characteristics. One characteristic is that the Department for Transport ranks relatively low in the wider political and departmental pecking order, which means that transport-related issues rarely receive high priority: 'Transport is rarely seen as a particularly important are of concern in relation to other policy matters' (Shaw and Docherty, 2014: 8). Consequently, transport has not received long-term sustained funding, as Shaw and Docherty (2014: 4) note: 'In four decades up to early 2000s, the UK spend on average 40% less as a proportion of GDP on its infrastructure each year than other leading countries in Europe.' Combined with the UK 'tendency of muddling through' (Shaw and Docherty, 2014: 6), the result is that UK transport has fallen behind leading European countries in the resilience and reliability of core infrastructure, comfort, and ease of using public transport, and the aesthetics of urban space.

Another characteristic is that UK transport governance 'is highly centralized, with very little power at the local level' (Wolmar, 2016: 106). Although local policymakers have become increasingly concerned about 'quality of life' issues such as air pollution, congestion, noise, and parking, their policy responses are constrained by their financial and regulatory dependence on Westminster. London is an exception because it has substantial policy discretion and because *Transport for London* (TfL) received dedicated funding from the Government (until 2018) and the Greater London authority.

A third characteristic is an increasing policy focus on technology and infrastructure projects rather than travel behaviour change or spatial planning. Successive transport ministers preferred 'large-scale road and (now) rail projects' over 'decidedly unsexy but very important local schemes' (Shaw and Docherty, 2014: 103). Wolmar (2016: 80) links this focus to the widespread use of cost-benefit analysis in transport decision-making and the emphasis in these analyses on time savings made by the users of new infrastructure: 'This tends to favour bigger schemes as the benefits can be presented as very large, and also results in ignoring schemes that deliver other benefits than time savings', such as traffic calming, safety, and environmental issues. The 2021 Transport Decarbonisation Plan (DfT, 2021a) and the new bus, rail, and cycling strategies, which are further discussed later, deviate somewhat from this characteristic because they introduced a new emphasis on modal shift, behaviour change, and local solutions besides the ongoing technological focus.

A fourth characteristic is that UK transport governance is relatively light touch and market-based, leaving it to consumers to choose their transport mode and vehicle (perhaps incentivised with some market-based instruments). Exceptions were the more interventionist 1998 *New Deal for Transport* plan (which was only limitedly implemented) and the recent phase-out policies of diesel and petrol cars. The 2021 Transport Decarbonisation Plan and the new bus and cycling strategies are also more interventionist because they hope to shift people from cars to public and active transport modes. The scope for change may be limited, however, by structural and cultural lock-in mechanisms that have created car dependence in many places: 'Decades of transport policies favouring road transport and the associated, established car culture mean that for many journeys people have little practical choice, or at least perceive that they have little practical choice, other than to drive to where they want to go' (Shaw and Docherty, 2014: 176).

5.3 The Railway System

5.3.1 Techno-Economic Developments

The railway system consists of extensive, costly infrastructures, which enable trains to travel uninterrupted between cities and villages, where stations enable passengers to enter and exit. Rail infrastructures include both trunk lines for long-distance, high-speed travel between major cities, and local networks with regional services that stop frequently. While trunk lines are electrified, many UK local networks still operate diesel trains. Railway signalling is a system used to direct railway traffic and prevent collisions.

Rail infrastructure length has decreased substantially since the Second World War, with particularly steep reductions following the 1963 Beeching plans, which by 1968 had reduced tracks by 7,000 km (Figure 5.19) and halved British Rail's supply of rolling stock and number of railway stations (Roberts and Geels, 2019). Only a third of UK railways are electrified, which is low compared to France (52%), Germany (58%), and Italy (71%). This is the result of the 'slow and piecemeal way in which successive governments chose to electrify the rail network' (Shaw and Docherty, 2014: 109).

Rail use declined slightly in the post-war decades, increased somewhat in the 1980s, and has more than doubled since the mid-1990s (Figure 5.20), despite rail fares increasing faster than the retail price index (Figure 5.11). Increased rail travel caused overcrowding in trains and congestion on the rail network, which led to increasing infrastructure investment. Many infrastructure projects focused on rail improvements in/out/across London, for example, Thameslink, Crossrail, high-speed railways to Europe (since 2007), and a second high-speed railway to Manchester and Leeds (called 'HS2'). These new projects (slightly) increased rail

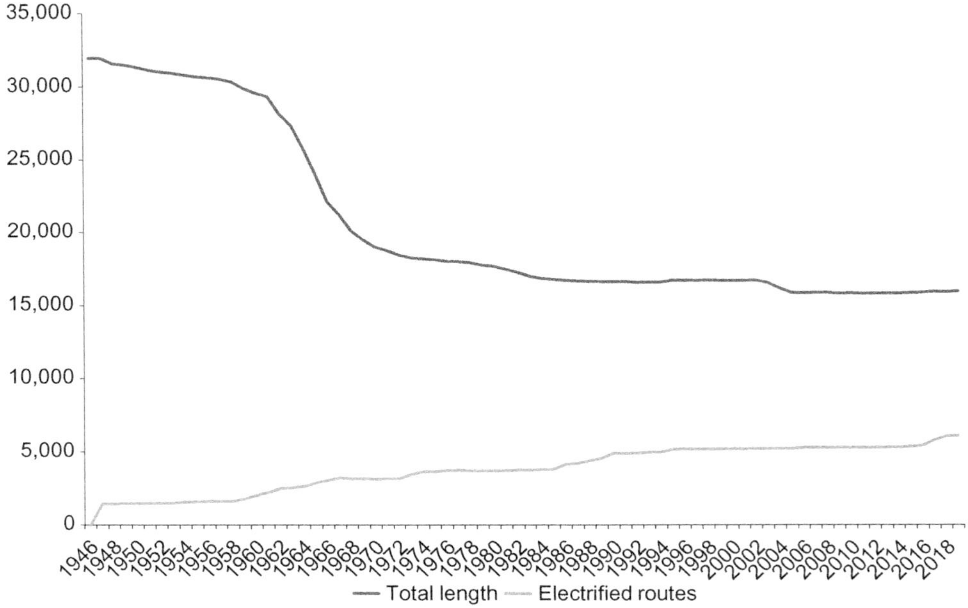

Figure 5.19 Length of railway infrastructure and electrified routes (in kilometres) in Great Britain, 1946–1919 (constructed using data from Department for Transport Statistics; Rail Statistics; Table RAI0101)

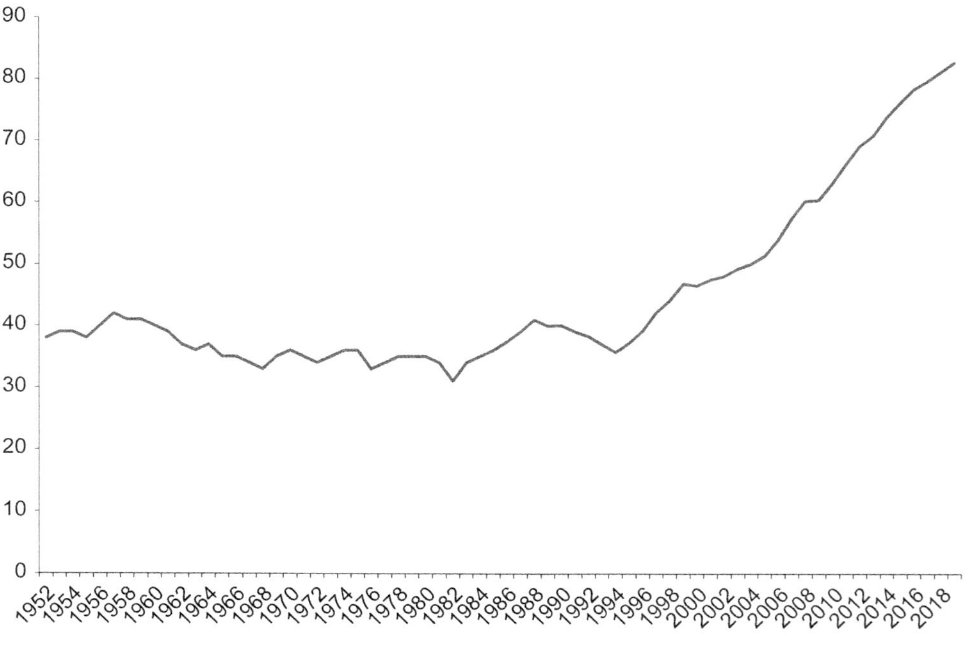

Figure 5.20 Passenger kilometres by rail, Great Britain, 1952–2019, in billion kilometres (constructed using data from Department for Transport Statistics; modal comparisons; Table TSGB0101)

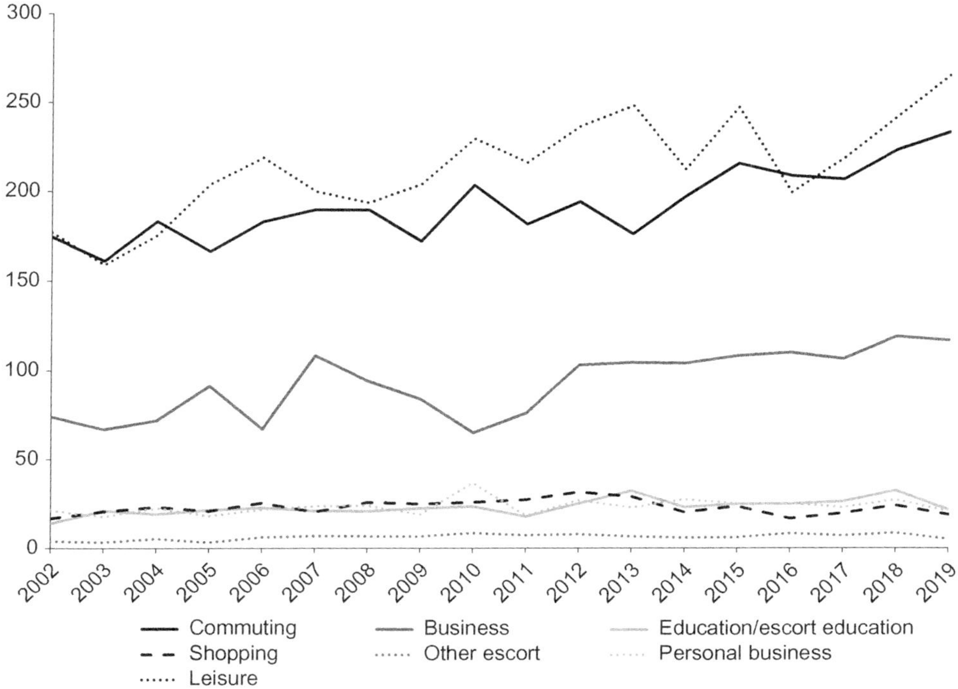

Figure 5.21 Average distance travelled by railways for different purposes (miles per person per year, England), 2002–2019 (constructed using data from Department for Transport Statistics; National Travel Survey; Table NTS0409b)

length after 2012. Services between other UK cities have remained less developed and are where 'the shortcomings of integrated network planning are most obvious' (Haywood, 2007: 210).

Rail travel collapsed due to COVID-lockdowns, reaching levels as low as 5% of normal use in April 2020 (Figure 5.4). Since then, rail travel has recovered slowly, but in July 2021 was still 50% below pre-pandemic levels.

5.3.2 Actors

Firms: British Rail, which was a vertically integrated state monopoly overseeing rail network development, maintenance, and operations, was privatised in 1994–1997. This created a complex network of actors, including: a) Train Operating Companies (TOCs) that operate 16 railway franchises for specific routes, b) rolling stock operating companies that lease trains to TOCs, c) Network Rail, which owns, maintains, and operates the railway infrastructure, d) the Office of Rail Regulation that oversees the industry, and e) the Department for Transport (DfT), which awards and sets general conditions for the railway franchises.

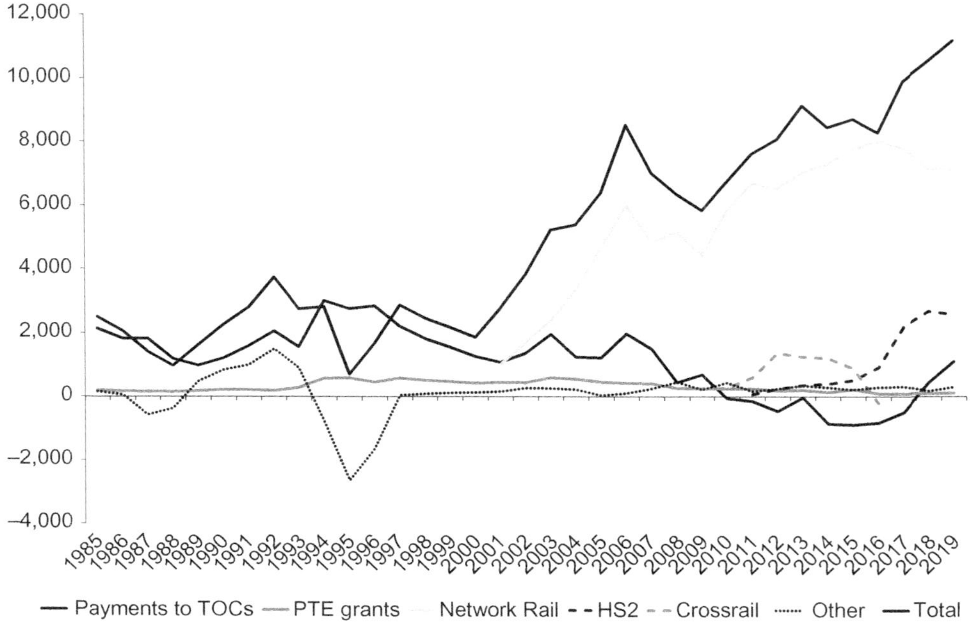

Figure 5.22 Government support to the rail industry, 1985–2019 in £million (constructed using data from Statistics at DfT; Rail Statistics; Rail finance tables; table RAI0302)[4]

Climate change is of limited importance to most of the industry actors, whose strategies mostly focus on financial gaming of the complex franchise and lease system, by privatising gains and collectivising costs.

TOCs compete for franchises, which they operate to optimise profits. TOCs earn income through: a) passenger revenues, which increased from £4.6 billion in 2000 to 10.2 billion in 2019 (through increased fares and passenger journeys), b) other train-related income (e.g., catering in trains and stations), which was \$1.2 billion in 2019, and c) direct government payments (through franchise awards and performance grants), which were £1.2 billion in 2019 (ORR, 2020); these government payments were negative between 2011 and 2017 (because of the way franchises were structured), but turned positive again in 2018 and 2019 (see also Figure 5.22). TOCs lease trains from rolling stock operating companies, pay 'track

[4] Government payments to train operating companies (TOCs) include franchise payments and performance payments. These payments turned negative after 2010 due to specifics of the franchise designs. Passenger Transport Executive (PTE) grants are payments to transport executives in large cities (e.g., Transport for London). Payments to Network Rail (which was renationalised after Railtrack went bankrupt in 2002) include operational funding, infrastructure investment, and loans. Funding for 'other' categories include British Transport Police, station improvement funds, security initiatives and research. The proceeds from the sales of rolling stock companies (ROSCOs) and British Rail non-passenger business in 1995 and 1996 (related to privatisation) are also included in the 'other' category.

access charges' to Network Rail for the use of rail infrastructure, and have several other operating expenditures, which have historically been lower than total income, allowing TOCs to make net profits of about 2–3% of revenue. But since they do not invest in trains or tracks, TOCs extract these profits at very low risk, leading to returns-on-capital that can be as high as 120% (Bowman et al., 2013).

TOCs also benefit from large *indirect* subsidies in the form of relatively low track access charges (which is a political decision), which are not enough to cover the investments by Network Rail in track maintenance and expansion, and these have therefore been paid through a mixture of government subsidy and debt accumulation, as we describe next.

The publicly owned Network Rail organisation became responsible for railway infrastructure when Railtrack (which was created after privatisation) went bankrupt in 2002. Because privately owned Railtrack made limited investments in track and signalling infrastructure, Network Rail had to address this backlog while also expanding and upgrading railways to accommodate growing rail traffic. Despite increasing rail traffic, TOC payments for infrastructure use decreased from £2.2 billion in 2008 to £1.6 billion in 2012 (Bowman et al., 2013). Although 'track access charges' increased again from £1.6 billion in 2015 to £2.5 billion in 2019 (ORR, 2020), TOC payments continued to fall short of Network Rail investments. Government payments to Network Rail have increased substantially over time, from £1.7 billion in 2002 to £7.1 billion in 2019 (see Figure 5.22). Additionally, Network Rail has borrowed money from the financial markets to finance rail infrastructure investments. By 2012, this had resulted in cumulative debt of £30 billion (Bowman et al., 2013), which by 2019 had ballooned further to £52 billion (ORR, 2020). This money thus forms an indirect subsidy to TOCs, which (with political consent) have managed to socialise infrastructure costs.

The COVID-related decline in passenger travel and fare incomes created major financial problems for the railway companies, which led policymakers to suspend the franchise model and provide emergency support to railway companies to prevent bankruptcy.

Users: Rail use increased despite soaring rail ticket prices, which increased by 102% between 1997 and 2014, making UK rail fares among the highest in Europe. People use trains for various purposes. Although leisure is the single largest end-use category, commuting and business-related travel together accounted for more than half of all railway passenger kilometres (51%) in 2019 (Figure 5.21).

Rail travel is dominated by the 'London effect': almost two-thirds of all rail journeys started or ended in London (DfT, 2019a). The most important explanation for increased rail travel growth is the socio-economic boom in Greater London and the South East, which stimulated more train travel in/out of London.

About half of the new rail travellers are drawn from the group of regular car users. Growth in rail use has thus led to some modal shift, which has been largest for 'men living outside London but travelling into London regularly for work-related purposes' (RAC, 2012: 84). Although there is ongoing debate about underlying causes of increased rail travel, proximate reasons are: a) rising London house prices forcing people to live elsewhere and commute to work, b) reduced company car use (related to fuel taxation changes), c) the London congestion charge and other car-hindering measures (RAC, 2012).

Railway usage decreased by 95% in April 2020 during the first national lockdown and has only slowly recovered since then (Figure 5.4), owing to ongoing COVID restrictions (such as the 2 metre distance rule) and health concerns that made people more reluctant to travel with others in confined spaces. It remains to be seen if this is a structural trend that will negatively impact public transport.

Policymakers: Policymakers privatised the railways with the stated aims of improving service quality, increasing competition and cost-efficiency, reducing public subsidies, encouraging investment, and improving environmental performance (Haubrich, 2001). Reality turned out differently, and aims have not been met with regard to service quality, cost-efficiency, and private investment (Haubrich, 2001; Jupe, 2013). Despite promises to the contrary, public subsidies also increased substantially (Figure 5.22), especially for Network Rail, which imply indirect subsidies to TOCs, who thus pay too little for the use of railway infrastructure.

To accommodate the growth in rail travel and address emerging problems, policymakers have substantially supported rail infrastructure expansion since the early 2000s (Figure 5.22). These investments have focused more on large-scale, eye-catching projects (such as HS2, Crossrail) than on improvements in existing tracks or the signalling infrastructure (Wolmar, 2016).

The various railway problems were investigated in the 2006 Eddington Review, the 2011 McNulty Review, and the 2013 Brown Review. The McNulty Review, for instance, criticised the lack of efficiency improvement, which created a 40% gap with European comparators. The review also found that the 'causes of GB rail's excessively high costs are many and complex', including 'fragmentation of structures and interfaces, roles of government and industry, ineffective and misaligned incentives, a franchising system that does not encourage cost efficiency sufficiently, management approaches that fall short of best practice, and a railway culture which is not conducive to partnerships and continuous improvement approaches' (p. 5).

Despite the critical reviews and public debates, policymakers did not show much appetite to substantially change institutional framework conditions (Jupe, 2013). Although Labour politicians repeatedly criticised existing arrangements and advocated railway renationalisation in the past 10 years, their proposals had limited direct effects because they were not in government. But the criticisms did have indirect effects in keeping problematic railway arrangements in the spotlight and shaping the discourse. The COVID-19 shock, which substantially disrupted the railway industry, then provided the conditions for Conservative politicians to introduce major institutional changes, which are further discussed later.

Wider Publics: Public debates about railways are mostly negative, focusing on overcrowding, delays, high train fares, and public subsidies (Taylor and Sloman, 2013). Although debates on the European continent often emphasise the public-good role of railways in providing mobility access for people without cars, this argument has been far less salient in UK debates in recent decades, where railways are mostly framed as a private enterprise (Roberts and Geels, 2018). There have, nevertheless, been ongoing critical debates about the dysfunctionalities of privatisation and liberalisation of the railways, which created the conditions for substantial institutional change (i.e., the Great British Railways reforms) when the COVID-pandemic disrupted the railway industry.

Public debate has focused on new infrastructure projects, which policymakers portray as economically necessary. Opponents, however, questioned the wisdom of large investments for HS2 (estimated at more than £100 billion) and criticised negative effects on the countryside. The COVID-pandemic reignited these debates because potential alterations in commuting and business travel erode the HS2 business case, leading the National Infrastructure Commission (NIC, 2020) to suggest that HS2 infrastructures beyond Birmingham should be 'reviewed' and potentially not go ahead.

5.3.3 Policies and Governance

Railway governance has been very London-centric and focused on large-scale projects (e.g., Crossrail, Thameslink, HS2) rather than on whole system improvement, including technical issues such as railway electrification and signalling. Instead of improving bottlenecks in existing railways, policy attention has mostly focused on new high-speed lines. But according to Wolmar (2016: 88), 'HS2 is a political project pushed through as a *grand projet* by politicians for reasons that do not stand up to political scrutiny. In effect, it is transport policy on a whim.' Although railway polices made some piecemeal improvements, there is a 'failure to follow through on an overarching vision for developing the railways' (Shaw and Docherty, 2014: 106).

Privatisation of British Rail in the 1990s was a major institutional change that under-delivered on the promises that were made in advance. Although the effects of privatisation were disappointing, successive governments were unwilling to substantially adjust the institutional framework, despite providing increasingly large subsidies (Figure 5.22). Repeated criticisms by the Labour opposition contributed to the erosion of the taken-for-granted legitimacy of existing institutional frameworks. In response to the COVID-pandemic, the government first provided substantial financial support to the railway companies, amounting to £12 billion in the first post-lockdown year (DfT, 2021b: 6), to prevent bankruptcy and ensure some continued service provision through the Emergency Measures Agreements (from March to September 2020) and the Emergency Recovery Management Agreements (from September 2020 to March 2022).

In May 2021, the government then introduced substantial institutional reform that aimed to simplify arrangements by creating a new arm's length public body, *Great British Railways* (GBR), which from 2023 'will own the infrastructure, receive the fare revenue, run and plan the network and set most fares and timetables' (DfT, 2021b: 7). GBR will also contract with train companies to operate trains to the timetable and fares it specifies. Train companies are expected to compete for the contracts, which will incentivise them on punctuality and efficiency rather than revenue raising (as in the franchise model). Although the government says it is committed to grow the railways, the institutional reforms also clearly aim to 'secure significant efficiencies' (p. 8) and reduce public subsidies because 'the current sums being paid to operate and maintain the railways are not sustainable' (p. 7). Many of the details are still to be decided, so it remains to be seen what the effects of these reforms will be.

5.4 The Bus System

5.4.1 Techno-Economic Developments

Buses share many road infrastructures with cars and cyclists, although some cities have started to introduce segregated bus lanes on some roads. Bus services are thus more flexible than (fixed) railways but do follow particular routes, which in many cities are determined by private bus companies. Some stops along routes are merely indicated by signs, whereas others also have shelters that protect waiting passengers from the weather. Although timetables indicate bus frequencies and arrival times, traffic congestion and other contingencies frequently cause delays, which create uncertainties for waiting passengers. Bus stops can be fitted with real-time passenger information displays that reduce these uncertainties by more reliably indicating estimated times of arrival. Buses mostly provide short- to medium-distance services within cities, although coaches also offer long-distance services between cities, and are often cheaper than trains.

Bus use is in long-term decline, with total bus passenger kilometres declining by 22% between 2010 and 2019 (Figure 5.3). In London, however, the number of bus

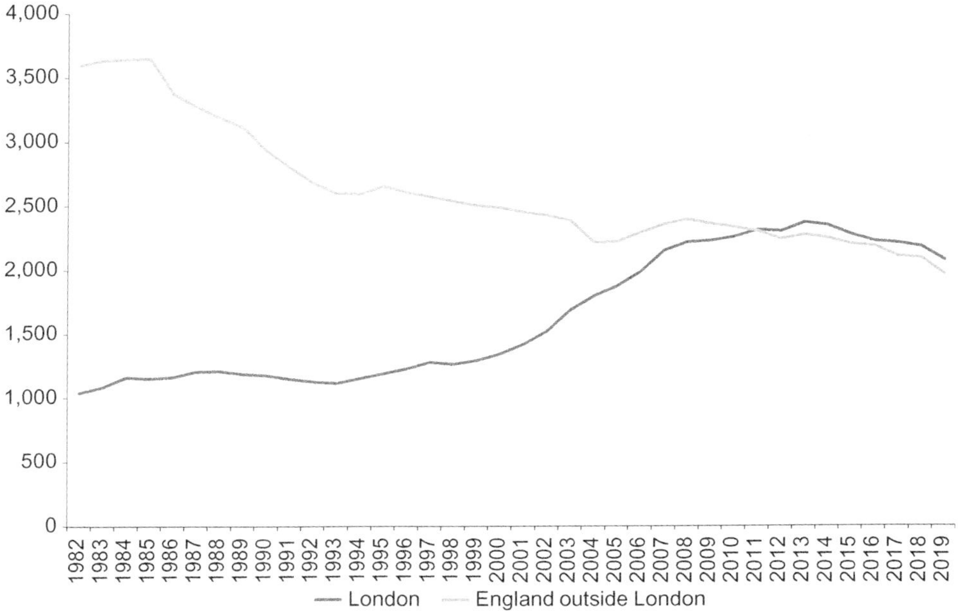

Figure 5.23 Number of English passenger journeys (in millions) on local buses, 1982–2019 (constructed using data from Department for Transport statistics; Bus Statistics; Table BUS0103)

journeys has increased substantially since 1995, accounting for about half of English bus passenger journeys in 2019 (Figure 5.23). Bus travel has become more expensive with fares rising faster than the retail price index, and faster than rail fares (Figure 5.11). The COVID-19 pandemic reduced bus travel by almost 90% in April 2020, which subsequently only partly rebounded to about 60% of pre-pandemic levels by July 2021 (Figure 5.4).

The great majority of buses in Great Britain (85%) still use diesel fuel (Table 5.3). Diesel-electric hybrid buses, which are 20–30% more fuel efficient, have also found some use, but mostly in London and far less in the rest of the country. Electric and gas-based buses are still relatively rare. Electric buses are mostly used in London, although a few city regions (e.g., Nottinghamshire, North Yorkshire, Greater Manchester) also experienced some deployment.[5]

5.4.2 Actors

Firms: The bus industry (except in London, and a few other places, including Nottingham and Reading) was deregulated in 1986, and subsequently privatised,

[5] See the website of the Zemo Partnership (formerly known as the Low Carbon Vehicle Partnership): www.zemo.org.uk/work-with-us/buses-coaches/low-emission-buses/areas-of-operation.htm.

Table 5.3. *Percentage of fuel consumption in buses operated by local bus companies in 2019/2020 (constructed using data from Department of Transport Statistics; Bus Statistics; Table BUS0609)*

	London	England	England outside London	Scotland	Wales	Great Britain
Electric (not hybrid)	4	2	1	1	0	1
Diesel-hybrid	40	14	3	6	0	12
Methane, bio-methane	0	1	2	0	0	1
Diesel	56	84	95	94	100	85

which initially resulted in a range of small companies, but subsequent take-overs created an oligopoly (Langridge and Sealey, 2000), dominated by five bus groups (First, Go-Ahead, Stagecoach, Arriva, National Express). The bus market is locally organised, with companies competing for specified tenders.

Privatisation allowed bus companies to decide on their routes, fares, and timetables. The promise was that increased competition between bus companies would result in improved services, lower fares, and increased bus use (Cowie, 2002). The results have been exactly the opposite: bus use has decreased, except in London (which did not privatise bus services); fares have increased faster than the retail price index (Figure 5.11); and service quality has decreased (Stradling et al., 2007). Bus companies mostly compete on costs, paying only moderate attention to innovation and quality improvement.

Despite being privatised, the bus industry is heavily subsidised. Decreases in public support, from 49% of total revenue in 2009 to 41% in 2019 (Figure 5.24), reinforced the industry's cost focus at the expense of other considerations. Although air pollution and climate change are not core considerations, bus companies have used government subsidies to gradually adopt hybrid and electric buses (especially in London). The COVID-pandemic substantially disrupted the bus industry, leading to emergency government support.

Users: Buses are the most frequently used and most accessible mode of public transport (McTigue et al., 2018). Bus use is primarily local. Most trips are between one and five miles. Buses are essential for people without cars, accounting for over 60% of all public transport trips. Bus use is most prevalent among lower income families, students, the young, and elderly people (DfT, 2014). People aged over 60 received concessionary bus travel passes in 2001 to enhance their access to mobility services. Local bus use is primarily for shopping, education, leisure activities, and commuting (DfT, 2017). Bus use has been declining, especially outside London, because of rising fares and declining service frequency and punctuality. Decline was further accelerated by the COVID pandemic and only partially rebounded (Figure 5.4), owing to lingering health concerns among the

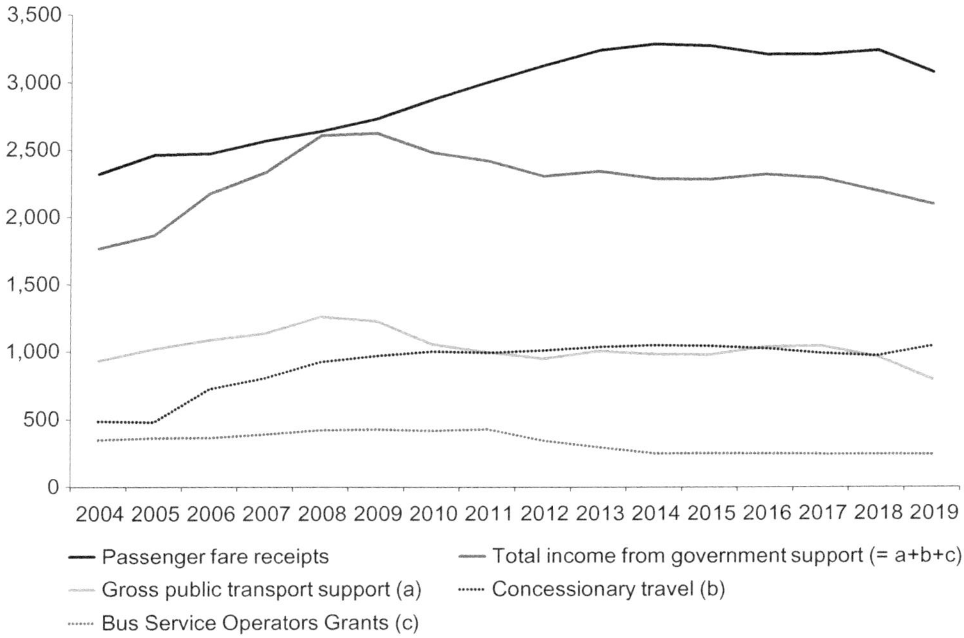

Figure 5.24 Operating revenue (in £millions at current prices) for local bus services in England, 2004–2019, through passenger fare receipts and (three forms of) government support (constructed using data from Department for Transport Statistics; Bus Statistics; Table BUS0501)

elderly and shifts to online teaching for students, which both may lead to long-term structural shifts.

Policymakers: Policymaking for bus transport has been fragmented since deregulation and privatisation, with national policymakers setting regulatory frameworks and controlling financial purse strings, and local authorities implementing policies and contract tendering. In response to the negative effects of privatisation, policymakers have tried since the late 1990s to improve coordination, innovation, and service quality in the fragmented bus industry. Despite various new policies, which are further discussed later, the problems have not been solved. One reason is that national funding for buses has decreased since the financial crisis, especially for bus operator grants and public transport support to local authorities (Figure 5.24).

Another reason is that local transport planning is hindered by limited policy discretion and dependence on national funding sources (Marsden et al., 2014). Local transport policymaking is also hampered by several barriers: shortage of skilled staff, limited political support from city councils, lack of funding, and insufficient strategic alignment between objectives and resources (McTigue et al., 2018).

A third reason is that local councils have 'insufficient political will to improve bus infrastructure where it involves reallocating road space away from cars, little desire to upset their local bus companies by interfering with their existing operations and a lack of in-house expertise capable of actually fashioning and managing a fully integrated transport network' (Shaw and Docherty, 2014: 90).

London, which privatised but not deregulated its bus services, is an exception. Bus use in London has expanded significantly since the early 2000s (Figure 5.23) for several reasons. First, successive mayors (Livingstone, Johnson) made buses a policy priority. Second, bus services remained regulated by Transport for London, which sets high standards in its bus tenders. Third, substantial financial support of around £600 million/year was made available to expand and improve London's bus system (Shaw and Docherty, 2014).

Wider Publics: The public image of buses is poor. They have become viewed as a 'last resort' means of transport (Shaw and Docherty, 2014). Public debates are mostly about negative issues such as rising fares and low service quality (Knowles and Abrantes, 2008; Stradling et al., 2007). There is no consensus about how to improve bus services (Currie and Wallis, 2008). It is surprising that air pollution problems have not led to stronger debates about dirty diesel buses, although this has begun to change in the last few years.

5.4.3 Policies and Governance

The negative unintended consequences of bus privatisation and deregulation gave rise to several policy initiatives. The 2000 Transport Act allowed local authorities to introduce 'quality partnerships' to stimulate innovation. In this partnership, local transport authorities would agree to improve bus stop infrastructure (e.g., raised kerbs, shelters, real-time passenger information), if bus companies invested in new vehicles and driver customer care training (White, 2010). These partnerships had limited success, however.

The 2008 Local Transport Act aimed to facilitate collaboration between operators and local authorities to co-ordinate ticketing and timetabling. The 2011 Local Transport White Paper further aimed to enhance the 'whole journey experience' and promoted integrated 'smart' ticketing. These national policies had limited effects (except for London), because they 'failed to compensate for the fragmentation installed in the horizontal level by the bus deregulation' (Sørensen and Gudmundsson, 2010: 14–15).

To address local air pollution and climate change, national policymakers also tried to stimulate the uptake of hybrid and electric buses via the Green Bus Fund (£90m between 2009 and 2015) and the Low Emission Bus Scheme (£41m between 2015 and 2017), which have delivered over 1,600 buses in service.

Wider diffusion was hampered, however, by the stop-start dynamic of national funding sources and changing priorities (McTigue et al., 2018). The Ultra-Low Emission Bus Scheme (£48million) aimed to drive further uptake between 2018 and 2021.

In response to the pandemic, policymakers introduced the COVID-19 Bus Service Support Grant (CBSSG), which provided about £1.4 billion support funding between March 2020 and July 2021.[6] This grant enabled bus companies outside London to provide service levels of up to 100% of pre-pandemic levels, which supported people relying on bus services, including many key workers. From September 2021, however, the government intends to reduce recovery funding from £27 million per week to about £8 million per week, which may substantially affect bus companies if travel patterns have not substantially increased.

Before the pandemic, in February 2020, the Johnson administration already announced a £5 billion bus and cycling fund to stimulate both transport modes, although it reportedly also aimed to stave off criticisms of the £100 billion HS2 go-ahead decision (Stewart and Walker, 2020). These plans were maintained and elaborated during the pandemic, which highlighted the importance of local transport and structural reform, culminating in the National Bus Strategy for England, launched in March 2021 (DfT, 2021c). Using £3 billion of funding over five years it aims to reform and improve bus services outside London and reverse their decades-long decline. Specific goals include: a) the introduction of simple, cheap flat fares (which enable multiple trips in one day), b) multi-modal tickets that can be paid with a contactless card, c) more frequent and reliable services, d) high service standards and improved digital information, e) new priority lanes for buses, and f) the introduction of about 4,000 electric buses (representing about 10% of the bus fleet). To achieve these goals, the National Bus Strategy also introduced the option of Enhanced Partnerships between Local Transport Authorities (LTAs) and bus companies, which give LTAs more influence over timetables, multi-operator ticketing, and bus services improvements, and made access to the additional £3 billion funding conditional on LTAs and bus companies implementing Enhanced Partnerships by April 2022.

For much of the studied period, the bus governance style had several characteristics that hampered policy effectiveness. First, buses had relatively low political priority, despite their importance for particular social groups. For many years, policymakers have given less attention (and resources) to buses than to trains or cars, because buses are less about big infrastructure projects and thus less eye-catching (Wolmar, 2016). Second, governance is fragmented between national

[6] www.gov.uk/government/speeches/supporting-vital-bus-services-recovery-funding

and local levels. Although local authorities are expected to develop Local Transport Plans, they have limited legislative or financial powers to enable strategic policymaking (McTigue et al., 2018). Third, national bus funding has not only diminished since 2009 but also frequently changed direction in terms of priorities, which further complicates local strategic policymaking. Because of these three characteristics, bus policymaking and implementation has been piecemeal, fragmented, and relatively ineffective in addressing persistent problems. Aiming for ambitious reform, the recent National Bus Strategy has increased the political priority of and funding for buses but remains a rather top-down policy. Its future success will therefore depend on local uptake and implementation and on the degree to which users will return to buses.

5.5 The Cycling System

5.5.1 Techno-Economic Developments

Basic bicycle technology has long stabilised, although innovations continue to be made, for example, folding bikes and electric bicycles. The average price of new UK bicycles was £427 in 2016 (Newson and Sloman, 2018). Sales of new bicycles fluctuated between 3 and 3.6 million per year between 2003 and 2016 (Conebi, 2017).

UK cyclists mostly share road infrastructures with cars, vans, buses, and lorries. Only a few UK cities (e.g., London, Brighton) have infrastructures with separate cycling lanes, although other cities have recently begun to follow suit.

The post-war transition towards cars was accompanied by a steep decline in British cycling from 23 billion passenger-kilometres in 1952 to 6 billion in 2019 (Figure 5.25). This represented a reduction from 11% of all passenger kilometres travelled in 1952 to 0.7% in 2019. Since the mid-2000s, British cycling has been on an upward trend, but it remains a very small regime with stabilised rules and practices, carried by relatively small social networks. UK cycling is low by EU standards (Pucher and Buehler, 2008), accounting for less than 1% of travel distance and 2% of trips (Table 5.1). The pandemic was an external shock that triggered a relatively rapid 46% increase in cycling in 2020 (DfT, 2021d), with particularly large expansion during and immediately after the first lockdown (Figure 5.26).

5.5.2 Actors

Firms: UK companies produce only a small quantity of bicycles, about 80,000 per year. The largest UK company is Brompton, which produces folding bikes. Most bicycles are imported from the Far East, principally Vietnam, Cambodia, China, and India. Bicycle sales account for about half of the total value of the bicycle

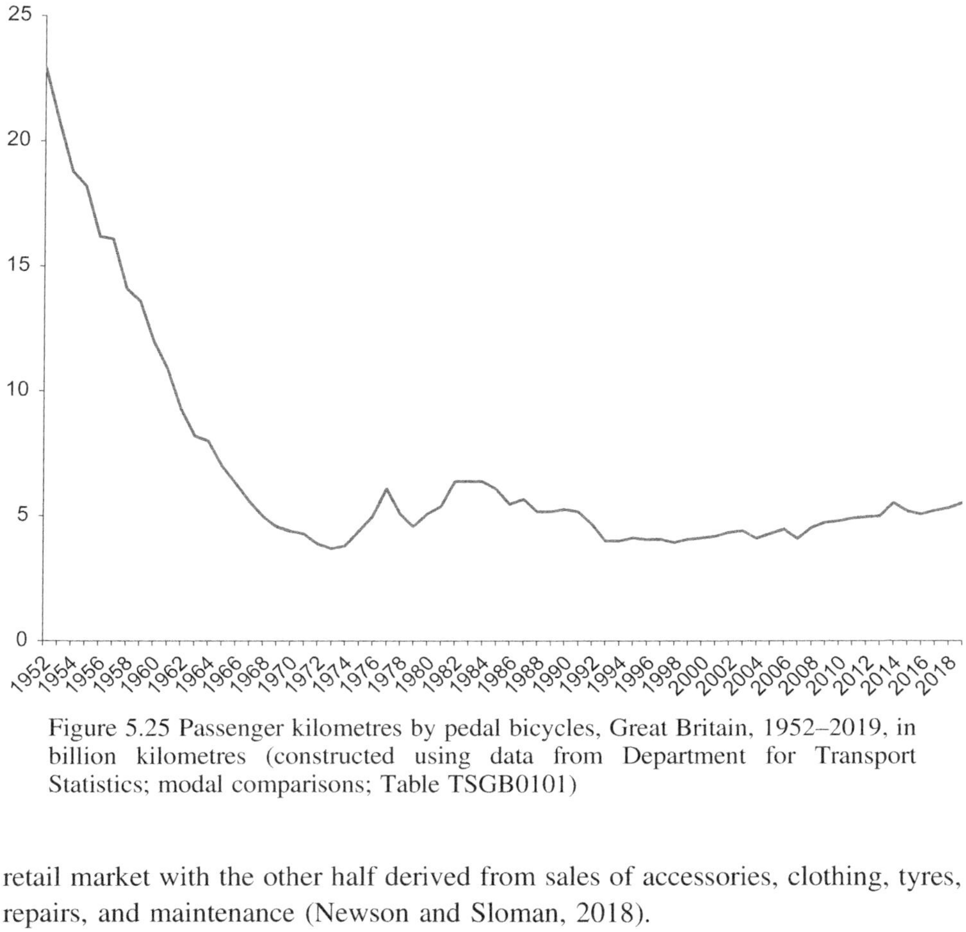

Figure 5.25 Passenger kilometres by pedal bicycles, Great Britain, 1952–2019, in billion kilometres (constructed using data from Department for Transport Statistics; modal comparisons; Table TSGB0101)

retail market with the other half derived from sales of accessories, clothing, tyres, repairs, and maintenance (Newson and Sloman, 2018).

Users: Cycling remains a marginal activity in the UK, perceived by most people as good for recreation or commuting daredevils but not for many other purposes. Bicycle use concentrates in cities with favourable conditions (e.g., London, Cambridge, Oxford, Brighton). In London, cycling has seen the fastest relative growth of all transport modes (Figure 5.18), especially for commuting purposes (Aldred and Jungnickel, 2014). Between 2001 and 2018, cycle flows across central London increased by 340% (Figure 5.27). Nevertheless, cycling still only accounted for 1.5% of all passenger-kilometres in London in 2018.

The majority (62%) of London's cyclists are male, 38% female. Most of London's cyclists are young urban professionals: 32% are from the 25–34 age group, 27% from the 35–44 age group, and 16% and 15% respectively from the 16–24 and 45–54 age groups (TfL, 2015). The main motivations for London's cyclists are increased fitness, and travel time and money savings. The main

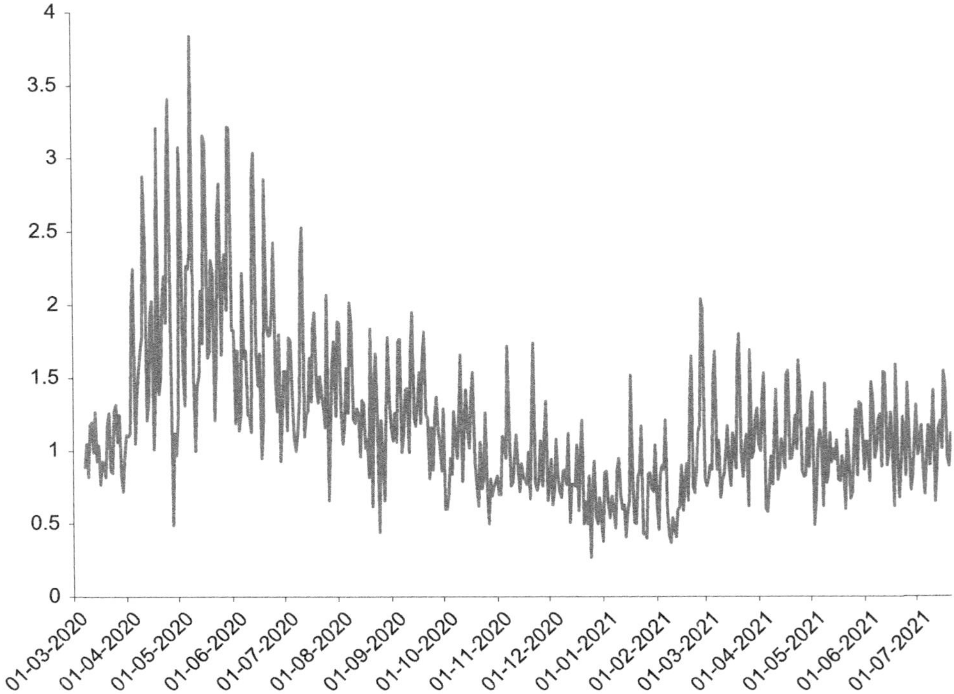

Figure 5.26 Daily use of bicycles in Great Britain between March 2020 and July 2021; figures are percentages of an equivalent day or week (constructed using data from Department for Transport statistics; Transport use during the coronavirus COVID-19 pandemic)

deterrents mentioned by non-cyclists are safety concerns, bad weather, health reasons, lacking accessibility, and limited confidence (De Boer and Caprotti, 2017).

The COVID-19 shock created opportunities for cyclists because the first lockdown in particular substantially reduced car traffic (Figure 5.4). On some days during and after the first lockdown cycling was more than 300% above pre-pandemic levels (Figure 5.26), as people cycled for their daily exercise on quiet roads. As lockdowns were lifted and car traffic returned to pre-pandemic levels (Figure 5.4), cycling also returned to pre-2020 levels in most places (Figure 5.26), which suggests that the cycling increase may not be a long-term structural trend. Some cities such as Manchester, however, report that cycling in July 2021 was still 20% above pre-pandemic levels (Laker, 2021).

Policymakers: Cycling policy was non-existent before the 1990s. The 1996 National Cycling Strategy was followed in 2000 by the 10-Year Transport Plan, which aimed to triple cycling trips in a decade. This goal was not reached,

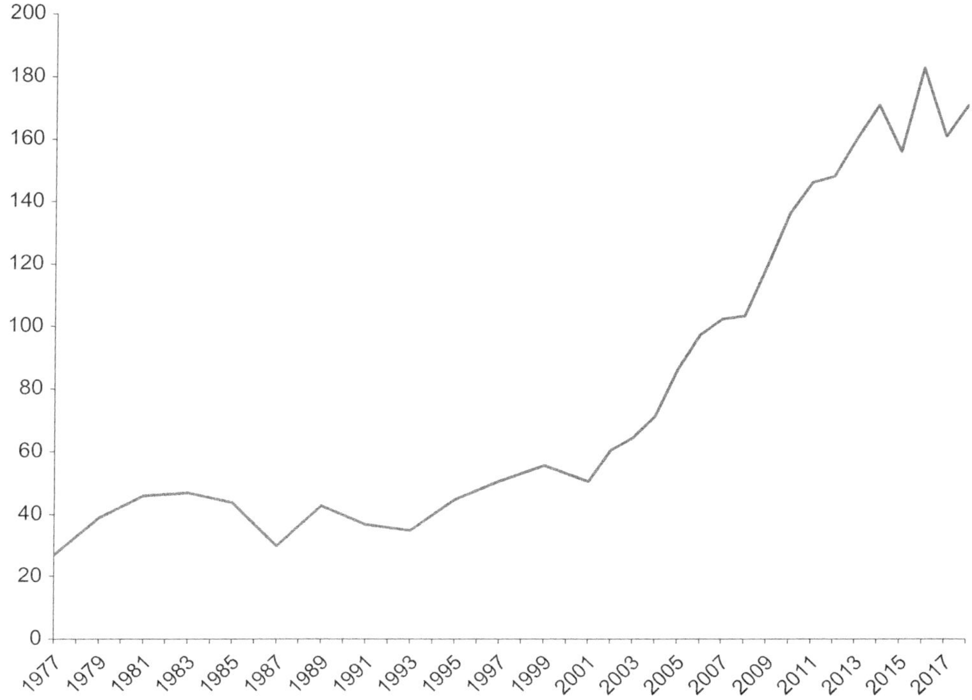

Figure 5.27 Trend in cycle flows (in thousand cycles per day) across central cordon, 1977–2018 (constructed using data from TfL (2019))

however, and cycling trips actually decreased in the subsequent decade. Reasons for this policy failure are the following: a) limited political will to systematically invest in cycle networks with separate lanes, b) perception of cycling as peripheral in transport policy, c) absence of specialised expertise in most local authorities, d) misalignment between national and local cycling policies, e) public perceptions of cycling as dangerous (Aldred, 2012; Pooley et al., 2011).

There has been some national funding for local cycle lanes and parking facilities (via the Local Sustainable Transport Fund and Cycle City Ambition Grants), but the stop-start dynamics of these schemes resulted in ad-hoc schemes rather than the building of interlinked cycling infrastructures. London has been an exception, because Mayor Boris Johnson (2008–2016) actively supported cycle hiring schemes and the development of London's Cycle Network. Several other cities (e.g., Cardiff, Newcastle, Bristol, Edinburgh, Leeds, Manchester) have also begun cycle infrastructure construction.

The COVID-19 shock, which disrupted most transport modes but boosted cycling, provided a window of opportunity for national policies and local initiatives to further support cycling. In May 2020, the Johnson government introduced a £250 million emergency active travel fund, which aimed to help city councils to reallocate road space to cyclists and pedestrians by creating dedicated cycle paths, pop-up bike lanes (using cones or flexible wands), and cycle and bus-only corridors.

In July 2020, the government introduced a new cycling strategy, which advanced the bold vision of a 'travel revolution in our streets, towns and communities' with 'half of all journeys in towns and cities being cycled or walked by 2030' (DfT, 2020b: 12). The vision, which was reportedly pushed by Johnson himself (Laker, 2020), was accompanied by 33 specific commitments for active travel and £2 billion funding over a five-year period to enable local councils to make on-the-ground changes. This top-down push resulted in many local initiatives so that by July 2021 sixty miles of new segregated cycle lanes on main roads have been constructed in London and forty miles in other cities, including active ones such as Manchester and Leicester (DfT, 2021e). But many other councils (including Liverpool, Brighton, Surrey, Kensington and Chelsea, Ealing, Wandsworth, South Gloucestershire, Trafford, Portsmouth) initially created (temporary) cycle lanes and low-traffic neighbourhoods but then later removed them in response to protests from car drivers, local residents, and other interest groups (Walker, 2021).

Frustrated about these local implementation problems, transport minister Heaton-Harris on 30 July 2021 wrote a letter to all local authorities with transport responsibilities, urging them to give active travel schemes a chance to bed in and warning them that they could lose future central government funding if they removed these schemes without sufficient evidence.[7] These tensions suggest that the success of the ambitious top-down cycling strategy is uncertain, depending not only on political struggles between national and local policymakers but also on social acceptance processes in cities and towns, which are likely to vary in terms of the actors involved and stakeholder consultation and governance processes.

Wider Publics: In most British towns 'cycling is not seen as normal' (Pooley et al., 2011: 1604). Many people perceive cycling as dangerous, because of the lack of dedicated cycle lanes. Bicycle accidents receive much attention in the media, which contributes to negative perceptions, despite impressive safety improvements since 1970. But although the fatality rate of cyclists (per billion passenger miles) decreased by 43% between 2006 and 2019 (DfT, 2020c), cycling is still fairly dangerous compared to other transport modes, with relatively high rates of casualties (deaths and injuries) and fatalities (deaths) (Table 5.4). People also tend to perceive cyclists as risk-takers and 'not like me', which is reinforced

[7] See www.gov.uk/government/publications/active-travel-schemes-supported-by-government-funding

Table 5.4. *Casualty and fatality rates per billion passenger miles by road user type in Great Britain (DfT, 2020c: 10)*

	Casualty rate	Fatality rate
Pedestrian	1,640	35.4
Cyclists	4,891	29.0
Motorcyclists	5,051	104.6
Car drivers	195	1.6
Bus passengers	141	0.6

by the wearing of materials such as lycra and high-visibility clothing (De Boer and Caprotti, 2017).

Nevertheless, public perceptions have improved in recent years due to positive storylines that frame cycling as good for the environment (e.g., climate change, local air pollution) and for health (as a form of exercise that may help mitigate increasing obesity problems).

5.5.3 Policies and Governance

Cycling has low policy priority and thus has long been marginalised. UK transport governance is fragmented between national and local levels. The national government published a cycling strategy in 1996 and a walking and cycling action plan in 2004 but did not follow up with sustained implementation activities. Cycling strategies thus remained little more than aspirations, leading to a pattern of over-promising and under-delivering.

One reasonably effective policy was the Cycle to Work scheme, introduced in 1999 and amended but maintained since then, which supported workers' bicycle purchases with tax exemptions and company-provided employee benefits. In 2016, about 150,000 bicycles were bought using this scheme (Conebi, 2017).

There has been some national funding for local cycling schemes and infrastructures, but this has remained ad-hoc and fragmented. 'Outside of London there is little sense of policy coherence and the amount of money available tends to be small. … The problem is that, at the national level at least, successive governments have failed to demonstrate the leadership necessary to raise the profile of walking and cycling above its current moribund state' (Shaw and Docherty, 2014: 70–71).

The exception has been London, where Mayor Boris Johnson introduced a comprehensive bicycle hiring scheme in 2010, modelled on the successful Lyon and Paris systems. And in 2012, he committed to spend £913 million over a nine-year period to improve cycling infrastructures. 'The one city to benefit from considerable cycling investment has been London, where a massive increase in cycle use, stimulated from the grassroots largely by young people commuting to the central zone, has almost forced local politicians to respond' (Wolmar, 2016: 99).

The government's new 2020 cycling strategy, introduced in response to the COVID-pandemic, aims to replicate London's relative success across other cities, using £2 billion funding over five-years to support local cycling infrastructure initiatives. Additionally, the government published an updated Cycle Infrastructure Design Guidance in July 2020 to provide technical design advice, issued new statutory Network Management Duty guidance that advises local authorities on reallocating road space to encourage cycling and walking, and published a Local Transport Note 1/20 with detailed design guidance for safe high-quality cycling infrastructure. The government also announced the creation of a new body for cycling and walking (*Active Travel England*), which from Autumn 2021 will oversee funding applications and inspect finished schemes.

While these new initiatives represent attempts at more active and interventionist governance for cycling, its top-down style and emphasis on financial carrots and sticks, combined with technical advice, may insufficiently address social acceptance problems that could scupper its success.

5.6 Niche-Innovations

This section analyses six niche-innovations, which have considerable carbon-reduction potential. Electric vehicles and biofuels are about the greening of cars. Tele-working, car-sharing, and ride-hailing are about technology-enabled changes in mobility practices. Intermodal transport innovations such as smart cards and mobility services create new linkages between transport modes. And self-driving cars are about changes in technology and behaviour, with potential transformative effects on how the automobility system functions. For each niche-innovation, we analyse techno-economic and socio-political developments.

5.6.1 Electric Cars

Techno-Economic Developments

Electric vehicles are modular innovations that substitute the internal combustion engine (ICE) with an electric motor and battery but leave many other parts of automobility systems unchanged. They do, however, require a new battery recharging infrastructure, and may also benefit from lighter car bodies (e.g., by replacing steel with composite materials).

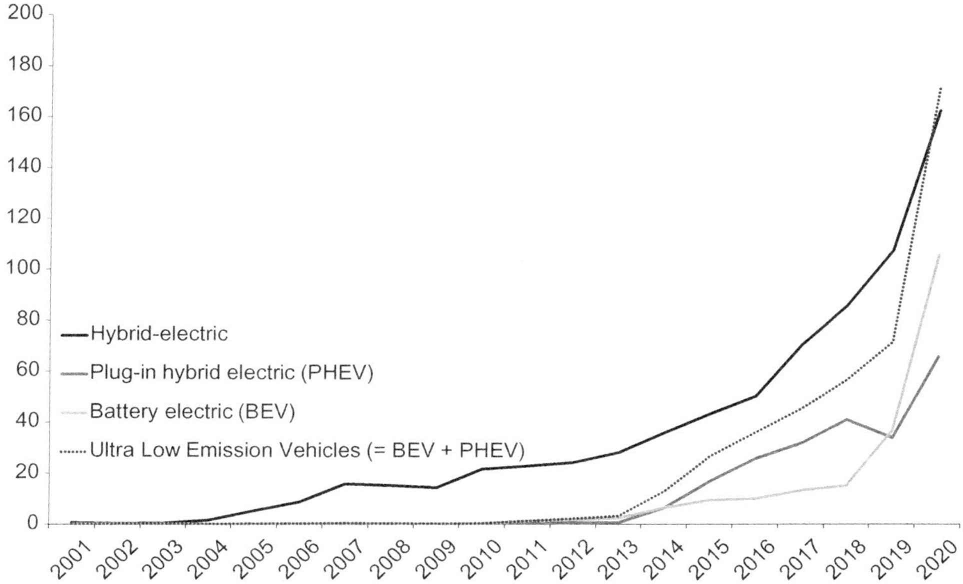

Figure 5.28 Annual electric car sales (new vehicle registrations) in Great Britain in thousands, 2001–2020 (constructed using data from Department for Transport Statistics; Vehicle Statistics dataset; table VEH0253)

There are different kinds of electric vehicles with varying technical and environmental characteristics. Battery-electric vehicles (BEVs) only use batteries and electric motors, as just described, which is a simpler configuration than ICE-vehicles as it removes the need for components such as crank shafts, pistons, and transmission. Hybrid-electric vehicles (HEVs) are more complex configurations involving both an ICE and electric motor, which can operate in serial mode (with a small ICE powering a battery that feeds the electric motor) or parallel mode (in which ICE and electric motor can both drive the wheels) to optimise fuel efficiency. A plug-in hybrid electric vehicle (PHEV) primarily uses its electric motor and battery, which can be charged from the grid, but switches to an ICE when the battery runs out of power. BEVs and PHEVs qualify as Ultra Low Emission Vehicles (ULEV) in the UK, because they emit less than 75 g of CO_2/km and can operate in zero-emission mode for at least 10 miles.[8]

Sales of different kinds of electric cars have steadily increased over the past two decades, cumulatively reaching 20.8% of all passenger car sales in 2020 due to increased electric vehicle sales (Figure 5.28) and decreased overall sales (Figure 5.10). HEV sales reached 163,700 in 2020, representing 10.1% of the UK market. Annual BEV and PHEV sales respectively increased to 106,700 and

[8] This may vary for other countries, depending on the carbon intensity of their electricity systems.

66,600 in 2020 (Figure 5.28), representing 6.6% and 4.1% of the market. In 2020, HEVs, PHEVs, and BEVs accounted for respectively 2.1%, 0.6%, and 0.6% of the total car fleet (based on data from DfT's Vehicle Statistics dataset, Table VEH0203).

While HEVs reduce CO_2-emissions by about 25% compared to ICE vehicles (Kay et al., 2013), BEVs and PHEVs promise deeper reductions (depending on the degree of electricity decarbonisation). Electric car purchase prices are still substantially higher than for comparable conventional cars: about £8,000–16,000 extra for BEVs, £7,000 for PHEV, £2,400 for HEVs (Kay et al., 2013; Nilsson and Nykvist, 2016). While a new Volkswagen Golf cost about £20,280 in 2020, the unsubsidised electric version (ID-3) cost around £29,990 (Jolly, 2020b). Electric vehicle purchase subsidies, discussed later, reduce but do not close the price gap.

Battery costs particularly push up electric vehicle prices, but power electronics and electric motors also contribute. The price of Li-ion battery packs has decreased by almost 90% between 2010 and 2020 (Figure 5.29), owing to learning-by-doing and scale economies. Once battery pack prices fall below $100/kWh, which is expected to happen between 2023 (BNEF, 2020) and 2025 (Nykvist et al., 2019), BEVs will become cost-competitive with conventional cars. But successful

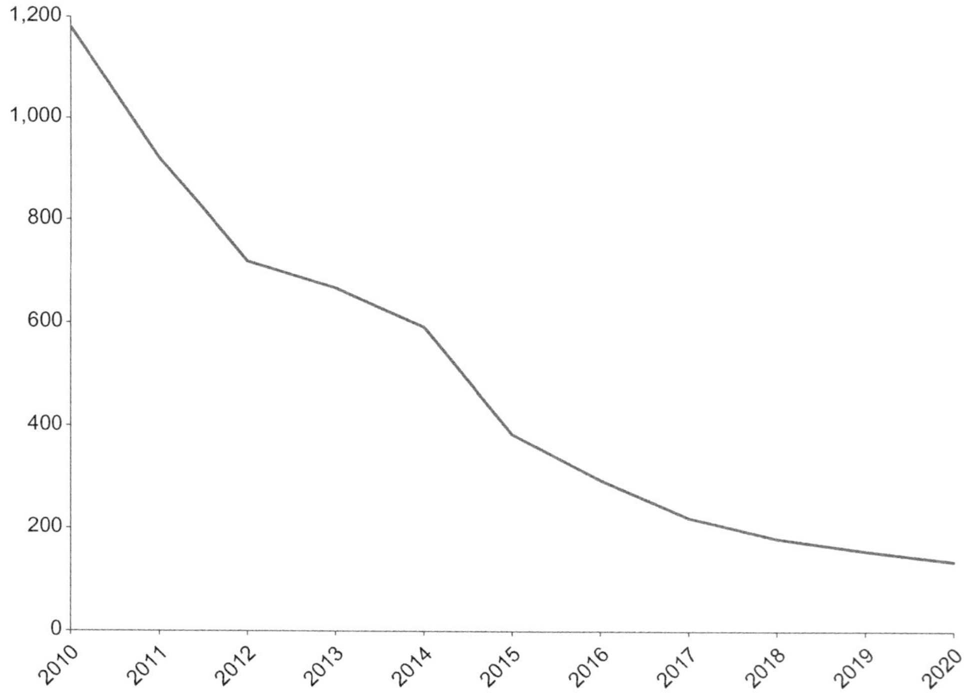

Figure 5.29 Battery pack price in real2020$/kWh (constructed using data from BNEF (2020))

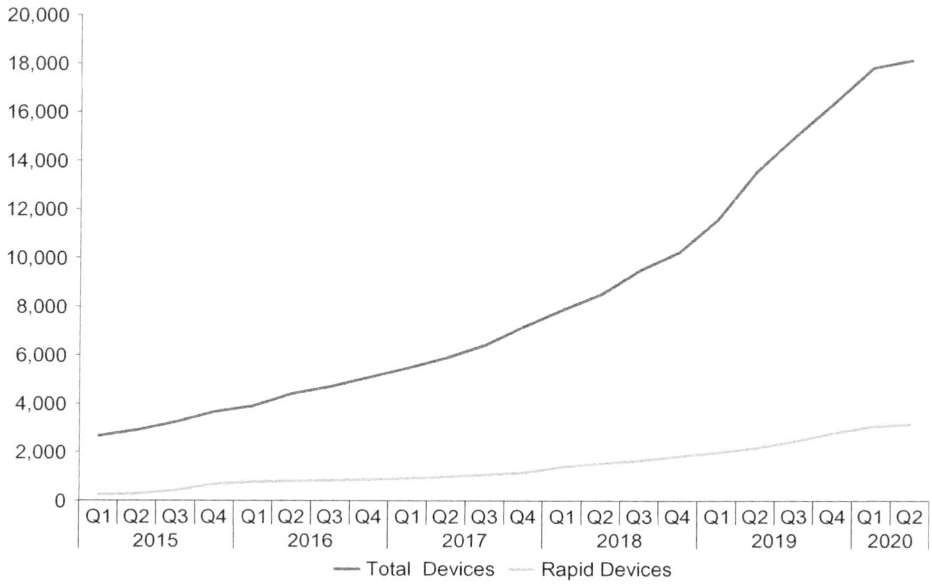

Figure 5.30 Number of public electric vehicle charging devices in the UK (constructed using data from Statistics at DfT; Electric Vehicle Charging Device Statistics; Table ECVD 02)

competition also requires improvements in other performance parameters: BEV-ranges should double to 300–500 km (which requires battery capacity of 50–100 kWh), while charging times should be reduced to 10–15 minutes.

An electric charging infrastructure is gradually being built in the UK, consisting of both public chargers (in communities, at supermarkets, along motorways) and private chargers at people's homes. The number of *public* electric vehicle charging devices has increased five-fold since 2015, reaching 18,265 in July 2020. The number of rapid public charging devices grew more slowly, reaching 3,206 in 2020 (Figure 5.30). Rapid chargers (which operate at 50–100kW) can recharge average electric vehicles from empty to 80% full in about 30 minutes. Slow chargers at home (operating at 3–7kW) take between 5 and 10 hours, depending on the car's battery size.

The geographical distribution of charging devices is uneven, which means that long-distance trips require careful planning. London has the greatest number and highest density (Table 5.5). Scotland also has relatively high density. Many other parts of the country have fewer and lower density charging devices. Densities are higher in cities such as Manchester, Milton Keynes, or the Midlands (SMMT, 2016). Further electric vehicle diffusion will require more rapid charge points and greater geographical coverage. Another problem that needs to be addressed is the

Table 5.5. *Geographical distribution of public electric vehicle charging devices in July 2020 (data from Statistics at DfT; Electric Vehicle Charging Device Statistics; Table ECVD 01)*

	Total devices	Devices per 100,000 of population
London	5,151	57
Scotland	1,910	35
North East	812	30
South East	2,478	27
South West	1,416	25
Wales	648	21
East Midlands	962	20
North West	1,476	20
East of England	1,119	18
West Midlands	1,030	17
Yorkshire and the Humber	951	17
Northern Ireland	312	16

lack of standardisation of public charging devices, which presently exist in a variety of sockets, power ratings, and payment methods.

Actors and Policies

The gradually accelerating diffusion of electric vehicles has co-evolved with car company strategies. Early developments were driven by diversifying incumbent automakers such as Toyota, which introduced the Toyota Prius (HEV) in the late 1990s, and new entrants such as Tesla, which in 2006 introduced the high-end Tesla Roadster that positively changed the public perception of BEVs (Penna and Geels, 2015).

Since then, perceptions and strategic orientations of global automakers towards electric vehicles have changed from reluctant engagement to a strategic innovation race (Bohnsack et al., 2020). Between 2006 and 2009, many automakers perceived electric vehicles (EVs) as a necessary evil they had to engage with (in the form of R&D, prototypes, and concept cars), partly to keep up with new entrants (e.g., Tesla) and first mover incumbents (e.g., Toyota, Mitsubishi, Nissan, and General Motors), and partly in response to increasing climate change concerns from policymakers and the wider public. Between 2010 and 2015, more incumbent car companies began to diversify towards EVs. Although industry perceptions of EVs became more positive, full commitment was delayed by doubts about whether EVs were developed for sustainability or commercial reasons (Bohnsack et al., 2020). After 2015, this perceived dilemma mostly disappeared because increasing sales and tightening climate regulations convinced the industry that EV diffusion was driven by both sustainability and commercial reasons. Between 2016 and 2018,

Table 5.6. *Examples of car manufacturers' commitments on electrification (CCC, 2018: 167)*

Manufacturer	Timing	Commitment
Nissan	2025	BEVs 50% of sales in Japan and Europe
Mercedes	2025	BEVs 15–25% of sales
Volkswagen	2025	EVs 25% of sales
Porsche	2030	EVs 100% of sales
Toyota	2030	EVs and conventional hybrids 50% of sales
Volvo	2030	EVs and conventional hybrids 50% of sales
Honda	2030	BEVs, PHEVs, and hydrogen 15% of sales

global automakers indicated their new strategic commitment with ambitious electric vehicle targets (Table 5.6), which were gradually increased in subsequent years. An indication of investors' future expectations is that by January 2021, Tesla's market valuation of $800bn was higher than the combined value of Toyota, Volkswagen, Daimler, General Motors, and Hyundai (*The Economist*, 23 January 2021).

The reorientation to electric vehicles is demanding for automakers and their supply chains. Car companies are investing large sums to build new factories and acquire new technical skills, which also leads to new collaborations. Additionally, a new supply sector is being created to produce enough batteries and electric motors for the expected electric vehicle boom.

The purchase costs of electric vehicles are still higher than for normal cars, even with subsidies. Early adopters therefore tend to be mostly middle-aged, male, well-educated, affluent urbanites with pro-environmental attitudes, a desire to save money on fuel costs, and an active interest in new technology (Nilsson and Nykvist, 2016). While it is sometimes claimed that consumers have 'range anxiety' regarding BEVs (fear of being stranded with empty batteries), research suggests that most users develop new skills and competencies to deal with this (Ryghaug and Toftaker, 2014). Increases in BEV ranges and charging points also increasingly alleviate this concern. HEVs and PHEVs have fewer range problems, because they can drive on normal fuels when their shorter electric-drive range (30–90 km) is exhausted.

UK policymakers have increasingly supported electric vehicles since the late 2000s, when biofuels encountered problems and electric vehicle expectations increased again. New organisations were created to support innovation and coordination between upstream actors. The Office for Low Emission Vehicles (OLEV) was created in 2009 to coordinate interactions between automakers, policymakers, and research organisations. The Automotive Council was created in 2009 to coordinate automakers and suppliers to develop an industrial strategy and roadmap. The Advanced Propulsion Centre was created in 2013 to position the UK as a global leader in low-emission powertrain research and development. To support this ambition, the government and automotive industry agreed to both invest £500 million, totalling £1 billion over 10 years (Skeete, 2019).

From 2011 to 2015, OLEV disbursed £400 million government funds to support R&D projects, build recharging infrastructure (via the Plugged-in Places scheme), and provide consumer adoption incentives for ULEVs. The government has allocated another £500 million for the 2015–2020 period to develop the ULEV market via a range of initiatives under the Go Ultra Low brand.

Various policies aim to stimulate electric vehicle adoption. In 2011, policymakers introduced the Plug-in Car grant, which paid 25% of BEVs and PHEVs up to £5,000. In 2018, this grant was cut to £3,500 for battery electric vehicles and removed for plug-in hybrid vehicles. In March 2020, the grant was further reduced to £3,000, followed by another reduction (to £2,500) in March 2021 (Jolly, 2021a). BEVs were also exempted from the Fuel Duty, Company Car Tax, and Vehicle Excise Duty.[9] In 2014, the government also introduced the Electric Vehicle Homecharge Scheme, which covered up to 75% of the installation costs of domestic recharging devices for BEVs or PHEVs, capped at a maximum grant of £900. The maximum cap was revised to £500 in 2018 and £350 in 2020 (OLEV, 2020). The typical cost of a home charger point in 2020 was around £800.

The government's phase-out plans of petrol and diesel cars, first announced in 2017 and sharpened in 2020, also support electric vehicles by creating mass markets. From 2030, all new vehicle sales will be hybrid or pure electric. By 2035, only pure electric vehicles will be sold. These phase-out policies not only aim to create mass markets for electric vehicles but also to signal to car companies that the UK is a good place to build new manufacturing plants.

The government's Clean Growth Strategy (BEIS, 2017a) also aimed to attract automakers to the UK for investing and building electric vehicle manufacturing plants. Results have been mixed because Brexit created trade-related uncertainties and complications and because other countries with similar aims may appear more attractive to automakers (Campbell et al., 2019). Tesla explored opening a new Gigafactory (for batteries and cars) in the UK, but in 2019 Brexit concerns made it opt for Germany instead. In 2017, Toyota invested 240 million into modernising its Derby car plant to produce the third generation Auris (a hybrid electric vehicle), but in 2020 announced that it would postpone further UK electric vehicle investments until 2027 (Jolly, 2020c). Jaguar Land Rover committed to building electric vehicle and battery manufacturing plants in the UK, and in 2021 announced that its Jaguar brand would only come as an electric vehicle by 2025 (Jolly,

[9] This exemption remained after the 2015 VED adjustments.

2021b). BMW decided that it would assemble the electric Mini in the UK but use batteries shipped from Germany (Campbell et al., 2019). Nissan announced in July 2021 that it would invest £1 billion in an electric battery plant and electric car manufacturing in Sunderland. Also in July 2021, Vauxhall's new owner Stellantis announced it would invest £100 million in its Ellesmere Port plant to build electric vans. Both of these investment decisions were facilitated by generous government subsidies, reportedly £100 million for Nissan and £30 million for Stellantis (Ambrose and Vinter, 2021).

5.6.2 *Biofuels*

Techno-Economic Developments

Biofuels are produced from biomass feedstocks, which can be grown specifically for processing into fuel or are waste products. The two main biofuel forms are biodiesel and bioethanol, which can be blended relatively easily with conventional fuels.[10] Although biofuels can be produced from many feedstocks and through many processes, it is common to distinguish three different generations (RAE, 2017):

- First generation biofuels are produced from food or animal feed crops such as sugar cane, sugar beet, corn, and wheat (for bioethanol) or palm oil, soybean oil, and rapeseed oil (for biodiesel).
- Second generation biofuels are produced from non-food related feedstocks, including dedicated lignocellulosic energy crops (such as Miscanthus or switchgrass), organic waste (e.g., agricultural residues, wood wastes), and other waste materials (e.g., used cooking oil and municipal solid waste).
- Third generation biofuels are produced from microalgae (RAE, 2017).

Biofuel use in the UK increased between 2004 and 2010, then stagnated, and increased again after 2017 (Figure 5.31), reaching 4.5% of road transport fuel (in real terms) in 2019 (Figure 5.32). This is relatively low by European and international standards (RAE, 2017). A 2011 revision of European rules allowed double counting for sustainable biofuels, which led to increasing divergence between *real* and *reported* transport biofuel use (Figure 5.32).

These new European rules triggered a substantial shift from first to second generation biofuels based on waste feedstocks, which qualified for double certification. Inputs for biodiesel shifted from soy and rapeseed towards used cooking oil, which by 2019 accounted for the great majority (79%) of biodiesel. Inputs for bioethanol shifted from sugar cane to corn, which by 2019 accounted for

[10] The use of 100% biofuels, which exists in some countries, requires some engine modification in existing cars.

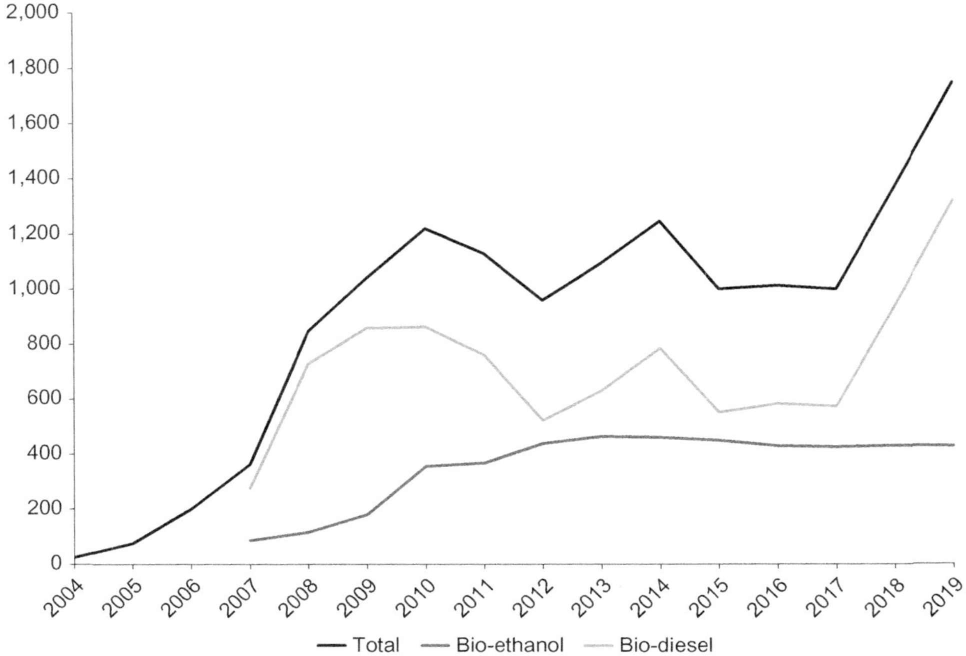

Figure 5.31 Biofuel use in UK transport 2004–2019 (excluding aviation), in million tons oil equivalent (constructing using data from Digest of UK Energy Statistics; Renewable Sources of Energy; Table 6.6)

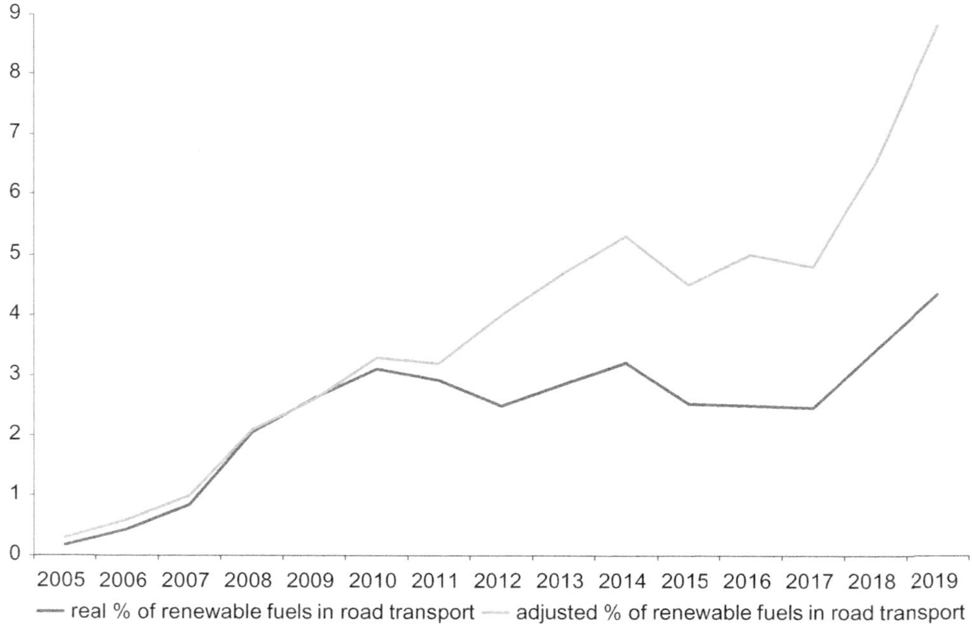

Figure 5.32 Real and adjusted percentage of biofuels in road transport fuel use, 2005–2019 (constructed using data from Department for Transport Statistics; Energy and Environment data tables; Table ENV0102)

43% of bioethanol (DfT, 2020d). The technologies for processing these inputs are well-known. Processing technologies for lignocellulosic biomass and algae are still under development, and not commercially used in the UK.

In 2019, the large majority (89%) of biomass feedstock was imported, mainly from China (especially cooking oil), United States, Spain, and France (DfT, 2020d). Biofuels are more expensive than fossil fuels, which is why diffusion has been stimulated by public policies.

Actors and Policies

The 2003 European Biofuels Directive stipulated that member states should replace 5.75% of all transport fossil fuels with biofuels by 2010. To implement this Directive, the UK introduced the Renewable Transport Fuel Obligation (RTFO) in 2005, which stipulated increasing biofuel targets that large transport fuel suppliers (like BP, Shell, Esso) had to meet: 2.5% in 2008/9, 3.75% in 2009/10, and 5% in 2010/11. The RFTO also created a trading system, in which biofuel suppliers could apply for Renewable Transport Fuel Certificates (RTFCs). Obligated fuel suppliers are required to acquire RTFCs in proportion to the volume of fossil fuel they supply (RAE, 2017). Until 2012, policymakers also stimulated biofuels with a 20 pence per litre fuel tax (duty) exemption.

In the absence of domestic biofuel production capacity, UK fuel suppliers initially imported biofuels, especially from Brazil and the United States. Domestic bio-refining capacity was gradually developed, with the first biofuel production plant opening in 2005. By 2013, there were 11 large-scale and more than 60 small-scale factories (Ecofys, 2014), with over 1,500 million litres annual production capacity, not all of which was used, owing to rule changes that restricted demand.

These rule changes related to a controversy over the negative environmental consequences of rapidly increasing first generation biofuels. Around 2007, UK environmental NGOs launched vocal protests against biofuels, because of unintended negative effects, including CO_2 emissions from indirect land-use change and competition with food production. In response, the UK government implemented the Gallagher Review, which in 2008 recommended a precautionary approach and pull-back from original targets (Harvey and Pilgrim, 2013). The government accepted the recommendations and, in 2009, down-scaled the RTFO-targets to 3.25% for 2009/10, 3.5% for 2010/11, 4% for 2011/12, 4.5% for 2012/13, and 4.75% for 2013.

The biofuel controversy also affected European policies. The 2009 European Renewable Energy Directive (RED) not only set new targets for 10% renewable transport fuels by 2020 but also introduced stronger sustainability criteria such as an increase of life-cycle CO_2 savings from 35% in 2011 to 50% in 2017. To stimulate a shift from first to second generation biofuels, the RED also allowed double counting of more sustainable biofuels such as wastes, residues, and non-food cellulosic material.

The lower than expected market demand and changing UK biofuel feedstocks created challenges for the UK's emerging biofuel manufacturing industry, leading to some factory closures (Ecofys, 2014). Smaller producers also struggled to sell RTFCs directly to obligated fuel suppliers and handle complex certification procedures (RAE, 2017). Consequently, the industry was increasingly dominated by large-scale operators.

After keeping the RFTO target at 4.75% for several years, UK policymakers introduced an upward increasing trajectory for the target to 7.25% in 2018, 8.50% in 2019, 9.75% in 2020, and further gradual increases until 12.40% in 2032. These policy changes, which boosted road transport biofuel use in 2018 and 2019 (Figures 5.31 and 5.32), also aim to further reduce first generation crop-based biofuels and stimulate advanced waste-based biofuels. So, despite an increasing policy focus on electric vehicles, discussed previously, there is also a new push for biofuels, aimed not only at passenger cars but also at aviation and freight transport, which are strategic but hard to decarbonise sectors. As part of its 2020 Transport Decarbonisation Plan, the government also decided to support four new biofuel plants. Two of these will be funded from the £20 million Future Fuels for Flight and Freight Competition and two from the £25m Advanced Biofuels Demonstration Competition.

5.6.3 Tele-work

Techno-Economic Developments

The emergence of an information society gave rise to the idea that tele-working (i.e., 'working at home' with the use of computers, internet, and other information and communication technologies) could reduce the need to commute to the office and back (Fu et al., 2012). The number of home-workers has gradually increased from 1.9 million in 1992 to 4.3 million in 2019, amounting to 13% of the UK workforce (Figure 5.33). The percentage of employees 'working at home' almost doubled from 2.6% to 5% in this period. 'Working from home', which refers to workers such as farmers, salespeople, or construction workers who use their home as a base but work in different places, increased from 4.7 to 8.1%.

In April 2020, the first national lockdown caused the 'working at home' percentage to increase to 46.6% of the working population.[11] When lockdown

[11] See www.ons.gov.uk/employmentandlabourmarket/peopleinwork/employmentandemployeetypes/bulletins/coronavirusandhomeworkingintheuk/april2020

Figure 5.33 Percentage of working population working 'at home' and 'from home', 1992–2020 (constructed using data from Felstead (2012) and Office for National Statistics databases: Labour Force Survey 1998–2014; Annual Population Survey 2012–2020)

restrictions were relaxed, many people returned to work, which is why the 'working at home' percentage for the whole of 2020 is only 8.5% (Figure 5.33).

Actors and Policies

Tele-working is a new social practice within the wider work regime, which is facilitated by mobile phones, laptops, and the Internet. Tele-working is particularly feasible for various kinds of knowledge workers who process data and information. Until 2020, the growth in 'working at home' has indeed come mostly from occupational categories with highly skilled roles in the economy, such as managers or senior officials, professionals, associate professionals, and skilled technical occupations (Figure 5.34). During the pandemic, it was also those highly skilled groups that were able to substantially shift their work activities online, while most occupational groups (except administrative and secretarial jobs) were far less able to do so.

Changes in organisational practices, increased use of video- and tele-conferencing, and the ability to create dedicated office space in their homes also

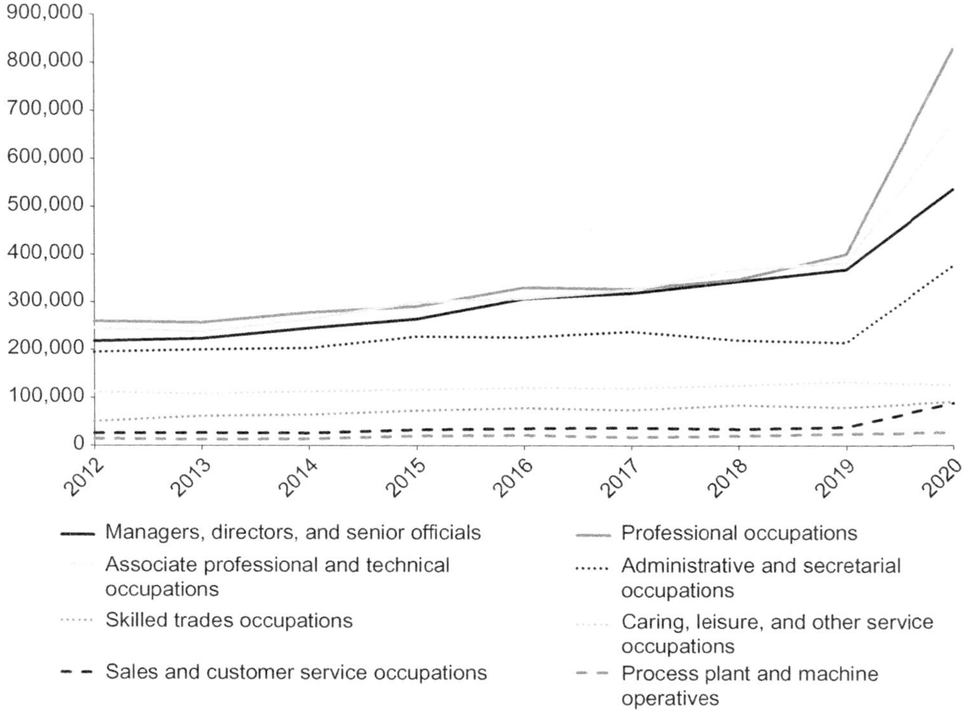

Figure 5.34 Major occupational groups (from ONS categorisation) in the 'working at home' category, 2012–2020 (constructed using data from Office for National Statistics; Labour Force Survey 1998–2014; Annual Population Survey 2012–2020)

enabled highly skilled occupational groups to keep 'working at home' (at least partly) once lockdown restrictions were lifted. Although these processes are still unfolding, it is likely that increased 'working at home' will represent a long-term structural change for these occupational groups.

Even before the COVID-pandemic, the effects of tele-working on mobility were not straightforward, because of the diversity in people's motivations (Haddad et al., 2009). Some people tele-worked a few hours in the morning, to avoid the rush hour, or in the evening, after a day in the office (Alexander et al., 2010), neither of which reduces the amount of travel. They also worked on their laptop *while* they travelled on public transport modes (Felstead, 2012). Homeworkers also made some *additional* daytime journeys (e.g., visiting friends) to compensate for a lack of social interaction. So, the relative importance of mobility substituting and inducing effects of tele-working is an ongoing debate in transport studies (Cohen-Blankshtain and Rotem-Mindali, 2016).

It is clear, however, that work temporalities have become more flexible than the standard 9-to-6 template (Alexander et al., 2010) and that the gradual increase in tele-working aligns with this trend. There has been no dedicated UK policy to stimulate tele-working. Before the pandemic, few employees wanted to work from home for most of the time, because the office was still seen to offer a range of benefits such as social contact, visibility, and influence on organisational politics (Fu et al., 2012). In the UK and Ireland, tele-work in the 2010s 'remained a marginal work practice for several reasons ... lack of legitimacy due to the absence of regulation, the ad-hoc arrangements that have ensued, the lack of support from management, the absence of training or direction, the dominance of the existing regimes of automobility and traditional ways of working' (Hynes, 2016: 25). So, while work regimes were becoming more fluid and spatially diverse, shifts towards tele-working were relatively gradual.

The pandemic clearly was an exogenous landscape shock that boosted the diffusion of the tele-working niche, which had gradually developed and stabilised in preceding years. Tele-working's long-term effects on mobility are not yet clear, and also depend on how structural its mass adoption will be, which is likely to vary between occupational groups. Effects on automobility, which has rebounded to pre-pandemic levels (Figure 5.4), are not yet observable. As video- and tele-conferencing have become more widely accepted, it is also possible that tele-working will reduce business travel via airplanes or trains. Since business travel is the third largest category (accounting for 17%) of rail travel (Figure 5.21), this would have implications for railway investments, including HS2, as noted in Section 5.3.2.

5.6.4 Car Sharing, Ride Hailing, and Ridesharing

Techno-Economic Developments

The use of smartphones, Global Positioning Systems (GPS), computers, internet platforms, and other digital technologies stimulated the emergence and expansion of multiple transport-related social and business model innovations in the last decade.

- Car sharing schemes (or car clubs) are short-term, membership-based car rentals booked online in advance. Users pay an (relatively low) annual membership fee and a fee depending on the duration of use. They collect the booked vehicle from a designated car parking space ('back-to-base scheme') or locate it in a circumscribed area using a smartphone app ('free-floating schemes').
- Ride-hailing services are provided by digital platforms (e.g., Uber) that connect licensed taxis or Private Hire Vehicles with passengers in real-time, often within minutes. Unlike traditional taxi operators, these platforms do not own taxi fleets but provide booking, payment, and rating services. For users, ride-hailing

Table 5.7. *Number of car club members and cars in 2018 (constructed using data from Steer Davies Gleave (2019a, 2019b, 2019c))*

	London	England & Wales	Scotland	UK total
Car club members	245,000	25,773	19,872	290,645
Car club cars	2,636	783	544	3,963

services are like taxis, except the digital platform makes their use more seamless and convenient.

- Ridesharing (or carpooling) services are provided by digital platforms (e.g., BlaBlaCar or UberPool) that connect potential riders with drivers going in the same direction, enabling them to share costs. These platforms charge a use fee.
- Recent platforms (e.g., Hiyacar) offer peer-to-peer (P2P) car rental services, enabling individual car owners to rent out their cars.

These new mobility options have grown rapidly in the last decade, especially in London, but they still constitute a small niche in the wider mobility system. Although UK *car sharing* organisations had 290,645 members in 2018, they only had 3,963 vehicles (Table 5.7), which is marginal compared to the 32 million UK car fleet (Figure 5.12). Membership growth in London (which accounted for 84% of members in 2018) has been impressive since 2010, but the number of vehicles has hovered around 2,500 (Figure 5.35).

Ride-hailing has also increased substantially in recent years, especially in large cities where higher passenger densities enable this business model to compete with traditional taxis. The number of Private Hire Vehicles (PHVs) in London, for instance, increased by 78% since 2008, up to 87,745 in 2018, while the number of PHV drivers almost doubled in this period (Figure 5.36) to 106,650 in 2018. Uber, which launched in 2012 in London, has been a major driver of this growth, resulting in about 45,000 Uber-related London-based drivers in 2018. Ride-hailing is competing with traditional taxi's (Cramer and Krueger, 2016), which in London declined by 10% between 2008 and 2018 (Figure 5.36).

Reliable numbers for *platform-mediated ridesharing* and *P2P car rental services* are hard to find. Despite growth in recent years, their contribution in wider mobility provision is, so far, likely to be very small.

Actors and Policies

Although *car sharing* initially emerged as community person-to-person grassroots initiatives (Ornetzeder and Rohracher, 2013), subsequent diffusion took the form of commercial ventures based on new business models. UK car sharing organisations purchase the vehicles, arrange insurance, and negotiate with local

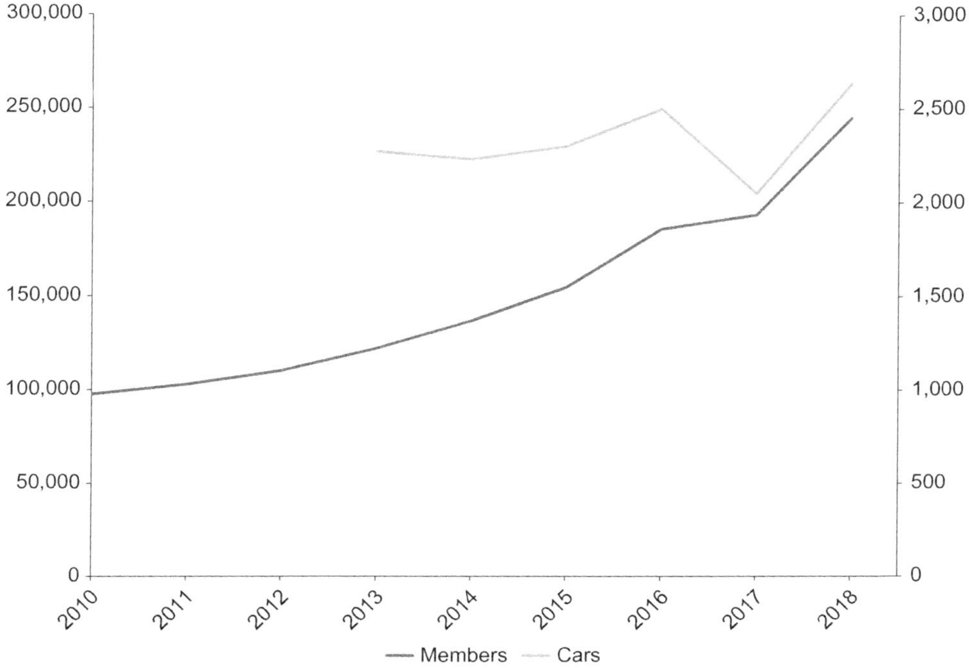

Figure 5.35 Number of members (left-hand axis) and number of cars (right-hand axis) in London car sharing clubs, 2010–2018 (constructed using data from Steer Davies Gleave (2019a, 2012, 2011, 2010))

authorities about parking spaces. These negotiations have been challenging in London, which is by far the largest car sharing market, because they involved separate interactions with 32 boroughs. A range of companies provide car sharing services in London and elsewhere. Although start-up companies (e.g., Greenwheels, Co-wheels) pioneered UK car sharing, the business has become dominated by organisations linked to incumbent car rental and automotive firms, for example, Zipcar (Avis Group), DriveNow (BMW), E-Car Club (Europcar), Enterprise Car Club (Enterprise Rent-A-Car), Bluecity (Bolloré group, which ended Bluecity in 2020), and Car2Go (Daimler, quit London in 2014). Several companies have failed, however, indicating tough competition and value extraction challenges in a relatively small market.

Although car club membership has grown rapidly, actual use of shared vehicles is limited, as indicated by relatively small car fleets (Table 5.7, Figure 5.35). The reason is that most members use car clubs as an infrequent back-up option. In 2015/16, 44% of members never used a car club vehicle in the past year; 36% used vehicles between 1–5 times per year; only 5% of members made more than 20 car club trips a year (Steer Davies Gleave, 2016). The user segment is also quite

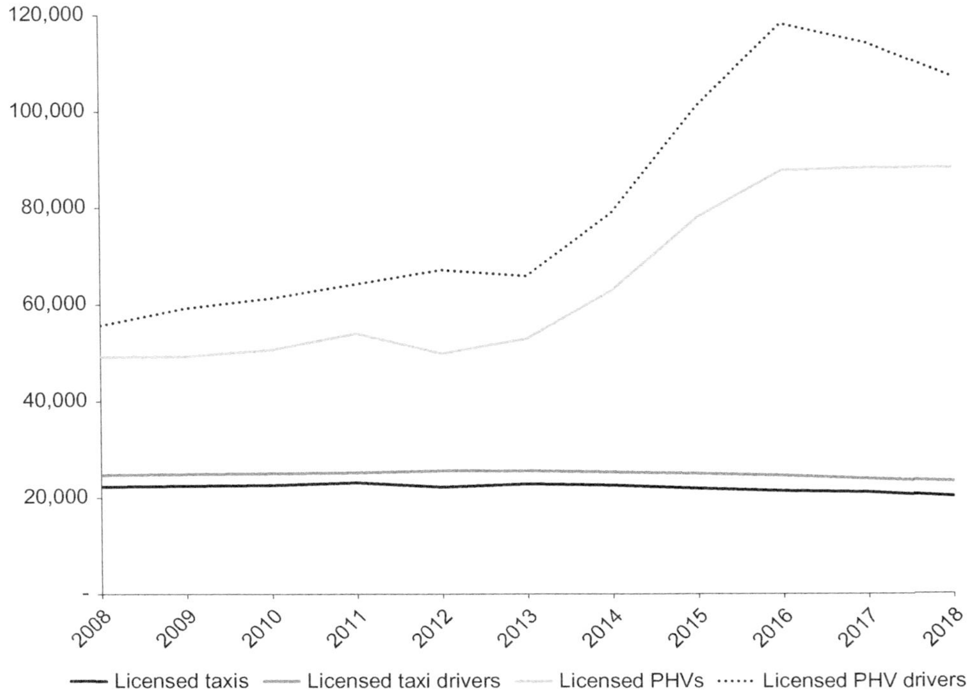

Figure 5.36 Number of licensed London taxis and private hire vehicles (PHVs) and drivers, 2008–2018 (constructed using data from TfL (2019))

specific, consisting mostly of highly educated, young professionals (between 25–35 years old) without children living in inner London boroughs (Akyelken et al., 2018). It remains to be seen if the appeal of car sharing will broaden in the future beyond this specific segment and if customers will use shared vehicles more than a few times per year.

London policymakers have welcomed car sharing and helped to create the Car Club Coalition in 2014 to facilitate interactions among companies, Transport for London (TfL), and London's councils. But because car sharing is perceived as a private sector initiative, local policymakers have, so far, refrained from stimulating it over other transport modes (e.g., by offering shared vehicles reduced parking fees or exemptions from the London congestion charge). Akyelken et al. (2018: 11) therefore conclude that policymakers have 'not yet formed a clear vision or a strategy to benefit from the opportunities of car sharing'.

Ride-hailing platforms have grown rapidly in recent years because they offer a more convenient, real-time service that also tends to be cheaper than traditional taxis. App-based ride-hailing is particularly popular with younger people living in inner London, who most frequently use it to travel to or from a night out and to or from airports (TfL, 2019). Although Uber is London's largest ride-hailing platform, others have emerged in recent years (e.g., Wheely, Hail, Bolt, Kabbee, GetTaxi). Uber has faced challenges from traditional taxi-drivers and has, since 2017, been involved in licensing struggles with Transport for London (TfL) over its business model, which does not include health insurance, sick pay, or accident and injury protection for drivers, whom Uber sees as independent entrepreneurs rather than employees. Although Uber made operational changes, legal struggles continued for several years, creating complications and uncertainties that led to some decline in Uber-related drivers (Figure 5.36). In September 2020, courts granted Uber a new license. But in February 2021, the UK Supreme Court ruled that Uber drivers should be treated as workers and receive minimum wages and paid holidays (Butler, 2021), which is likely to substantially affect Uber's financial business model.

Although new platform mobility services such as car sharing or Mobility-as-a-Service (which is discussed later) hold potential, they have, so far, not been transformative due to their small-scale, narrow user segment, and ambiguous position between public and private transport regimes. They presently fill and exploit gaps between these regimes, but it remains uncertain if they will trigger wider reconfigurations: 'The current wave of lean platforms are largely interstitial and ephemeral, rather than transformative, realized in the gaps between the scale of central city walking and that of region-serving commute infrastructure' (Stehlin et al., 2020: 1263).

5.6.5 Intermodal Transport

Techno-Economic Developments

Innovations to stimulate intermodal transport, which refers to the use of two or more transport modes in the same journey, can take many forms, but generally aim to create new linkages between transport modes that facilitate easy transfer. Well-established options such as *park-and-ride* facilities are car parking lots with good public transport connections. They can enable car drivers to park at the edge of cities and use public transport to travel to city centres. Likewise, *kiss-and-ride* parking spots enable car drivers to easily drop friends or relatives off at public transport modes.

Intermodal ticketing or *smartcards* (which use a microchip to store, process, and write data) is a more recent innovation, which can be used to pay electronically for different public transport modes, thus contributing to a seamless intermodal travel experience. Smartcards can also accelerate payment, especially in buses where current cash payments delay travel.

Smartcards, which are credit-card sized cards, diffused substantially from less than 50,000 in the 1990s to more than 3.7 billion public transport journeys in 2013 (Slavin, 2014). Most smartcard journeys are made in London, where the Oyster card became the dominant fare medium, until contactless credit cards took over. Outside London, only 700 million smartcard journeys were made in 2013 (Slavin, 2014).

Mobility-as-a-Service (MaaS) is a more recent technical and commercial package that integrates different mobility services (e.g., public transport, car sharing, bike sharing, taxi) using a single digital platform that provides intermodal journey planning, booking, payment, and real-time information (Kamargianni et al., 2016). Members can use MaaS either as Pay-as-You-Go or buy advance mobility packages to suit their travel needs. The promise of smooth and efficient intermodal mobility services has generated policy and academic enthusiasm, because of the hope that MaaS may enable a shift away from ownership and use of private cars (Finnish Transport Agency, 2015; Shaheen and Chan, 2016; Transport Systems Catapult, 2016). Although MaaS systems have been trialled since the mid-2010s in a few dozen, mostly European, cities (e.g., Stockholm, Amsterdam, Birmingham, Helsinki, Basel, Vienna), 'the cases of MaaS are, thus far, few as well as limited in terms of numbers of users' (Smith et al., 2020: 163). 'The performed pilots have thus far been small-scale. Moreover, very few of the services have been systematically evaluated' (Smith and Hensher, 2020: 54). UK demonstration projects are taking place in Scotland and Birmingham to explore technological dimensions of MaaS (e.g., smartcard payment mechanisms, online booking systems), social acceptance, and business models (Government Office for Science, 2018).

Actors and Policies

The 1998 UK Transport White Paper expressed an interest in the benefits of *smartcards* for transport. In 1999, the *Integrated Transport Smartcard Organisation* (ITSO) was created to develop interoperable standards and stimulate uptake. Diffusion proceeded more slowly than anticipated, because of technical challenges, difficult negotiations between a wide range of actors (ICT companies, data-processing firms, financial organisations, transport operators, local transport authorities), and the time it took to persuade actors to join smartcard schemes (AECOM, 2011; Blythe, 2004). Developments progressed fastest in London where Transport for London had governance capabilities and greater financial resources. The Oyster Card was launched in 2003 as a £1.2 billion private finance initiative scheme (Blythe, 2004). Developments in other UK cities have largely been piecemeal.

A 2009 national government document developed a national smartcard vision: 'The Government's vision is for smart and integrated ticketing across public transport in England, with the ITSO specification allowing for seamless travel, potentially across the country, using the same smartcard' (DfT, 2009: 3). Its strategy was to first establish ITSO-compliant smartcards in the nine largest urban areas (by 2015) and then link these 'urban islands' of local ticketing schemes with corridors of intercity transport. Local Transport Authorities were meant to drive local smartcard implementation, acting as coordinating bodies between transport operators and other organisations. By 2013, however, the government recognised the scale of the task and the difficulties involved, in particular the need for 'new business processes and commercial agreements. These may be complex and difficult to negotiate' (DfT, 2013: 27). Although some cities (e.g., Manchester) have since introduced smartcards, other integrated payment options such as smartphones and contactless credit cards are now more dominant.

The concept of *Mobility-as-a-Service* (MaaS) started in Finland and Sweden in the early 2010s, supported by innovation agencies, transport ministries, city transport agencies, and IT companies, with the aim of developing new digital products and services (Pangbourne et al., 2020). The concept then spread to other countries such as the Netherlands, Germany, and the UK, accompanied by visions that promised multiple benefits such as enhanced freedom and choice for citizens (i.e., the availability of multiple travel modes), increased comfort and convenience for citizens (i.e., door-to-door mobility services), optimised network flows and higher efficiency for transport authorities (i.e., better utilisation of public and private transport assets), improved sustainability for city authorities (e.g., less congestion, less emissions), and new business opportunities (Pangbourne et al., 2020). Long-term visions suggested that people might shift from privately owned cars to MaaS-type systems, enabling them to seamlessly, quickly, and efficiently use multiple transport modes, perhaps even including automated vehicles, to travel to their destinations (Sperling, 2018).

Optimistic MaaS visions were disseminated in the UK by the think tank CIVITAS (2016) and the Transport Systems Catapult (2016), whose report was commissioned by the Department for Transport (DfT). MaaS Scotland, a network of 75 public and private sector organisations, was formed in 2017 to facilitate discussions between public and private stakeholders such as digital platform providers, IT and telecom companies, public transport providers, car sharing and ride-hailing companies, and national and local policymakers. MaaS Scotland started its first pilot projects in early 2020 (https://maas-scotland.com/).

Another pilot project was launched in 2017 by Birmingham's public transport authority, which used the Whim smartphone app (from the Finnish start-up Maas Global) to combine public transport, taxis, and rental cars. Multiple public and private stakeholders were involved in the project (Table 5.8), leading to complex negotiations to address issues such as specific roles, mandates, business models,

Table 5.8. *Range of MaaS packages in Birmingham and the West Midlands region (Pangbourne et al., 2020: 41)*

	Pay-as-you-go (PAYG)	Basic (£99/month)	Unlimited (£349/month)
Buses, trams, and trains within West Midlands county (National Express West Midlands; Transport for West Midlands)	PAYG	Unlimited	Unlimited
Taxi 3-mile radius (Gett)	PAYG	PPR (per person round trip)	Unlimited
Car (Enterprise)	PAYG	Max £49 per day	Unlimited
Bikeshare (nextbike)	Coming soon	Coming soon	Coming soon

data-sharing, and risk and profit sharing (Hirschhorn et al., 2019). Users could initially choose from three packages with different prices (Table 5.8). The project struggled to recruit participants, however, and in 2019 removed the monthly packages, reverting to pay-as-you-go only (Pangbourne et al., 2020). Evaluations of Finnish and Swedish projects also indicated that users often struggled to understand and appreciate the subscription model, in which they would buy mobility services in advance (Karlsson et al., 2020).

The Department for Transport monitored the project but was not directly involved. Although DfT has positive orientations, it is presently still an open question 'whether and how the UK will regulate MaaS' (Hirschhorn et al., 2019: 184). A research report commissioned by the UK Government Office for Science (2018) acknowledged the potential for a MaaS-related transport paradigm shift, but also noted that 'MaaS does not yet have a major presence in either the research arena or in society more generally' (p. 14). There are also uncertainties about consumer interest and economic viability of MaaS, especially outside dense city centres where public transport and car sharing are less frequent and available (Pangbourne et al., 2020). Two high-profile international failures (Kutsuplus in Helsinki, Bridj in the United States) underline that the MaaS business case may be more challenging than the optimistic visions about an imminent paradigm shift suggest.

5.6.6 Self-driving Personal Cars

Techno-Economic Developments

There have been several waves of attention and R&D programmes about automated and self-driving cars (in the 1930s, 1960s, and 1980s), which failed to materialise (Geels and Smit, 2000). Since the 2010s, progress in information and communication technologies has led to another wave of attention as well as many collaborative R&D and real-world demonstration projects. The reason for this renewed enthusiasm is that proponents of self-driving personal cars envisage many *potential* benefits such as the following (Parkhurst and Lyons, 2018): a) elimination of vehicle collisions (and associated deaths and mutilations), b) smoothed traffic flow and greater effective road capacity, c) using travel time for other (economically productive) activities in the vehicle, d) reduced energy use and greenhouse gas emissions (because of smoother rides), and e) enhanced mobility for those unable to drive themselves (e.g., impaired people and elderly). Deeper transformations are possible if self-driving cars align with car sharing and mobility services, leading to a future in which people do not own and drive their own cars, but instead use phone apps to smoothly access shared automated vehicles (Sperling, 2018).

Efforts to monetise potential benefits have generated high numbers. The McKinsey Global Institute (2013) estimated that potential global economic benefits could range between \$0.2 and 1.9 trillion by 2025 with an uptake of 5–20%. Assuming a 90% penetration rate, Fagnant and Kockelman (2015) suggest that net direct benefits could be \$196 billion per year for the United States, mainly due to reduced collision and congestion costs. Arbib and Sebat (2017) estimate global productivity gains from better used travel time alone at \$1 trillion by 2030. Combined with technical progress, these optimistic benefit estimates generated high enthusiasm in the 2010s, which will be discussed further here.

There are different technological levels of vehicle automation (Table 5.9), but only higher levels (4 and 5) qualify as driverless or self-driving cars. Some digital

Table 5.9. *Five levels of driving automation specified by Society of Automotive Engineers (SAE, 2016)*

L0: Driver only	Driver in complete control
L1: Assisted ('hands on')	Driver in charge, but some small steering or acceleration tasks can be performed by the car without human intervention (e.g., lane centring, parking assistance)
L2: Semi-automation ('hands off')	The system controls steering, acceleration, and breaking in relatively simple, specific circumstances. Adaptive cruise control, for instance, can maintain speed and safe distance to other vehicles in motorway driving.
L3: Conditional automation ('eyes off')	The system steers, accelerates, and brakes under specified conditions (e.g., traffic jam chauffeur); driver must take over, when the system indicates.
L4: High automation ('mind off')	The system has full control in situations deemed suitable by driver (e.g., urban driving)
L5: Full automation ('steering wheel optional')	The autonomous system drives the vehicle under all circumstances; no human driver needed.

technologies with level 1 or 2 functionalities (e.g., parking assistance, lane centring, cruise control) are already used in higher-end personal vehicles. Fully automated self-driving vehicles also already exist in specific application domains, for example, underground metro-lines and trucks in open-cast mining operations.

Deployment of self-driving personal cars in cities or on highways is more complicated, however, because these environments are more dynamic and unpredictable. To monitor these environments, essential technologies include cameras (to recognise and 'read' traffic signs, traffic lights, road lanes, etc.), laser sensors such as lidar (which measure distances to objects by sending laser light and measuring the returning reflection), radar sensors (which send out and receive radio waves that bounce off objects), and ultra-sonic sensors (which measure distance using ultrasonic waves) (Mora et al., 2020). To assess the car's location within environments, important technologies include GPS, digital road maps, cloud servers, and big-data-based vehicular networks (Soteropoulos et al., 2019; Sperling, 2018). To interpret and process incoming information, driverless cars also need ultra-fast computers and multiple software packages to build models, interpret patterns, assess possible outcomes of actions (e.g., braking, steering, accelerating), and make decisions about the best course of action. These packages include Artificial Intelligence (AI) and machine-learning algorithms that enable computer systems to gain experience and learn from mistakes.

Although technical developments in many areas have been fast, 'most of the research is [still] experimental in nature' (Mora et al., 2020: 12). There are also real-world demonstration projects such as Google's *Waymo* autonomous vehicle, which by January 2020 had driven more than twenty million miles on public US roads. Increasing real-world experience indicates that self-driving cars perform reasonably well in stable environments, but may struggle in bad weather (e.g., rain, snow), when pedestrians, dogs or plastic bags make sudden movements, or when traffic signs are less visible (e.g., owing to graffiti, mud, or snow). Accidents, including fatal ones such as a 2016 crash with Tesla Autopilot technology (Stilgoe, 2018), led engineers and researchers to increasingly focus on safety and reliability issues.

Technical challenges for high safety and reliability performance relate to sensors, which need to become more accurate, and software packages, which need to improve their ability to interpret information and make the right decisions in any possible scenario (Faisal et al., 2019). A recent edition of *The Economist*'s Technology Quarterly on 'Artificial intelligence and its limits' (13 June 2020) noted that engineers are beginning to doubt if current AI and machine-learning technologies are able to achieve this level of reliability for self-driving cars:

> Many of the grandest claims made about AI have once again failed to become reality, and confidence is wavering as researchers start to wonder whether the technology has hit a wall.

> Self-driving cars have become more capable, but remain perpetually on the cusp of being safe enough to deploy on everyday streets. ... Machine-learning systems are not 'intelligent' in the way that most people understand the term. They are powerful pattern-recognition tools, but lack many cognitive abilities that biological brains take for granted. They struggle with reasoning and generalising from the rules they discover. ... Without another breakthrough, these drawbacks put fundamental limits on what AI can and cannot do. Self-driving cars, which must navigate an ever-changing world, are already delayed, and may never arrive at all. (p. 4)

Costs are difficult to assess in this early development stage. Google's self-driving cars presently cost more than $150,000, with the lidar laser system being particularly expensive at $70,000 (Wolmar, 2018). Scale economies and learning-by-doing will drive costs down, but it is uncertain by how much. Volvo's CEO estimates that full autopilots may cost $10,000 by the early 2020s, which would restrict initial introduction to premium vehicles (Fagella, 2020).

Actors and Policies

Established automakers and ICT firms as well as research groups, consultancies, and specialised suppliers are crucial actors in developing self-driving cars. Large amounts of money have been spent on R&D, collaborative projects, mergers, and acquisitions (Fagella, 2020). In 2016, General Motors spent $581 million to acquire Cruise Automation as well as $500 million to buy 9% shares of Lyft as part of its strategy to first introduce autonomous vehicles in the ride-hailing market niche. In 2017, Ford invested $1 billion into start-up Argo AI, and £15 million into creating an Autonomous Vehicle Research Centre at Carnegie Mellon University. In 2015, Toyota invested $1 billion in the Toyota Research Institute to develop robotics and AI technology. In 2016, Volvo entered into a $300 million joint venture with Uber to develop autonomous vehicles. Tesla and Google have developed their own autonomous vehicles (Autopilot and Waymo). Microsoft has collaborated with car companies such as Renault-Nissan (Fagella, 2020).

Self-driving cars have experienced a hype–disappointment cycle, in which product champions initially made optimistic promises to attract attention, funding, and partners, which then subsequently do not materialise as quickly as predicted, owing to technological or other setbacks (van Lente et al., 2013). Sperling (2018: 91) identified 2015/16 as the starting point of the hype-cycle, because this is 'when the business and popular press began overflowing with adulation and enthusiasm'. In 2015, Tesla's CEO predicted the arrival of 'complete autonomy' by 2018. GM's 2016 acquisition of Cruise was accompanied by promises to commercially launch self-driving cars in San Francisco by 2019. In 2017, Ford's CEO announced plans to introduce Level 4 automated vehicles by 2021. CEO's of Volvo, Fiat-Chrysler, and Toyota also envisaged that self-driving cars would be commercially available by 2021 (Fagella, 2020).

By 2019, however, Ford's CEO acknowledged that 'We overestimated the arrival of autonomous vehicles', adding that initial applications will be in targeted niches: 'Its applications will be narrow, what we call geo-fenced, because the problem is so complex' (cited in Fagella (2020)). Other companies, except Tesla, have similarly downscaled expectations, because of serious technical challenges to achieve the high safety and reliability standards associated with introducing new cars (Topham, 2021).[12]

Important actors in UK driverless car developments are research groups (e.g., Bristol Robotics Group, Mobile Robotics Group at the University of Oxford), university spin-offs and start-up companies (e.g., Conigital, Oxbotica, GoBotix), incumbent ICT and engineering companies (e.g., Telefonica, Thales, BAE Systems, ARUP, RDM Group, Bosch), and existing car companies (e.g., Jaguar Land Rover, Ford, Tata Motors, Williams) (Hopkins and Schwanen, 2018).

The UK government has also been highly supportive of driverless vehicle technology, which it saw as having 'the potential to be a real game changer on the UK's roads, [...] delivering major benefits for road safety, social inclusion, emissions and congestion' (DfT, 2015a: 5). The government would like the UK to capture parts of the global market for self-driving cars, which could be as large as £907 billion by 2035 according to a UK Transport Systems Catapult estimate (TSC, 2017). It therefore aims to position the UK as 'a world leading centre for vehicle research and technology' (DfT, 2015b: 5), which it hopes will attract investment in UK-based driverless car development.

To stimulate these driverless car developments, the government allocated £19 million funding in 2015 to real-world trials in Bristol, Greenwich, Coventry, and Milton Keynes (Hopkins and Schwanen, 2018). In 2016, 2017, and 2018, it provided further funding (£21m, £31m, and 51m respectively) for research, development, and demonstration projects in three rounds of Connected and Autonomous Vehicle (CAV) competitions, which helped to create new partnerships between companies, universities, local authorities, and technical consultancies.

The government also developed Codes of Practice for real-world testing in 2015 and 2019 (DfT, 2019b, 2015b) and created the Centre for Connected and Autonomous Vehicles in 2015 within the Department for Transport to oversee and coordinate technical developments, invest in innovation, and bring together expertise from across the public, private, and academic sectors. The government also developed supportive legislation such as the 2018 *Automated and Electric*

[12] 'Unaccustomed to dealing with stringent regulations regarding safety, [tech companies] are learning that bringing a car to market is very different from launching tech gadgets and software' (Sperling, 2018: 94).

Vehicles Bill, which proposed that accidents with self-driving cars should be addressed by adjusting insurance frameworks rather than through product liability frameworks (which would hold manufacturers responsible). More than other countries, UK policymakers have also started to discuss cyber-security issues, particularly data theft, hacking, and resilience to foreign attack (Mladenović et al., 2020).

Governance discussions are dominated by experts and specialists and do not include civil society groups or wider publics (Hopkins and Schwanen, 2018). Policymakers and experts deterministically present self-driving car developments as positive and inevitable rather than as something that can be discussed with wider stakeholders in terms of desirability and potential negative side-effects. A report from the House of Lords Science and Technology Select Committee (2017) is one of the few policy documents that raises critical questions, emphasising deep uncertainties about the potential impacts of driverless cars: 'There is little hard evidence to substantiate the potential benefits and disadvantages of CAVs because most of them are at prototype or testing stage. Furthermore, as with any new technology or advancement, there may be unforeseen benefits or disadvantages that have not yet presented themselves' (p. 6). It also concludes that the government has focused too much on upstream R&D and dedicated 'inadequate effort on thinking about deployment . . . and the main social and behavioural questions relating to CAV' (p. 6).

Deployment issues have, however, been investigated by social scientists, who suggest that social and environmental impacts will vary widely, depending on vehicle operating profiles (e.g., will they be shared, hailed, or operated individually), settlement patterns (e.g., will people decide to live further away when automated vehicles drive them to work), and travel choices (e.g., will automated vehicles make public transport more efficient or will people shift from public transport to self-driving cars). Wadud et al. (2016: 1) suggest that 'automation might plausibly reduce road transport GHG emissions and energy use by nearly half – or nearly double them – depending on which effects come to dominate'. Based on a comprehensive review of modelling studies, Soteropoulos et al. (2019: 29) conclude that automated vehicles (AVs)

> are mostly found to increase vehicle miles travelled and reduce public transport and slow modes share. This particularly applies to private AVs, which are also leading to a more dispersed urban growth pattern. Shared automated vehicle fleets, conversely, could have positive impacts, including reducing the overall number of vehicles and parking spaces.

Depending on user behaviour and wider knock-on effects (on public transport and residential patterns), self-driving cars may thus *increase* rather than decrease traffic.

Self-driving vehicles may also threaten the jobs of drivers of heavy goods vehicles, buses, coaches, taxis, and private-hire vehicles, which in the UK amounts to about 1.6 million people (Wolmar, 2018).

Safety benefits are also uncertain. First, it is technologically challenging to develop 100% reliable, fail-safe automated systems, as noted at the beginning of the section. Engineers can see technological options that could bring driverless cars to 96–97% of this goal but assess that the final 3–4% will be exponentially harder to achieve (Turley, 2019). Reliable fail-safe CAVs may thus require further technical breakthroughs, as noted previously. Second, a fully reliable and functional system may still be vulnerable to hacking, which would provide intruders with control over thousands of vehicles. Sperling (2018: 103) suggests that 'the industry is years away from solving the cyber security problem'. Third, safety benefits will be modest until all cars are fully automated, because 'mixing driverless and non-driverless cars will have uncertain safety benefits' (Sperling, 2018: 99). Fourth, liability issues in case of accidents are challenging to address, because responsibilities for well-functioning self-driving cars are distributed between manufacturers, software providers, owners, city councils (who have responsibilities for traffic signs or road paint), and regulators who set standards (Wolmar, 2018).

In sum, powerful private and political actors are committed to developing self-driving cars. The technology is, however, still in early developmental phases and faces substantial techno-economic and socio-political challenges. Early high expectations about the technology's imminent real-world deployment are being replaced by more balanced understandings of challenges and timescales: 'Under almost any scenario, it will take many decades for the full transition to driverless cars to be realized' (Sperling, 2018: 100). The positive and negative effects of this transition are potentially large and open-ended, and dependent on manifold actions and choices. While this creates deep uncertainties, it also means that self-driving car developments are not autonomous, but can still be shaped by actors, including policymakers.

5.7 Low-Carbon Transition through Whole System Reconfiguration

Pulling together information from the system and niche-innovation descriptions, this section first analyses low-carbon whole system reconfiguration through the three lenses (techno-economic, actors, institutions) and then addresses speed, scope, and depth of change.

5.7.1 Low-Carbon Innovations Driving GHG Emission Reductions

GHG emissions from the land-based passenger mobility system (which includes cars, railways, buses, bicycles) decreased by 14% between 2007 and 2019, despite an 11% increase in total passenger-kilometres in the same period. While this does not yet represent a decisive low-carbon transition, it does show that relevant reconfigurations are underway in the passenger mobility system. Based on the analysis in preceding sections, we conclude that the GHG emission reductions resulted from multiple landscape, system, and niche developments.

- The financial-economic crisis (and subsequent recession) and high oil prices were two landscape pressures that decreased socio-economic activities in the 2008–2013 period, which, in turn, decreased passenger mobility demand by cars and associated fuel use by 2%. Both these landscape pressures weakened after 2013, boosting car mobility (Figure 5.3).
- GHG emissions were also reduced by incremental engine innovations that improved the fuel efficiency of *new* petrol cars by 25% and *new* diesel cars by 27% in the 2007–2016 period (Figure 5.13). One qualifier is that emission performance numbers refer to laboratory test results, which can deviate substantially (up to 30–40%) from real-world driving performance. Another qualifier is that the fuel efficiency improvements refer to *new* cars. A precise assessment of their contributions to emission reductions thus also requires information about vintages of the whole car fleet, which is beyond the chapter's scope.
- Another climate-relevant incremental change in the automobility system was the gradual shift towards more fuel-efficient diesel cars after 2001, when the government introduced CO_2-banding in its Vehicle Excise Duty. This trend was reversed by the 2016 Diesel-gate scandal, which triggered major decreases in diesel car sales.
- A climate-relevant incremental change in the railway system was gradual expansion, as railway passenger-kilometres increased by 46%, from 57 billion in 2007 to 83 billion in 2019 (Figure 5.3). Some of this passenger expansion resulted from a moderate modal shift from cars to trains, especially for commuting in and out of London, as discussed in Section 5.3.2.
- The expansion of cycling (in some cities) is beneficial for health, air quality, and congestion, but has not contributed much to CO_2-reduction because it is such a small system. Despite substantial expansion in some cities, including London, cycling's contribution to overall passenger mobility increased only from 0.5% of passenger-kilometres in 2007 to 0.7% in 2019 (Figure 5.3). Although cycling receives much attention in public media and academic sustainability articles, it is likely to remain of marginal relevance for climate mitigation. Even a doubling of cycling, which would be a challenge in the UK, will have limited climate mitigation effects.
- One niche-innovation that contributed to GHG emission reductions was biofuels, which increased from 0.8% of fuel blends in 2007 to 4.3% in 2019 (Figure 5.32).
- Other climate-relevant niche-innovations were electric vehicles (HEVs, BEVs, PHEVs), which increased from 0.3% of the passenger car fleet in 2007 to 3.3% in 2020 (Section 5.6.1), and 20.8% of new car sales.

- A third niche-innovation, tele-working at home, increased from 2.7% of the working population in 2007 to 5.1% in 2019 (and jumped to 8.5% in 2020). The influences on GHG emissions are uncertain, however, because tele-working has both mobility substituting and inducing effects, as discussed in Section 5.6.3.
- The emission reduction effects of driverless cars, car sharing, and mobility services have been small, because these niches are still marginal, except for ride-hailing, as discussed in Sections 5.6.4, 5.6.5, and 5.6.6.

This summary clearly demonstrates that GHG emission reductions between 2007 and 2019 resulted from multiple exogenous landscape influences, system developments (incremental improvements and relative size changes), and niche-innovations that together reconfigured the passenger mobility system.

5.7.2 Techno-Economic Reconfiguration

To further interpret mobility systems reconfiguration, we use Summary Table 5.10, which positions the various innovations and changes in the techno-economic mapping framework that we developed in Section 2.2.1. Combining this table with the dynamic analyses in the previous sections, we identify the following pattern. *Incremental innovations* in system modules (e.g., engine improvements, shift from

Table 5.10. *Mapping system reconfiguration opportunities in the UK passenger mobility system*

<table>
<tr><th colspan="6">Core elements</th></tr>
<tr><td rowspan="5">Linkages (coupling) between system components</td><td></td><td>Reinforced</td><td colspan="3">Substituted</td></tr>
<tr><td rowspan="2">Unchanged</td><td rowspan="2">Modular incrementalism
- Efficiency innovation in internal combustion engines
- Shift from petrol to diesel cars (2001–2016)
- Motorway extension
- Railway extension, better signalling</td><td colspan="3">Modular substitution</td></tr>
<tr><td>System–system switching:
Limited modal shift (from cars to other modes)</td><td>Niche-system hybridisation:
Biofuels, HEV, PHEV</td><td>Niche-system replacement:
BEV, ride-hailing, tele-working</td></tr>
<tr><td>Changed</td><td>Architectural stretching
- High-speed rail</td><td colspan="3">Architectural reshaping
- Mobility-as-a-Service, car sharing
- Driverless cars
- Intermodal transport</td></tr>
</table>

petrol to diesel cars, motorway and railway extension) were prominent in the early period and also remained relevant in later periods.

During the last 5–10 years, they have increasingly been complemented, however, with *modular substitutions*, which replace system components but do not require deeper changes in the system architecture. These modular substitutions took different forms. First, some limited system-to-system switching occurred due to a small modal shift from automobility to railways. Second, substantial hybridisation occurred within the automobility system, due to biofuel blending (which increased markedly since 2017), hybrid-electric vehicles, and plug-in hybrid-electric vehicles (which use both electric motors and combustion engines). Third, substantial niche-to-system component replacement is underway in the form of battery-electric vehicles and ride-hailing (which increasingly substitutes traditional taxis). Tele-working can also be seen as providing (partial) niche-to-system replacement, although this niche-innovation is less about changing a technical component and more about reducing the demand for a certain kind of mobility purpose (commuting to work), which indirectly affects the (auto)mobility system.

There has also been some *architectural stretching* in the last 5–10 years, particularly in the form of big railway projects such as high-speed railways, the strategic relevance of which is still under debate, as discussed in Section 5.3.3.

There are also seeds for deeper *architectural reshaping*, but the associated innovations have remained relatively small and have uncertain growth paths. Car sharing and Mobility-as-a-Service could deeply reconfigure mobility systems if they would lead to a shift from private to collective car ownership or to a service model. Presently, however, both innovations are marginal and mostly an occasional add-on to existing mobility practices for particular user groups (see Section 5.6.4).

Driverless cars could also deeply transform existing transport systems, both in terms of functionality (with people trusting computers to do essential driving tasks) and performance (e.g., less accidents, better time use, smoother ride). When combined with sharing or ride-hailing, future mobility systems could even consist of robo-taxis driving around continuously in search of customers. Such high-tech visions are unlikely to materialise any time soon, however, given the technological, social, and legal barriers driverless cars face (discussed in Section 5.6.6).

Intermodal transport innovations can also reshape system architectures, particularly at the local level, by creating new linkages between transport modes. There are examples of such deeper system reconfiguration in London, and to a lesser extent in smaller cities such as Bristol, Brighton, and Oxford (Schwanen, 2015). In London, many radical niche-innovations (electric car schemes, car sharing, bike-sharing, intermodal ticketing) have come to fruition, and existing systems have changed significantly through the expansion of bus services, new train schemes, and construction of dedicated bicycle infrastructures. The alignment of systems has resulted in an effective intermodal transport system (Table 5.2) and a pronounced modal shift, with car use declining by 35–40% between 2002 and 2018 (Figure 5.18). Apart from London and some small cities, such deep reconfigurations are, however, rare in the UK.

Although there are multiple system reconfiguration options, modular substitution is currently the dominant pathway, especially in the form of a transition towards electric vehicles and, somewhat less, towards biofuels, which have also received a new policy impetus in recent years.

In the past five years, electric vehicles have received increasing support from policymakers, automakers, consumers, and wider publics (as discussed in Section 5.6.1). Electric vehicle sales (HEV, PHEV, BEV) have rapidly increased in recent years, reaching 20.8% of the passenger car market in 2020. While automakers still defended internal combustion engine vehicles in the early 2010s (Bohnsack et al., 2020; Penna and Geels, 2015), they have in the past few years begun to seriously reorient towards electric vehicles.

Critics rightly argue that an electric vehicle transition still maintains the centrality of the automobility system and private passenger cars (Bergman et al., 2017; Holden et al., 2020), and in that sense is not as deep a reconfiguration as some of the architectural reshaping options because it leaves most elements unchanged. Electric vehicles may also not be sufficient to reach deep decarbonisation targets (Milovanoff et al., 2020).

On the other hand, the focus on electric vehicles is understandable considering that cars are deeply locked-in (economically, culturally, socially, technologically), and accounted for 85% of all passenger-kilometres in 2019 (Figure 5.3). It is thus very likely that cars will remain the dominant transport mode in the next 10–20 years, and perhaps even longer, particularly for mid- to long-range travel. In most parts of the UK, except in some large cities, alternative transport is infeasible or insufficiently developed to enable a substantial modal shift. Furthermore, the majority of total motor vehicle traffic (62%) in 2019 was on major, interlocal roads (Figure 5.7). Most of the associated journeys simply cannot be replaced by cycling, mobility services, or local integrated transport systems. So, if passenger cars are likely to remain dominant in the next 10–20 years, it does make sense to stimulate a shift towards electric vehicles.

This obviously does not mean that deeper reconfiguration options should not also be pursued. In fact, we agree that policymakers should give more support to these options (Holden et al., 2020). But we do suggest that it is rather unlikely in the medium- and probably long-term that everyone will give up cars, start cycling, use public transport, or use sharing and mobility service options.

A final conclusion is that the technology-oriented focus on modular substitution options tends to come at the expense of systemic improvement efforts. The creation (and standardisation) of public battery-recharging, for instance, lags behind the development and purchase of electric vehicles. Basic infrastructure issues (such as local road maintenance, pothole filling, parking, traffic-calming schemes) similarly receive little attention. And railway electrification or signalling improvements, which could help reduce delays, also receive relatively little attention, compared to big rail schemes such as HS2.

5.7.3 Actor Reconfiguration

Focussing on actors and social networks, we conclude that there has been substantial flux and increasing reconfiguration due to the appearance of new actors and changes in the views, interests, and strategies of existing actors, particularly in the automobility system.

New actors have especially appeared with regard to technology development and mobility supply options, as the following summary indicates:

- In car manufacturing, Tesla has been a successful new entrant, whose high-end electric vehicles boosted market demand and created competitive pressure on incumbent automakers to reorient towards EVs.
- New public and private organisations (such as the Office for Low Emission Vehicles, the Automotive Council, and the Advanced Propulsion Centre) have been created to facilitate coordination and collaboration in the development and manufacturing of electric vehicles, leading to a new innovation system.
- A new biofuel manufacturing industry has emerged in the UK to process (mostly) imported biomass feedstocks and supply biofuel to mainstream fuel providers.
- Ride-hailing companies (such as Uber) are successful new entrants, whose rapid, convenient, and cheap mobility services have disrupted the taxi market and are generating new options for multimodality in urban settings.
- Car- and bike-sharing organisations have appeared as new actors providing new mobility services. The uptake has, so far, remained relatively small, focused on a narrow user segment (e.g., highly educated, tech-savvy, young people in big cities). Initial start-up companies have increasingly been replaced by the entry of incumbent automakers and car rental companies.
- Mobility-as-a-Service (MaaS) providers are new organisations that use platform-based business models to provide new intermodal mobility services. Many participants in these new organisations are incumbent actors (e.g., taxis, public transport). The uptake is still very small and the business model fragile.
- Incumbent IT companies such as Google, Microsoft, and Apple have moved into the automotive sector, focusing especially on driverless cars, often in collaboration with existing automakers. New R&D organisations and university spinoffs are also working on this new technology, which is in an early developmental stage but supported by highly positive expectations.

Reconfiguration of existing actors and social groups occurred through changes in interpretations, interests, and strategies in response to new (technological and economic) opportunities and (environmental and social) problems. These incumbent actor reconfigurations include the following:

- Incumbent automakers have very substantially changed their views and strategies about (various forms of) electric vehicles, in response to increasing consumer demand, CO_2 regulations and phase-out targets, public debates, and competitive pressures. Their reorientation towards electric vehicles is generating tremendous turbulence and flux in the automobile industry because it requires major investments and changes in factories, supply chains, and technological capabilities.

Incumbent automakers have also started to explore driverless cars, investing billions of dollars globally in research and technological development. And they have moved into car sharing and ride-hailing ventures to learn about new business models and the mobility services market.

The large investments for these reorientations and diversifications are financially challenging for automakers, whose profits are under pressure from declining passenger car sales in the UK and globally since 2017.

- The preferences, views, and mobility practices of some consumers were also changing before the pandemic, mostly through diversification processes leading to increasing fragmentation of mobility practices. Driver's licenses and car ownership were becoming less prevalent among young people, compared to previous generations, but there is an ongoing debate about the relative importance of cultural reasons (e.g., changing attitudes towards cars) and economic reasons (e.g., increasing car operating costs and lower post-recession incomes).

Young people in large cities such as London partially shifted towards ride-hailing, car sharing, and cycling, but it is not yet clear if these new mobility practices are long-lasting or if they change when people marry, buy houses, and have children.

More people were taking trains, leading to some modal shift. Most of the rail travel growth came from journeys in relation to leisure, business, and commuting (Figure 5.21). The latter two purposes accounted for 51% of rail travel in 2019.

Another change has been the increased purchase of electric vehicles (which accounted for 20.8% of passenger car sales in 2020). Despite purchase subsidies, electric vehicles are still more expensive than normal cars. Early adopters were mostly middle-aged, affluent urbanites with an interest in new technology and environmental issues, but other social groups also increasingly buy electric vehicles.

Although the aforementioned changes are highly relevant, we should not forget, however, that most consumers still showed strong attachments to cars.

The percentage of households with two or more cars has gradually increased since the early 2010s (Figure 5.16). Passenger-kilometres by cars remained at 85% of total passenger mobility between 2007 and 2019. And the percentage of Sports Utility Vehicles in passenger car sales increased from 6.6% in 2009 to 21.2% in 2018. So, although the various new mobility behaviours are important and underpin various technical innovations and industry reorientations, a major cultural and behavioural revolution does not yet appear to be underway.

- The COVID-pandemic was an exogenous shock that substantially affected consumer preferences and user practices, although it remains to be seen which changes will be long-lasting and structural. Car travel was heavily affected by lockdowns but has rebounded to pre-pandemic levels (Figure 5.4). Rail and bus travel decreased even further during lockdowns but subsequently rebounded to only 50% and 60% of pre-pandemic levels, respectively, owing to lingering health concerns, which suggests that a shift away from public transport may be longer lasting. Cycling increased drastically during and immediately after the first lockdown but seems to have returned to pre-pandemic levels in 2021 (Figure 5.26) in most though not all cities. Tele-working increased substantially in 2020 (Figure 5.33), especially for highly skilled professional workers (Figure 5.34), for whom some of the behaviour changes are likely to be structural, with potential consequences for commuting and business travel (via railways or air).

The pandemic also affected the perceptions and practices of policymakers, who not only provided emergency support funding for railways and buses but also used the disruption to introduced substantial reforms and new policy initiatives for railways, buses and cycling.

The four subsequent tables more systematically summarise and interpret degrees of actor reconfiguration in the four systems. The two columns in the tables address both the main actor changes that support low-carbon transitions and the lock-in mechanisms and competing issues that constrain their engagement with low-carbon transitions.

Table 5.11 shows that actor reconfiguration in the automobility system has been large for automakers and policymakers, and medium for consumers and wider publics. All actors also face medium or large lock-in mechanisms or competing issues that constrain the speed of their reorientation. Tables 5.12, 5.13, and 5.14 show that actor reconfigurations in the railway, bus, and cycling systems have been small or medium, while lock-in mechanisms and competing issues have been high or medium. Changes in actors' views, interests, strategies, and capabilities have thus been more limited in these three systems.

Table 5.11. *Changes and lock-ins for actors in the automobility system*

	Actor changes supporting low-carbon transition	Actor lock-ins and competing issues constraining low-carbon transition
Carmakers	LARGE - Shifting from internal combustion engines to electric vehicles (EV) - Also exploring car sharing, mobility services, and driverless cars	MEDIUM - Protect sunk investments in car manufacturing (e.g., capabilities, factories) - Immediate economic pressures (e.g., declining markets and profitability, under-utilisation of factories) more important than climate change - Deal with post-Brexit trade problems
Policymakers	LARGE - Increasing climate mitigation pressure from local, national, and European policymakers, but mostly with technological focus (especially EVs) - Increasing car-restrictive measures in urban areas (e.g., rising parking fees, one-way roads, reduced city centre access)	MEDIUM - Other non-climate issues also prioritised at national level (e.g., support domestic car industry, congestion, safety), leading to a £27 billion road building programme (2020–2025) - Local policymakers concerned about parking, air pollution, congestion, but have limited policy and financial scope for action
Users	MEDIUM - Increasing EV purchase and ride-hailing in big cities - Young people less likely to own cars	LARGE - Ongoing attachment to cars, leading more cars per household - Traditional purchase criteria (e.g., price, size, reliability, comfort, safety, appearance, brand) more important for most consumers than climate mitigation, leading to more SUV sales
Civil society organisations, public debate	MEDIUM - More public debate about climate change, air pollution, and cars - Public debate portrays electric vehicles as the central climate solution	MEDIUM-LARGE - Cars remain associated with positive values (e.g., freedom, success, excitement) - Other issues (e.g., fuel prices, congestion, road conditions, air pollution) more prominent in public debates than climate change

Table 5.12. *Changes and lock-ins for actors in the railway system*

	Actor changes supporting low-carbon transition	Actor lock-ins and competing issues constraining low-carbon transition
Train operating companies (TOC)	SMALL - Climate change not a major consideration - Nevertheless, rail expansion is positive for climate mitigation	MEDIUM-LARGE - Other issues (e.g., financial profitability, punctuality, overcrowding) more important than climate change
Policymakers	SMALL-MEDIUM - Policymakers have supported railway expansion but abandoned explicit modal shift policies - Railway investments mostly focused on large-scale, London-centric projects rather than whole system improvement (e.g., electrification, signalling) - Large COVID-related emergency support prevented TOC bankruptcies	MEDIUM-LARGE - Prolonged reluctance to change institutional frameworks until the COVID-pandemic provided the conditions for substantial reform and the creation of Great British Railways from 2023 - No overarching policy vision beyond accommodating railway growth
Users	SMALL-MEDIUM - Increased railway travel, mostly for economic purposes (commuting, business) and leisure	MEDIUM - Many people depend on railways to travel in and out of London (which accounts for majority of rail journeys), for which they have limited alternatives - High fares and delays frustrate users - Large COVID-related decrease in rail travel, and lingering health-related fears to travel with others in confined space.
Civil society organisations, public debate	SMALL - Limited public debate about railways and climate mitigation, beyond occasional mention of modal shift	MEDIUM - Other issues (e.g., rising train fares, overcrowding, delays, and public subsidies) more important than climate change - Criticisms of privatised railway arrangements created context for COVID-related institutional reform.

5.7.4 Policy Reconfiguration

Formal Policies

The unfolding low-carbon reconfiguration has been supported by different kinds of policy instruments. Some innovations and transport systems received more policy support than others, which helps explain uneven developmental patterns.

Table 5.13. *Changes and lock-ins for actors in the bus system*

	Actor changes supporting low-carbon transition	Actor lock-ins and competing issues constraining low-carbon transition
Bus companies	SMALL - Climate change not a major consideration - Some subsidised purchase of hybrid-electric buses in London	MEDIUM-LARGE - In declining markets, bus companies mostly compete on costs - Other issues (e.g., quality improvement, better bus stops with electronic information) receive more attention than climate mitigation
Policymakers	SMALL-MEDIUM - Some policy support for hybrid and electric buses - Stop-start dynamics and changing priorities of funding schemes lead to piecemeal and fragmented improvement - The 2021 National Bus Strategy aims to reform and improve bus services and reverse decades-long decline	MEDIUM-LARGE - Fragmented bus governance (between national and local levels) - Other issues (mitigating decline, maintaining some service provision, stimulating some innovation) more important than climate change
Users	SMALL - Climate mitigation unimportant consideration for most bus users	LARGE - Some social groups (low-income families, students, elderly) dependent on bus services - Bus use is declining in most of country, except London - Increasing prices and low quality frustrate bus users (potentially leading to downward spirals)
Civil society organisations, public debate	SMALL - Little public debate about buses as climate solution	LARGE - Negative public image of buses - Other issues (e.g., rising fares, low service quality, air pollution) more pertinent than climate change

- Electric vehicles have been particularly well-supported with public finance for R&D and technology development, subsidies for consumer purchase, and investment in recharging infrastructure. Policymakers also created a dedicated organisation (the Office for Low Emission Vehicles), which strengthened the innovation system by stimulating interactions and collaborations between upstream actors. Tightening European emission regulations and a UK phase-out policy for diesel and petrol cars also stimulated development and demand for electric vehicles.

Table 5.14. *Changes and lock-ins for actors in the cycling system*

	Actor changes supporting low-carbon transition	Actor lock-ins and competing issues constraining low-carbon transition
Bicycle manufacturers and shops	SMALL - Majority of UK bikes are imported - UK bicycle shops sell bikes and accessories, but limitedly engage with climate change	WEAK - Limited manufacturing lock-ins
Policymakers	SMALL-MEDIUM - Active policy support in London and a few other cities - Ambitious but top-down 2020 cycling strategy aims to increase cycling infrastructure and use	MEDIUM-LARGE - Ad-hoc and fragmented national bicycle governance with low priority level (prior to 2020)
Users	MEDIUM - Increased uptake by some social groups in large cities (mainly young urban professionals) - Cycling boost due to pandemic lockdowns in 2020, but limited evidence of structural effect - Increased fitness, travel time, and money savings more important motivations than climate change	LARGE - Most citizens uninterested in cycling
Civil society organisations, public debate	SMALL - Cycling not seen as 'normal'	LARGE - Cycling widely perceived as dangerous and unappealing

- Driverless cars also received substantial and increasing support, mainly through subsidised research, development, and demonstration (RD&D) projects as well as emerging regulations.
- Biofuels have received some domestic RD&D support but were mainly driven by EU and UK bio-blending targets, which were substantially increased in recent years.
- The railway system was mainly supported through large-scale infrastructure investment and increasing operational support for Network Rail. In response to the COVID-pandemic, the government not only provided about £12 billion emergency support but also in May 2021 introduced substantial institutional reforms that from 2023 will create a new arm's length public body (*Great British Railways*) in charge of rail infrastructure, fares, timetables, and contracting with train companies to provide operational services. These reforms aim to

simplify organisational arrangement and responsibilities, improve efficiency, decrease the need for public subsidies, and further grow rail networks and travel.

- The bus system has been supported with small and decreasing bus service operator grants (Figure 5.24), some HEV adoption subsidies, and bus travel passes for the elderly. In response to the pandemic, the government not only provided about £1.4 billion emergency support but also in March 2021 introduced a National Bus Strategy for England, which aims to reverse the decades-long decline of bus travel outside London and revitalise bus use in cities and towns. The strategy proposes institutional reforms that give Local Transport Authorities more influence over timetables, fares, and service quality, and provides £3 billion funding over five years to introduce changes such as bus priority lanes, 4,000 new electric buses, contactless intermodal payment systems, and improved digital information.
- Limited policy support has been given to cycling (except in a few cities), teleworking, car sharing, and mobility services. These are typically options with a substantial behavioural change component, which suggests that UK transport policy has focused more on technology than on mobility practices. The new active travel strategy, introduced in July 2020, represents a substantial policy shift because it aims to boost walking and cycling in cities and towns and provides £2 billion funding over a five-year period to enable local councils to make on-the-ground changes such as dedicated cycle lanes and low-traffic neighbourhoods. While this strategy and funding led to many local initiatives, numerous councils also removed or downscaled them in response to protests and social acceptance problems, which means the strategy's success is not guaranteed.

Governance Style

Neoliberalism has long been the dominant guiding principle in UK transport policy, leading to 'the encouragement of competition, privatization and a light touch from government' (Wolmar, 2016: 106–107). Neoliberalism affected public and private transport systems differently. In the bus and railway systems, it underpinned privatisation and liberalisation policies, which fragmented industry structures and led firms to focus on cost-competition. Neoliberal notions resonated well with the individualism and freedom associated with private cars, although significant regulations and public investments imply that the automobility system does not operate as a 'free market'.

The increasing emphasis on free markets and consumer choice also led to the decline of explicit modal shift and behaviour change policy. Although the 1998 *New Deal for Transport* plan had ambitious and interventionist goals in this regard, the 1997–2010 Labour government gradually abandoned the idea that transport policy could be about intervening to change the share of mobility accommodated by each mode. Towards the end of its term of office, Labour's policy had become one of 'modal agnosticism'. ... The narrative shift from promoting 'sustainable' transport, to 'integrated' transport and then 'choice' can be readily discerned in Labour's copious transport policy statements from 1997 to 2010. (Shaw and Docherty, 2014: 123)

The very recent active travel strategy (2020) intends to signal a new direction, because it aims to boost cycling and walking so that these transport modes account for half of all journeys in towns and cities by 2030. Although this top-down strategy represents a transformative push, it remains to be seen to what degree it will drive change or form another instance of 'overpromising and underdelivering'.

Attempts at more interventionist governance are also visible regarding railways and buses, but since the new 2021 plans and strategies focus more on vision than on implementation detail, the transformative effects are difficult to assess. Regarding electric vehicles, policymakers have more demonstrably adopted an increasingly interventionist governance style in the last few years, introducing a raft of market-shaping policies such as purchase subsidies, recharging infrastructure investment, tightening emission regulations, and a phase-out policy of diesel and petrol cars. This shift towards an interventionist governance style stems from the alignment of transport, climate, and industrial policy considerations, which aim to help the UK car industry in its repositioning in a global innovation race.

One stable characteristic of the UK transport governance style has long been the focus on technologies (e.g., electric vehicles, driverless cars, biofuels) and infrastructure. This technological focus was further strengthened by increasing alignments between transport policy, industrial policy, and climate policy, which led policymakers to prioritise those solutions that not only address transport and climate problems but also have the potential to galvanise new industries, jobs, and economic growth. With regard to infrastructure, policymakers have focused more on *new* large-scale road and rail projects than on less eye-catching bottleneck or 'whole system' improvements, such as railway electrification or better signalling, or local schemes, such as 'junction safety improvements for pedestrians and cyclists, repairing local pavements, building cycle paths and implementing traffic-calming schemes' (Wolmar, 2016: 103). If the recent active travel strategy succeeds in the coming years, it would represent a substantial shift that would complement the (ongoing) technological strategies with more behavioural and locally oriented policies.

Another characteristic is a preference for financial incentives, which leave decisions to the market, over more interventionist policies. 'Politicians are surprisingly wary of doing things that they judge might be unpopular with the public, and this has had massive implications on the direction of British transport policy, especially in the last two decades' (Shaw and Docherty, 2014: 8). Since some of the recent bus and cycling policies aim to change people's travel practices, it remains to be seen how policymakers will respond to public protests, which are beginning to emerge in places.

A third characteristic is the fragmentation between national and local jurisdictions, which has hampered local transport policies for buses, cycling, road maintenance, and local transport schemes. Wolmar (2016: 107) describes British transport policy as 'highly centralized, with very little power at the local level'. Marsden et al. (2014: 630) note that: 'The local authorities have no legislative powers and remain highly dependent on national government for resources for new projects (capital) and funding ongoing activities (revenue). . . . Our interviews with local government stakeholders frequently turned to the hollowing out of technical capacity and of reductions in core financial resources'. Schwanen (2015: 7086) therefore concludes that: 'The autonomy of small- and medium-sized cities as agents in bringing about transformational change toward low-energy urban mobility should not be overestimated.' In the coming years, this characteristic may well form a substantial problem for the local implementation of the new top-down bus and cycling strategies.

The exception to the fragmentation problem is London, which has managed to deeply reconfigure its passenger transport system in the past two decades, leading to a marked modal shift away from cars. This reconfiguration was enabled by a dedicated local governance structure (notably *Transport for London*, which was created in 2000 with substantial budgetary and policy discretion), political leadership by successive mayors, and targeted policies, including investments to improve public transport and cycling systems, and the London congestion charge that dis-incentivised automobility. London's system reconfiguration thus shows that substantial change is possible in favourable governance contexts, including political will and resources.

5.7.5 Scope, Depth, and Speed of Reconfiguration

The scope of the unfolding techno-economic reconfiguration is relatively limited in the automobility system because most change has come to centre on electric vehicles, which is a modular component substitution. Other low-carbon options receive less attention, support, and investment, including incremental fuel efficiency improvements, other modular component substitutions (e.g., biofuels, ride-hailing), and architectural reshaping options (e.g., driverless cars, car sharing, intermodal transport).

The scope of actor reconfiguration in the automobility system has been rather substantial for electric vehicles (especially for automakers and policymakers and somewhat less for users and wider publics) but more limited for other low-carbon options, so that we evaluate the overall scope as moderate.

The scope of policy reconfiguration in the automobility system has remained limited in terms of overall goals, which have increasingly focused on the 'greening of cars', particularly through electric vehicles. In terms of policy instruments, however, electric vehicle policies had broad scope, using multiple types of instruments, including financial incentives (for consumer purchase and automaker investments), regulations (e.g., CO_2 emission performance standards, diesel and petrol car sales bans), and direct infrastructure investments (in recharging facilities).

The depth of the unfolding techno-economic reconfiguration in the automobility system is moderate, because biofuel blending and various forms of electric vehicles represent modular substitution changes that do not require deeper changes in the system architecture (although the creation of an electric recharging infrastructure creates some new linkages to the electricity system with potential future knock-on effects via vehicle-to-grid innovations). Deeper architectural reshaping options (such as car sharing, Mobility-as-a-Service, driverless cars, and intermodal transport) have remained relatively limited, except in London where an effective intermodal transport system has been created.

The depth of actor reconfiguration in the automobility system is substantial for electric vehicles, where all major automakers have reoriented investment strategies and technical capabilities, while maintaining a mass production-focused business model. Policymakers remained supportive of cars and the car industry, but substantially shifted their support from petrol and diesel cars to electric vehicles, leading to a range of new instruments, including a phase-out policy. But actor reconfiguration has been low to moderate with regard to most of the other car-oriented innovations, which are still small (car sharing), in early developmental stages (driverless cars), or not very radical (fuel efficiency, biofuels, ride-hailing).

Policy reconfiguration in the automobility system has become deep for electric vehicles but is of limited and moderate depth for other innovations.

The speed of reconfiguration is substantial for electric vehicles and will likely increase in the coming decade (driven by phase-out policies, technological and cost improvements, and competitive dynamics). But speed is relatively limited for most of the other innovations, except perhaps for ride-hailing in big cities. Cumulative effects of multiple changes have been substantial, however, leading to 12% GHG emission reductions from automobility between 2007 and 2019, while total passenger car mobility increased by 10%.

Phrased in MLP terms, the automobility system is experiencing destabilisation in internal combustion engine technology, industry strategies, policies, and mobility practices (especially of young urbanites), while multiple niche-innovations have emerged, some of which are gaining momentum (EVs, ride-hailing, biofuels). But the automobility system also continues to have substantial degrees of obduracy related to extensive road infrastructures, established consumer preferences, the economic importance of car manufacturing, and ongoing political support, which is why electric vehicles are receiving more support and resources than other radical innovations.

In the railway, bus, and cycling systems, the scope of the unfolding techno-economic reconfiguration has been limited because of the limited breadth of innovation. Recent bus and cycling strategies aim to increase the reconfiguration scope by stimulating electric buses, intermodal smart tickets, improved information provisioning, and dedicated bus and cycle lanes, but it remains to be seen if these succeed.

The scope of actor reconfiguration in these three systems was also limited until 2020 when the pandemic led to a substantial reduction in rail and bus travel, with potential longer-term structural effects due to lingering health concerns, and a temporary increase in cycling. The scope of policy reconfiguration in these three systems was also limited until 2020 when new strategies changed the policy instruments and incentives in rail, bus, and cycling systems.

The depth of techno-economic reconfiguration has been limited in these systems because change is mostly about increased use of well-established technologies, except for (the intended expansion of) electric buses, which represents moderate-depth modular substitution. The depth of actor reconfiguration in these three systems has also been limited since increased use of existing technologies builds on existing capabilities. The depth of policy reconfiguration was similarly limited until 2020 when new strategies intend to substantially alter the goals, governance styles, and/or institutional arrangements in rail, bus, and cycling systems.

The speed of change has been substantial in the railway and cycling systems in terms of expansion but not in terms of transformation: railway use by passengers has more than doubled since the mid-1990s, while cycling increased by 40% between 2007 and 2019, and 46% in 2020. Despite this expansion, railway and cycling remained relatively small systems, accounting for, respectively, 9.5% and 0.7% of passenger-kilometres in 2019, compared to the automobility system, which accounted for 85% of passenger-kilometres in 2019.

5.7.6 Future Outlook

Although GHG emissions have not yet declined sufficiently, there are some encouraging developments, such as rapidly increasing market shares of electric vehicles in car sales, which are likely to substantially reduce emissions in the coming decades. Although the replacement of diesel and petrol car fleets will lag behind, the diffusion of electric vehicles is likely to constitute an important future pathway for passenger mobility system reconfiguration, even though it mostly represents a modular component substitution rather than a Great Reconfiguration. The speed of this change is likely to increase in the coming years because of strong policy support, including phase-out policies for petrol and diesel cars, strategic reorientations by automakers, and further decreases in the cost of batteries and electric vehicles.

Deeper reconfiguration of passenger mobility systems is possible in the coming years, especially at the local level, where buses, cycling, car sharing, ride-hailing, intermodality, and mobility services have substantial transformative potential. Recent national policy strategies have recognised this potential and introduced some policies to stimulate developments in this direction, but it remains to be seen to what degree local implementation will succeed or be transformative. One reason for doubt is that local policymakers have limited technical capacities, financial resources, and policy discretion. Another reason is that national policymakers, automakers, and many users remain committed to personal cars, which complicates a deep reconfiguration of local transport systems.

Tele-working, and the long-term consequences of COVID-19 more generally, may also contribute to future reconfiguration of passenger transport by altering mobility patterns. While increased tele-working is likely to be a permanent trend (especially for white collar professionals), there are uncertainties about the scale of this change and its effects on mobility patterns. So far, the effects on car travel, which has rebounded to pre-pandemic levels, seem to be limited, while the effects on rail travel, which has rebounded much less, may be larger. It also remains to be seen how employers' and workers' attitudes towards tele-working, and COVID-19 more generally, will develop in the coming years.

In sum, electric vehicles are likely to substantially reconfigure parts of the passenger mobility system in the coming decades and there are additional options for deeper reconfiguration. It is presently uncertain, however, if these additional options will be further developed or if modular substitution will remain the dominant decarbonisation pathway.

6

Heat System

6.1 Introduction

The heat system is relatively localised because heat cannot easily be transported over long distances. The heat system is therefore closely linked to buildings, where heat is used to warm up rooms (space heating) or water (e.g., for showers, baths, laundry). While there are many types of buildings (e.g., homes, offices, factories, schools, shops), the residential sector uses most heat-related energy (Table 6.1). This chapter therefore focuses on the residential sector, and particularly on space heating, which is the largest segment.

UK domestic space heating is dominated by gas-fired boilers, which are used in 85% of homes. Electric storage heating is used in 5% of homes, oil central heating in 4%, other options (e.g., coal) in 4%, and 2% of homes are linked to heat networks (BEIS, 2018a: 20). The UK's gas-dominated residential heat system has a hybrid centralised-decentralised form, because most heat conversion is done on-site (using individual gas-fired appliances), but fuel is supplied by a national gas distribution infrastructure.

We conceptualise heating and buildings as two 'orthogonal' or intertwined systems. On the one hand, energy infrastructures (for gas) or supply chains (for coal, oil, biomass) feed into conversion devices (e.g., gas boilers, coal or biomass stoves, electric heaters, heat pumps) to generate heat. The schematic representation in Figure 6.1 focuses on the dominant gas-based heating system. On the other hand, heat demand is shaped by the building shell (including walls, roofs, windows, doors, floors) and design choices made in the construction industry and its supply chains. Many UK buildings are poorly insulated and draughty, leading to relatively high heat demand (and heat losses).

Despite increasing numbers of homes, greenhouse gas emissions from residential buildings decreased by 29% between 2001 and 2014

Table 6.1. *UK heat consumption (in thousand tonnes of oil equivalent) in different building types in 2018 (constructed using data from BEIS Statistics; Energy Consumption in the UK; energy consumption by end use, Table U1)*

	Space heating	Water heating
Domestic	27,144 (58%)	7,040 (15%)
Services (public administration, and commercial & miscellaneous)	9,531 (20%)	1,349 (3%)
Industrial	1,872 (4%)	0
Total	38,547	8,389

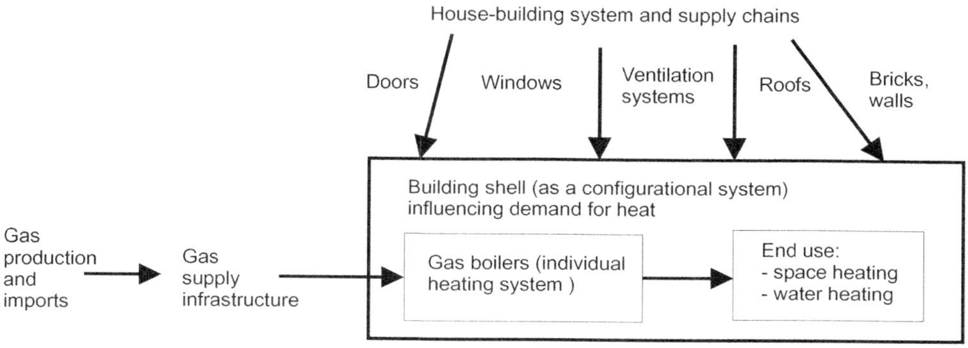

Figure 6.1 Schematic representation of the material elements and flows in the dominant UK (domestic) heat system (involving heating and buildings)

(see Figure 6.2).[1] These reductions mostly came from incremental efficiency improvements in gas boilers and the thermal performance of buildings (e.g., various insulation measures). Since 2014, emissions have stagnated and slightly increased, leading the Committee on Climate Change to warn that progress is not on track to meet climate change targets: 'The progress made on buildings remains insufficient even to meet the previous target for an 80% reduction in emissions relative to 1990 levels ... In order to go further to meet net-zero ambitions, bold and decisive action is urgently needed from Government' (CCC, 2019a). Further decarbonisation will require deeper and more fundamental changes in heating and buildings systems (e.g., low-carbon heat sources, very energy efficient buildings).

To further explain heat-related emission trends and assess potential options for deeper system reconfiguration, we will analyse the heating system (including fuel supply, distribution, on-site conversion, and heating practices) in Section 6.2 and the buildings system in Section 6.3. These systems are related to different actors,

[1] The number of UK dwellings increased by 14.9% between 2000 and 2018 (*Figure 6.12*).

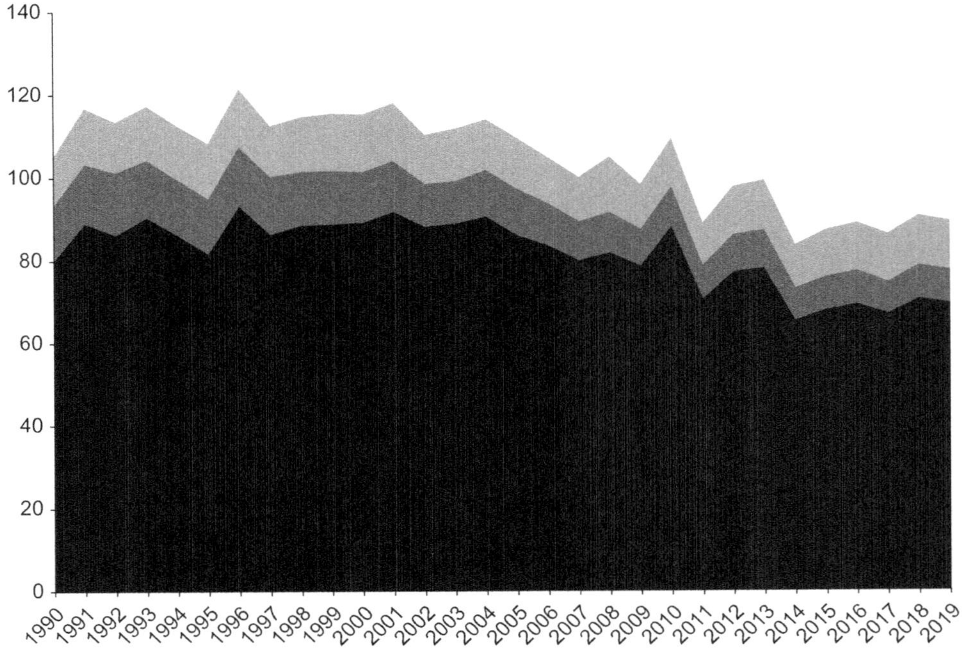

Figure 6.2 UK greenhouse gas emissions (in $MtCO_2e$) from buildings 1990–2019 (constructed using data from BEIS, 2020 Final UK greenhouse gas emissions national statistics)

technologies, and policies. We will then address several niche-innovations (heat pumps, biomass heating, solar thermal, greening the grid, heat networks and passive housing) in Section 6.4, and finally assess the speed and depth of low-carbon heat reconfiguration in Section 6.5.

6.2 Heating System

6.2.1 Techno-Economic Developments

Energy supply for UK heating is dominated by natural gas, which is distributed through a dense, centrally operated, piped network established from the 1960s through the integration of local networks and systematically expanded as the UK 'dashed' for gas (Arapostathis et al., 2014). Relevant technical operations include gas production (exploration, drilling), delivery and processing (at coastal terminals), gas transport, storage, shipping, supply, and third-party activities. The UK gas network involves (high-pressure) transmission and (medium- to low-pressure) distribution to end-users, though ownership and transport operations have been unbundled.

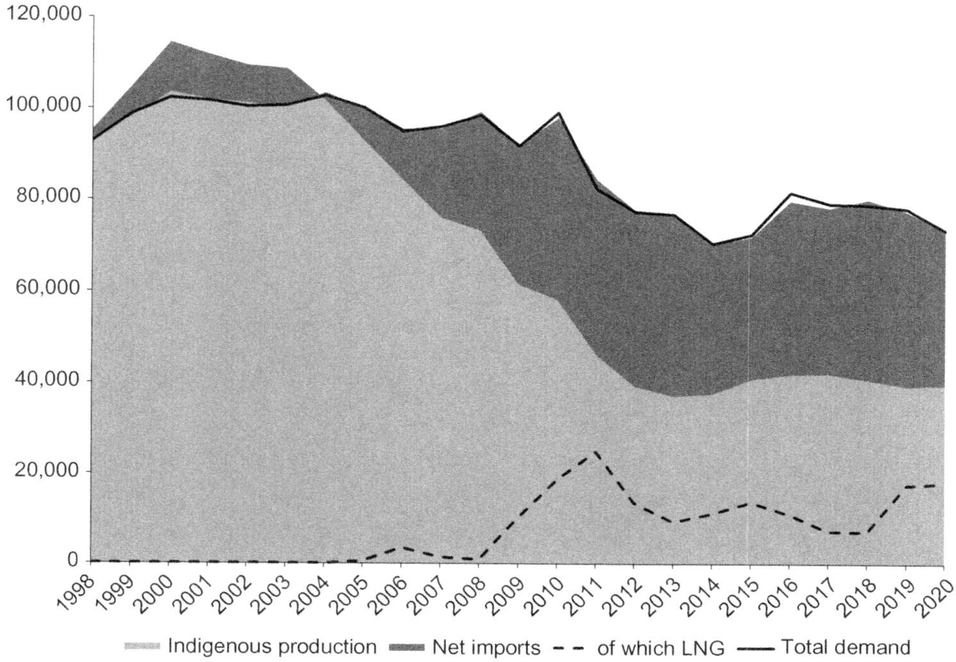

Figure 6.3 Gas production, imports, and demand in the UK in million cubic metres, 1998–2020 (Constructed using data from BEIS, Energy Trends, ET Table 4.1 Natural gas supply and consumption)

Domestic (North Sea) gas production peaked in 2000. Since then, gas imports (largely from Norway, but also LNG from further afield) have steadily increased (Figure 6.3). This was accompanied by investments in storage and (LNG) import facilities (Kopp, 2015), and a trend towards the diversification of sources to ensure energy security (Bradshaw et al., 2014), shorter contracts (for LNG), but also increasing dependence on global market fluctuations. The costly expenses associated with the creation of new facilities (storage, pipelines, controls) were passed on to users, but rising gas prices (Figure 6.4) were mainly caused by rising oil prices (to which gas prices are linked).

Heat conversion in the UK is highly decentralised and located at the point of consumption in individual home-based boilers and stoves. Most households are self-contained heating generation and consumption units. Gas-based central heating systems have been installed in 85% of homes, with oil, coal, electric storage, and heat networks making up the remainder (BEIS, 2018b). Biomass (pellets and woodchips), solar-thermal, and heat pumps are slowly emerging as alternatives, but mainly as additional sources of heat for environmental or aesthetical motives (see Section 6.4.2).

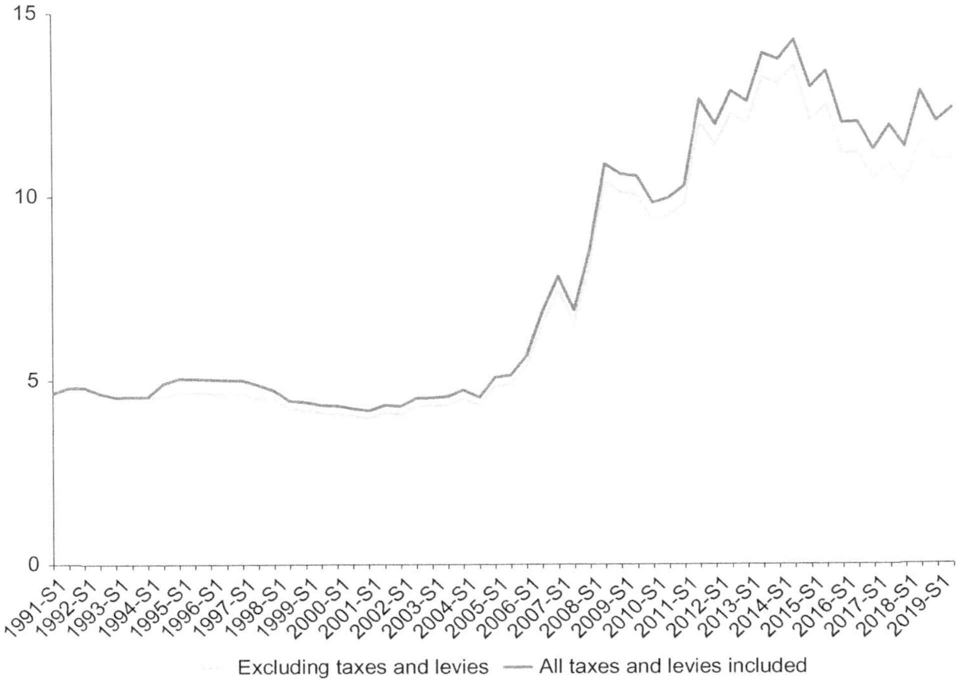

Figure 6.4 Evolution of average gas price (in GPB/GJ) for households in the UK, 1991–2019 (Constructed using data from Eurostat: Gas prices for domestic consumers – bi-annual data (until 2007) [nrg_pc_202_h], Gas prices for household consumers – bi-annual data (from 2007 onwards) [nrg_pc_202])

The number of homes without a boiler has decreased substantially since the 1970s, which was linked to the diffusion of central heating. Since the 1990s, there have also been shifts towards more fuel-efficient boilers, first from standard to combination boilers, and then towards condensing and condensing-combination boilers (Figure 6.5). These incremental boiler improvements helped to reduce domestic gas use (see Figure 6.6) and greenhouse gas emissions.

Energy use for ***residential space heating*** increased from the 1970s to the mid-2000s (Figure 6.6) because of the switch to gas-fired central heating systems (which enabled heating of multiple rooms in houses) and because of increases in the average internal temperature in UK homes (Figure 6.7), which was driven by desires for higher thermal comfort and the heating of multiple rooms (Chappells and Shove, 2005). From the mid-2000s, energy use for space heating started to decline, with the exception of an unusually cold winter in 2010 leading to a heat consumption spike (Figure 6.6). This decline was driven by a switch to more efficient boilers (Figure 6.5), stabilisation and slight decrease of internal temperatures (Figure 6.7), and domestic insulation improvements (discussed

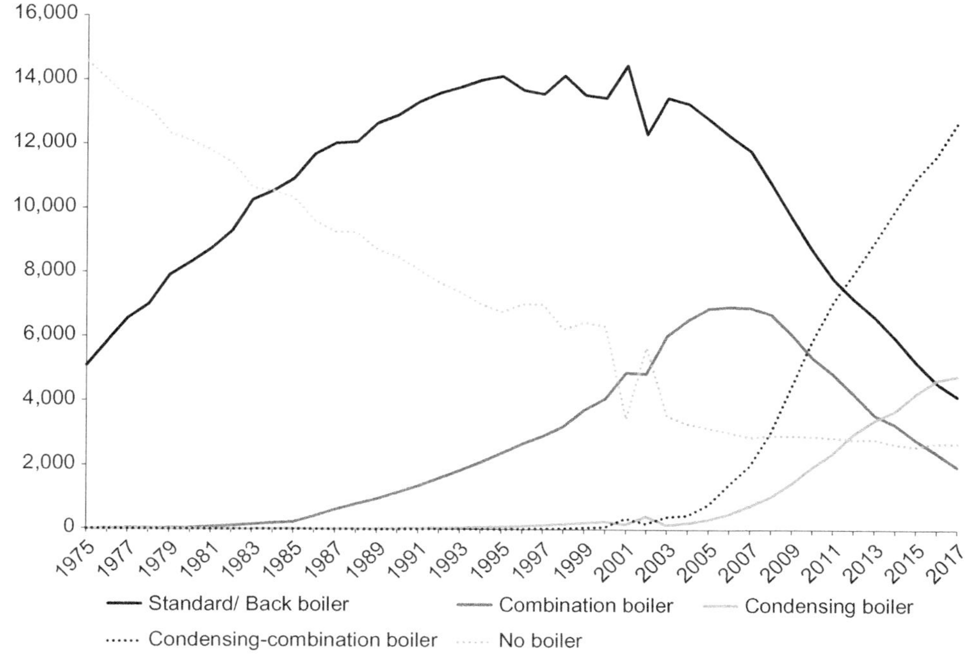

Figure 6.5 Boiler types in the UK, 1975–2017 (constructed using data from DUKES; Energy Consumption in the UK; Supplementary tables; Table S6)

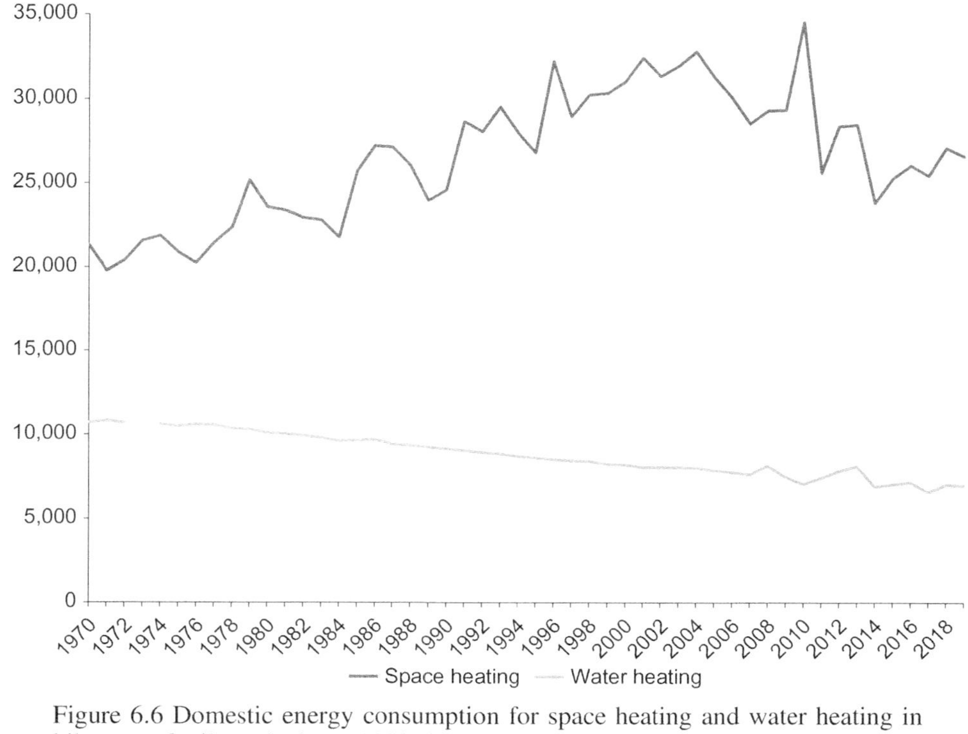

Figure 6.6 Domestic energy consumption for space heating and water heating in kilotons of oil equivalent, 1970–2019 (constructed using data from DUKES; Energy Consumption in the UK; energy consumption by end use; Table U3)

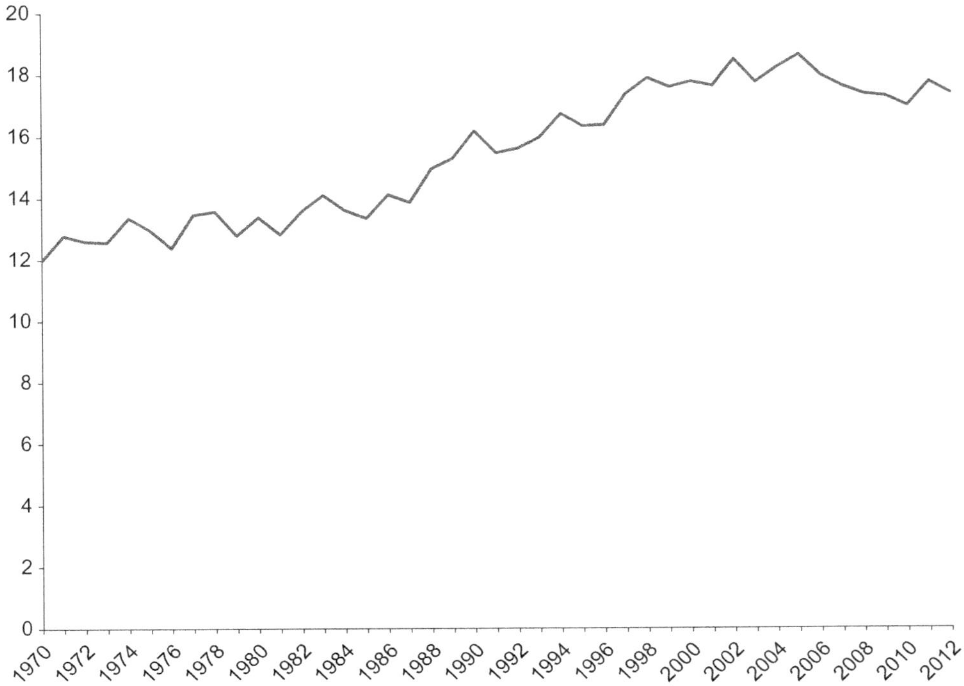

Figure 6.7 Average internal temperature in UK homes, degrees Celsius, 1970–2012 (constructed using data from DUKES; Energy Consumption in the UK; Supplementary tables; Table S3)

under buildings system). Energy use for water heating has declined steadily since the 1970s, as hot water storage tanks were replaced by other heating appliances.

6.2.2 Actors

Energy Supply Actors: The actor configuration for gas distribution is like that of electricity distribution; it is fairly concentrated around a large technical infrastructure and a regulated market. National Grid Gas owns the National Transmission System (NTS), while eight regional Gas Distribution Networks (GDNs) are owned by four operators (Cadent, Northern Gas Networks, SGN, and Wales & West Utilities) who, for a fee, distribute gas to households on behalf of gas supply companies. Privatisation and liberalisation processes in the 1990s culminated in the emergence of the 'Big Six' suppliers (Centrica/British Gas, E.ON, NPower, SSE, Scottish Power, and EDF), which dominated the domestic gas market (Figure 6.8) in a vertically integrated oligopoly until 2014. Since then, the emergence of new players (e.g., Bulb, Co-operative Energy, Green Star

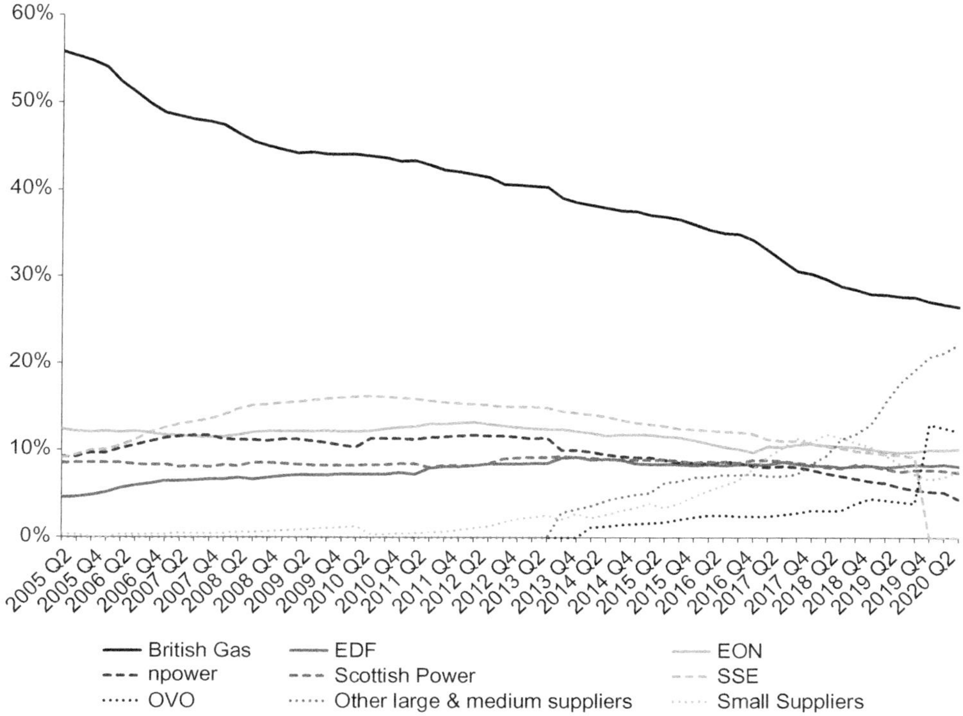

Figure 6.8 Domestic market gas supply market shares in GB, 2005–2020 (Source: Ofgem retail market indicators)

Energy, Octopus Energy, OVO Energy, Shell Energy) has led to increased competition and shrinking market shares of the Big Six (Figure 6.8), notably for British Gas.[2]

Gas industry actors face increasing economic pressures from recent decreases in gas demand (for heating and for power generation), substantial infrastructure costs, operational challenges to maintain supply security, and alternative low-carbon heating options that may increase to address climate change (National Grid, 2019). In this context, gas industry actors strategically seek to maintain their position of centrality with regards to heating. Indeed, most of the energy supply actors are strongly attached to the gas-based heating system, because of sunk investments in pipeline infrastructures, business models, market positions, and technical capabilities. Their historic centrality is also proving an important advantage for shaping heat decarbonisation pathways amid significant uncertainties and policy indecision (Lowes and Woodman, 2020).

[2] In 2020, SSE Energy Services and OVO merged, with OVO taking over gas supply activities.

For a long time, these actors paid limited attention to climate change. But since the mid-2010s, they have 'woken up' to potential threats of climate change and heat decarbonisation (Pearson and Arapostathis, 2017), because many low-carbon transition scenarios (by the government, Committee on Climate Change, or researchers) envisaged a smaller role for gas in UK heating and a greater role for low-carbon alternatives such as heat pumps, heat networks, or biomass (discussed further later). The Energy Networks Association (ENA), which represents interests of transmission and distribution operators, has therefore started to use lobbying and framing strategies that propose alternative visions of the future, which advocate for a continued role for gas infrastructures, notably around arguments of security and affordability in the face of uncertainties, claiming that 'Decarbonising gas – reducing the carbon emissions linked to its use – is the least disruptive way of delivering a cleaner, greener future' (ENA, 2018: 10). These visions and scenarios include continued use of natural gas (via hybrid heat pumps or energy efficiency improvements) but also large-scale distribution of biomethane (which is currently practised in demonstration facilities and requires little to no adjustments to infrastructure and appliances, but poses significant scaling challenges) or hydrogen (which is less developed and more technically challenging) through repurposed gas grids (Lowes et al., 2020; Richards and Al Zaili, 2020; Speirs et al., 2018, see also Section 6.4.1). An ENA-commissioned report by KPMG (2016) suggested that a heat transition based on the 'evolution of gas' relying mainly on conversion of the gas grid to hydrogen would be as much as three times cheaper than other scenarios. These defensive strategies have increased the uncertainties that policymakers face with regard to potential low-carbon heat transition pathways (Lowes and Woodman, 2020).

Appliance Manufacturers and Installers: Concerning heat conversion, there are many appliance manufacturers (e.g., boilers, radiators, controls) and installers (e.g., plumbers, builders, heat technicians). More than twenty companies supply gas boilers, but four international companies (Baxi, Worcester Bosch, Vaillant, and Ideal) dominate the UK market (DECC, 2013a: 74). While appliance manufacturers have remained committed to gas-dependent heating,[3] producers of other heat-related products (e.g., cylinders, radiators, controls) are more versatile and hence less resistant to significant changes in the sector (Lowes et al., 2018). The Energy and Utilities Alliance (EUA), which represents the appliance

[3] Because heat pumps rely on technical components issued from the cooling and ventilation sectors, they are primarily produced by specialised companies or within electronics conglomerates. Thermal heating appliance manufacturers in the UK have to date not shown much interest in this technology and remain a separate manufacturing segment. This is likely to change over time, as prefigured by the Vaillant group's involvement in thermal appliances and heat pumps.

manufacturers, ran an advocacy campaign on ‘Green Gas’ that also advocated continued use of existing gas infrastructure:

> Rather than rip out heating systems and make the grid obsolete, it makes sense to decarbonise the gas we use; using so called green gases such as Biomethane and bio SNG, in addition to hydrogen, will deliver affordable and sustainable solutions to the challenges the UK faces. (EUA website www.eua.org.uk/green-gas, accessed December 2020)

There are also many heating equipment installers in the UK (e.g., builders, plumber, heating engineers) whose skills are mainly tied to conventional gas-fired boilers and central heating systems. While there are 74,000 registered Gas Safe registered businesses (representing over 100,000 gas engineers) in the UK (Gas Safe Register, 2017), the number of installers certified under the Microgeneration Certification Scheme (MCS) never exceeded 4,000, most of which were specialised in the installation of solar PV rather than heat pumps or biomass boilers (Hanna et al., 2018).

The community of gas heating installers is concerned that a transition towards renewable heat or heat networks may disrupt their livelihood and knowhow (Gas Safe Register, 2017) and would require significant re-training to address the skills gap:

> Heating engineers are naturally familiar with [conventional gas-fired] technology, meaning that a domestic boiler can be bought and installed in under a day. Low carbon heating technologies do not have these advantages, being relatively new to the market. With greater complexity and longer installation time, such technologies tend to be far more expensive. The high price of alternative heating systems is a clear barrier to uptake. (DECC, 2013a: 81)

The existing skills of installers thus help to lock-in the gas-based heating system, while a transition to low-carbon heat systems is hampered by the lack of technical installation skills:

> Growth in the low carbon heat sector will lead to new installer jobs, and upskilling of existing jobs as existing gas installers cross-train. (DECC, 2013a: 74)

As ‘middle actors’ between appliance manufacturers and users, heating installers play important mediating roles through the advice they provide to users (Wade et al., 2016), often in periods of stress when the existing boiler has broken down. Because they know more about gas-fired boilers, and have their own views on the reliability and performance of different brands, they tend to advise households to purchase conventional systems (Wade et al., 2016).

Policymakers: Interventionist supply-side heat policies of the 1960s and 1970s, which focused on the transition from coal to gas and central heating (Pearson and Arapostathis, 2017), were followed by privatisation, liberalisation, and a more hands-off approach in the 1980s and 1990s, focused on increasing competition and lowering prices. British Gas was privatised in 1986 and broken up in 1997 to create Centrica and Transco. Freedom of supplier choice was established for all customers from 1998, leading to more competition, overseen by the regulator (Ofgas, later Ofgem).

Affordability became a salient political issue in the mid-2000s, when rising gas prices and higher heating bills led to critical debates about affordability and energy company pricing strategies. These debates led Ofgem to emphasise its role as maximising access to affordable energy for all (Ofgem, 2019).

Energy security also became an important issue in the mid-2000s (Kuzemko, 2014), when increasing reliance on gas imports strengthened concerns about vulnerability (especially when Russia closed gas supplies to Eastern Europe in 2005), price stability, and so on. This led to 'considerable debate as to whether the UK Government should incentivise new storage, but the current position is to leave it to the market. The net result being that very little new storage capacity has been built' (Bradshaw, 2018: 3).

Energy poverty reduction has been a policy goals since the 1990s, because many vulnerable social groups (e.g., elderly, benefit claimants) struggled to afford decent heating. Winter fuel payments for elderly people were introduced in 1997. Policymakers also introduced the Warm Home Discount (for low-income pensioners) in 2008 and increased Cold Weather Payments (for benefit claimants) in 2009.

Heat remained a relatively unproblematic and invisible part of climate policy until 2008, when the Climate Change Act triggered debates about the decarbonisation of heat. Since then, the strategic policy visions on preferred options and transition pathways have changed several times (Winskel, 2016), focusing initially on heat pumps, then on heat pumps and heat networks, and recently also on hydrogen and biomethane in gas grids.

The various visions were not translated, however, into concrete strategies or plans for action, leading the Committee on Climate Change to lament in 2016 about a lack of direction and delivery: 'Progress to date has stalled. The Government needs a credible new strategy and a much stronger policy framework for buildings decarbonisation over the next three decades' (CCC, 2016: 7). In 2019 the Committee still concluded that UK 'buildings and heating policy continues to lag behind what is needed' (CCC, 2020: 19) to reach emission reduction targets.

For policymakers, climate change thus appears to have been of less importance than other heat-related issues such as affordability, energy poverty, and energy security.

Users: There are around 29 million homes in the UK, corresponding to a slightly lower number of heating installations, given that a small number of homes are supplied by block- or district-level heat networks. Domestic heating has relatively low cultural visibility and consists of a relatively undifferentiated product, with largely routinised and taken-for-granted practices and low levels of user engagement. The operation of heating systems requires little competence, and basically consists of pushing a button or adjusting a thermostat. Conventional heating practices are passive (Hope et al., 2018), and domestic residents primarily demand comfortable, hassle-free, and affordably heated homes regardless of the heat source (CCC, 2016). From a user-perspective, there are few current incentives to shift away from gas:

> Surveys report that customer satisfaction with natural gas boilers is extremely high. Gas-fired central heating is affordable, provides very high levels of thermal comfort, responds quickly when customers adjust the required temperature, and is convenient and familiar to both householders and installers. (Gross and Hanna, 2019: 358)

Users give relatively little thought to their heating systems, which are linked to routine behaviour. DECC (2013a: 82) reports that 'the majority of homeowners would only consider replacing their heating system if it needed significant repairs or services'. Homeowners who rent their properties to tenants also have limited incentives to upgrade their heating equipment. Nevertheless, households have steadily adopted more energy-efficient boilers (Figure 6.5), which is partly due to public policies that have phased out inefficient boilers (discussed in Section 6.2.3), and partly due to anticipated cost savings associated with reduced gas use. Combined with insulation measures (discussed in Section 6.3.1), this has resulted in average reductions in household gas use.

Concerns over rising costs have led to more user switching between energy providers, which can lead to substantial savings. Since 2014, annual gas supplier switching rates have increased, reaching over 20% of customers in 2019 (Ofgem, 2019).

Most users have few motivations to adopt more radical low-carbon innovations such as heat pumps, biomass boilers, or solar thermal options (Balcombe et al., 2014). The Renewable Heat Incentive (RHI) sought to encourage the uptake of these technologies but has had only limited effect and largely benefited more affluent users and homeowners. Furthermore, user trust with regards to the quality of installers is an additional structural barrier to wider uptake of radical innovations. The industry quality insurance scheme (the Microgeneration Certification Scheme) does not appear to have substantially addressed these concerns (Hanna et al., 2018), notably due to poor inspections and the recurrence of sub-optimal installations.

Wider Publics: Heating is not a highly visible issue in public debates, because it remains rather abstract and technical. If there are societal debates about heating, these are mainly about costs (because of rising energy prices) and energy poverty. There is little discussion about the environmental implications of heating, which is remarkable because households use much more energy for space heating than for electricity.

Civil society activity on heating is primarily oriented towards raising awareness, putting issues on the policy agenda, and maintaining pressure for more ambitious policies. NGOs have been most vocal on the issue of fuel poverty, which is a particular form of inequality and injustice. NGOs dedicated to fuel poverty include National Energy Action, National Right to Fuel Campaign, Fuel Poverty Action, Energy Bill Revolution, and End Fuel Poverty Coalition.

Several more generalist organisations seek to promote low-carbon heating by influencing policy, supporting the development of supply chains, or enabling informed user choices. For instance, the Centre of Sustainable Energy manages sustainable energy projects, and the Association for the Conservation of Energy represents the energy conservation industry and lobbies for more stringent energy efficiency policies. The Green Alliance is a green think tank that published a Manifesto for Sustainable Heat (Green Alliance, 2007), which contributed to raising policy awareness of the need to develop a long-term heat strategy and supporting a wider portfolio of options. The Energy Saving Trust offers information and user advice about low-carbon heating and home insulation options.

6.2.3 Policies and Governance

Formal Policies

Since the 1990s, successive governments have introduced a range of policy instruments to address fuel poverty and financially assist vulnerable groups in cold periods (Table 6.2).

Table 6.2. *Direct financial support for users in vulnerable situations (Ofgem, 2019: 107)*

Policy	Eligible	Recipients (winter 2017–2019)	Payments to individuals, nominal (£)	Total cost, 2018 prices (£m)	Funding source
Winter Fuel Payments	All pensioners	11.8 million individuals	£100 to £300	£2,055	Central government
Warm Home Discount (core group)	Low-income pensioners	1.2 million individuals	£140	£173	Energy bill payers
Warm Home Discount (broader group)	Consumers on a low income and vulnerable to fuel poverty	0.6 million individuals	£140	£161	Energy bill payers
Cold weather payment	3.8 million benefit claimants	4.7 million payments	£25 for each cold week of weather	£121	Central government

European and UK policymakers also played an active role in supporting the diffusion of more efficient gas boilers and appliances, which led to substantial yet incremental changes in existing technologies (Figure 6.5). The European Boiler Efficiency Directive (1992), for instance, mandated minimum performance standards, which led to the phase-out of boilers with efficiencies under 70% (G-rated), though it excluded back boilers. Efficiency standards were ramped up in the Building Regulations revision (2000), requiring a phase-out of boilers with efficiencies under 78% (D-rated) by 2002. The 2005 Building Regulations, brought forward by the Energy White Paper of the same year, further raised efficiency standards, requiring a phase-out of boilers with efficiencies under 86% (B-rated). This effectively mandated a switch to condensing boilers. In 2018, the Boiler Plus Standard further mandated new boilers to comply with a minimum efficiency of 92% as well as to include time and temperature controls.

Besides tightening standards, policymakers also used other instruments to stimulate the diffusion of more efficient gas boilers. Between 2009 and 2010, the Boiler Scrappage Scheme provided a £400 incentive for 125,000 households to upgrade from the least efficient, G-rated boilers to new high efficiency boilers (Dowson et al., 2012). The government also ran several programmes that placed energy savings obligations on energy suppliers, which required them to make energy efficiency improvements in households (Rosenow, 2012). The third Energy Efficiency Standards of Performance programme (2000–2002), the Energy Efficiency Commitments (2002–2008), the Carbon Emissions Reduction Target (2008–2012), the Community Energy Saving Programme (2009–2012), and the Energy Company Obligations (from 2013 onwards) all stimulated energy suppliers to assist the diffusion of efficient gas boilers as well as other measures (e.g., insulation, lighting, appliances).

Policy engagement with more radical heat innovations has been much more fragmented, patchy, and short-lived, however. The Renewable Heat Premium Payments (2011–2014) offered a single payment to assist households with the purchase of renewable heating technology (e.g., solar thermal panels, heat pumps, biomass boilers). It was replaced by the domestic Renewable Heat Incentive (RHI) (from 2014 onwards), which provides varying subsidies for alternative heating systems based on renewable sources (e.g., biomass boilers, heat pumps, deep geothermal, solar thermal). This scheme has been criticised as ineffective, narrowly targeted, and under-delivering on emission targets (NAO, 2018). It also 'provided a disproportionate incentive for domestic biomass boilers compared to other technologies' (CCC, 2020: 100). Although the RHI-scheme offered generous returns, roll-out remained limited and less than £100 million was spent in 2018 (CCC, 2019b: 29). This was because the RHI was 'not supported by a package of measures to encourage and enable customers to make changes easily' (CCC, 2020:

96). The reliance on a single (financial) instrument has thus remained ineffective in boosting renewable heat options.

These limited and relatively ineffective instruments have created major discrepancies between actual policy delivery and the far-reaching visions of heat decarbonisation transition that have been advanced since 2009. Already in 2016, the Committee on Climate Change (CCC) warned that: 'The existing set of policies is not an effective overall package for decarbonising heating' (CCC, 2016: 13). In 2018, the CCC commended the vision of the 2017 Clean Growth Strategy but also diagnosed that it contained 'few new specific policies to deliver real emissions reduction' (CCC, 2018a: 16). And in 2020, the CCC welcomed extensions of RHI and plans for a Green Gas Levy (to support green gas) but also assessed that 'the current plans are far too limited to drive the transformation required to decarbonise the UK's existing buildings' (CCC, 2020: 20). It therefore urgently suggested that 'policy needs a step change in ambition and delivery this year' (CCC, 2020: 21).

There have been some recent policy announcements, but these remain relatively limited and do not yet add up to a step change in ambition and delivery. In 2019, the government announced plans for a ban on gas boilers in *new* homes from 2025, but this does not apply to replacement of boilers in *existing* homes, which is a far greater market, and has since then been reformulated as a rather vague ambition in the Energy White Paper (HM Government, 2020a):

> We will [. . .] consult in early 2021 over new regulations to phase out fossil fuels in off-grid homes, businesses and public buildings, including a backstop date for the use of any remaining fossil fuel heating systems. (p. 110)

In June 2020, the government awarded £14.6 million for the Electrification of Heat Demonstration Project, but the heat pump trial remains limited to 750 homes. There are also plans to replace the *non-domestic* RHI in 2021 with a Green Gas levy to further support the deployment of biomethane in the existing gas grid. More substantially, perhaps, the government launched the £320million Heat Networks Investment project in 2018 that supports local governments, firms, and third sector organisations in building and operating heat networks in areas of denser heat demand. The first projects to receive funding were announced in February 2020, but overall funding in this round was limited to £40m.

Although incremental boiler improvements in the 1990s and 2000s were achieved with a mix of policy instruments (e.g., efficiency standards, supplier obligations, financial incentives), the governance style with regard to low-carbon heat transitions seems to have narrowed in the 2010s to a technology-neutral, market-based approach (relying mainly on incentives and information): 'Across all the different heating strands, the Government wants to make progress without prescribing the use of specific technologies. Instead, information for market players, including households and businesses, should be improved to enable effective decision-making' (DECC, 2013a: 79).

Governance Style

In terms of governance style, heating has largely been approached as a demand-side technical performance problem to be addressed through efficiency improvements at the point of heating. Mandated performance levels and minimum standards have been rather effective in improving the efficiency of fossil heating appliances, but the potential for further incremental improvements is limited. Aside from clarity on performance standards, low-carbon heat policy has been marked by hesitancy and a lack of coherence.

The reliance on isolated market modulation mechanisms (e.g., the RHI) has, so far, been ineffective in driving low-carbon heat transitions, because instruments tended to be fragmented, short-lived, and not part of a wider policy mix. Transition governance has been weak and the successive changes in long-term visions have not provided clear directionality to support innovation, market formation, supply-chain development, or the build-up of relevant skills and competences on the required scale. Strategic visions (e.g., 2012, 2013, 2020), although multiplying in recent years, were also not translated into concrete policies and instruments, leading to lack of delivery. These strategic visions were also shaped by various influence groups, leading to a lack of coherence concerning technological preferences (Broad et al., 2020).

A new Buildings and Heat Strategy was expected in 2020 but has been delayed, and it remains to be seen if this will provide more clarity about future directions of travel and back these up with effective policy instruments. The general heat policy objectives that such a strategy needs to meet have been spelled out in the latest Energy White Paper (HM Government, 2020a) and Ten Point Plan (HM Government, 2020b), but many uncertainties remain concerning delivery on these announcements. Repeated delays in publishing the new Buildings and Heat Strategy are indicative of disagreement about implementation and feasibility.

6.3 Buildings System

6.3.1 Techno-Economic Developments

The demand for domestic space heating is strongly influenced by the insulation properties of the building shell, which are poor in the UK compared to other countries. The UK housing stock is relatively old: most houses were built before 1960 (Figure 6.9). Houses also have high material obduracy and are deeply

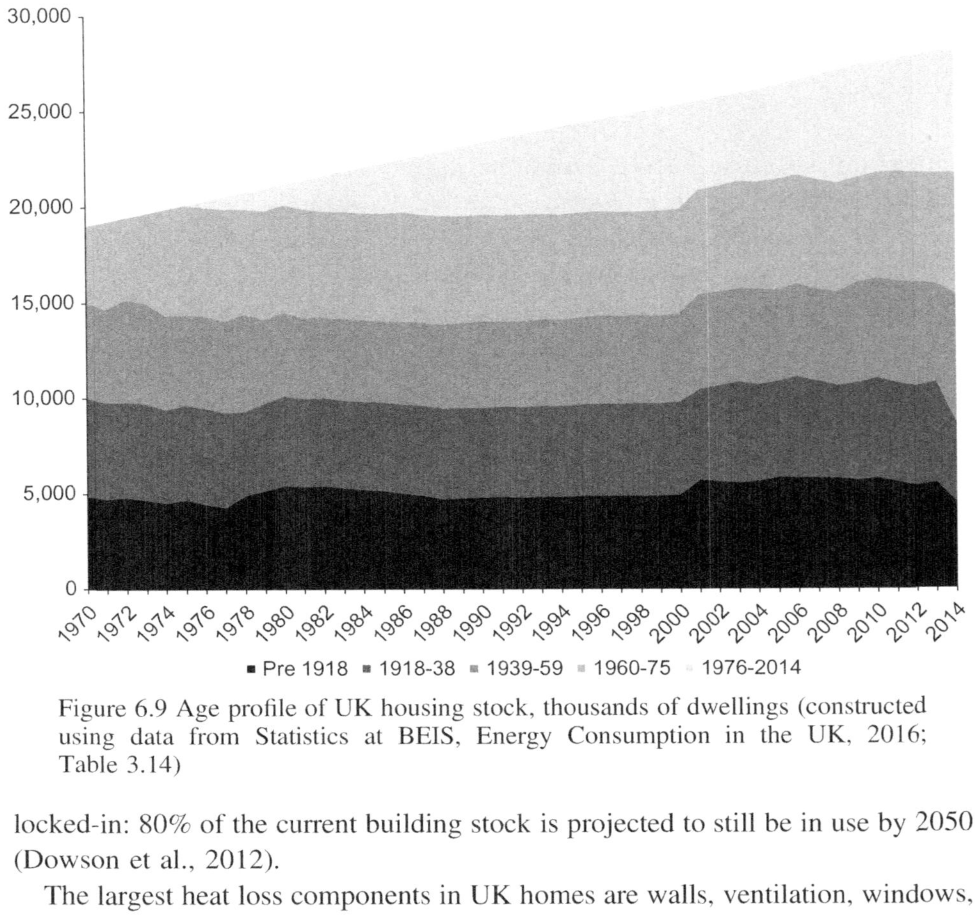

Figure 6.9 Age profile of UK housing stock, thousands of dwellings (constructed using data from Statistics at BEIS, Energy Consumption in the UK, 2016; Table 3.14)

locked-in: 80% of the current building stock is projected to still be in use by 2050 (Dowson et al., 2012).

The largest heat loss components in UK homes are walls, ventilation, windows, and roofs. Between 1970 and 2008, the average heat loss per dwelling was reduced by about 33% (Figure 6.10) through a range of incremental improvements in existing houses. Most houses with lofts now have some degree of loft insulation (although the depth of insulation varies). Since the 1970s, the percentage of houses with some degree of double glazing also increased very substantially, to 96% in 2016 (Figure 6.11), although this does necessarily imply that all windows are double glazed. The percentage of houses with some degree of cavity wall insulation increased from close to zero in the 1970s to 69% in 2016 (Figure 6.11). Solid wall insulation, which is expensive and 'hard to treat', has been limitedly applied, however. While the diffusion rates of these insulation techniques are impressive, these have often been applied in piecemeal fashion rather than with a whole-house approach, which presents significant limits as to the efficiency improvement potential.

The number of UK dwellings has increased steadily to 28.9 million in 2018 (Figure 6.12). Between 2014 and 2019, 1,095,870 new dwellings were built

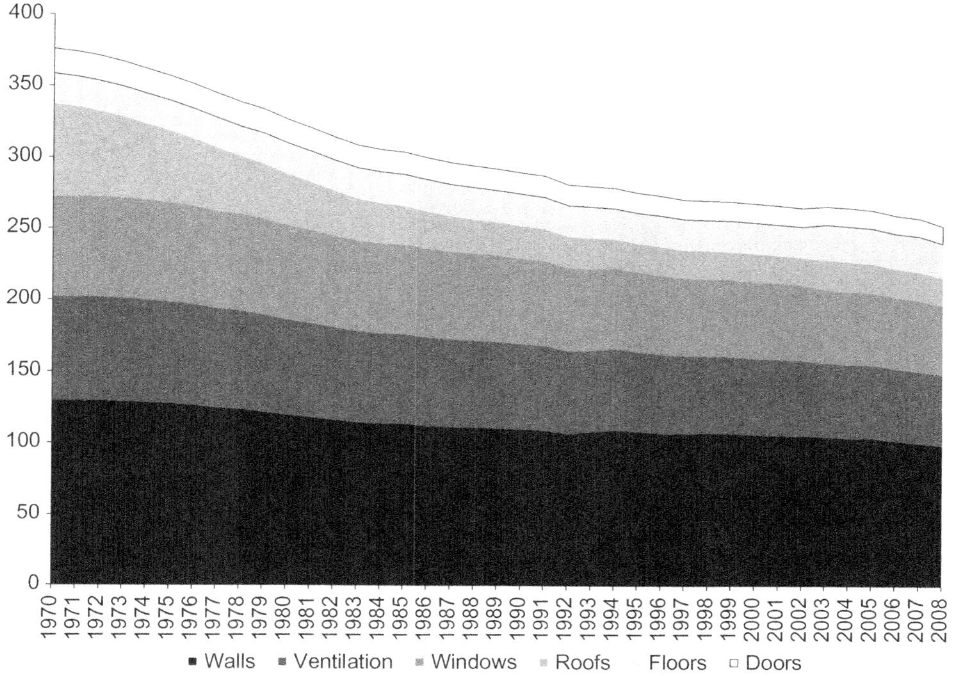

Figure 6.10 Average heat loss per dwelling in W/°C (constructed using data from DECC (2012a))

(Figure 6.15), representing a 4% building stock increase. Most UK homes are owner-occupied (outright owned and owned with mortgage). Social renting (from local authorities or housing associations) has declined strongly since the 1980s' Right to Buy policy (Figure 6.12).[4] Private renting has increased steadily since the turn of the century. Penetration rates of insulation measures vary greatly according to tenure type, with privately rented homes ranking poorer on average. Privately rented homes are less likely to benefit from cavity wall insulation, loft insulation, or double glazing (Figure 6.13), owing to split incentives problems.[5]

The deployment of incremental insulation measures in the existing building system (Figure 6.11) helped to reduce heat demand from the early 2000s (Figure 6.6) and contributed to decreasing greenhouse gas emissions from UK buildings (Figure 6.2). Most of the thermal efficiency improvements have, so far, been piecemeal and incremental rather than leading to radical changes in building methods or building stock (e.g., whole-house retrofits). They have also focused on

[4] The Right to Buy policy gave eligible people who live in council properties in England the right to buy the home they live in at a large discount.

[5] Homeowners have limited incentive to insulate rented properties since they bear the costs, while tenants enjoy the benefits.

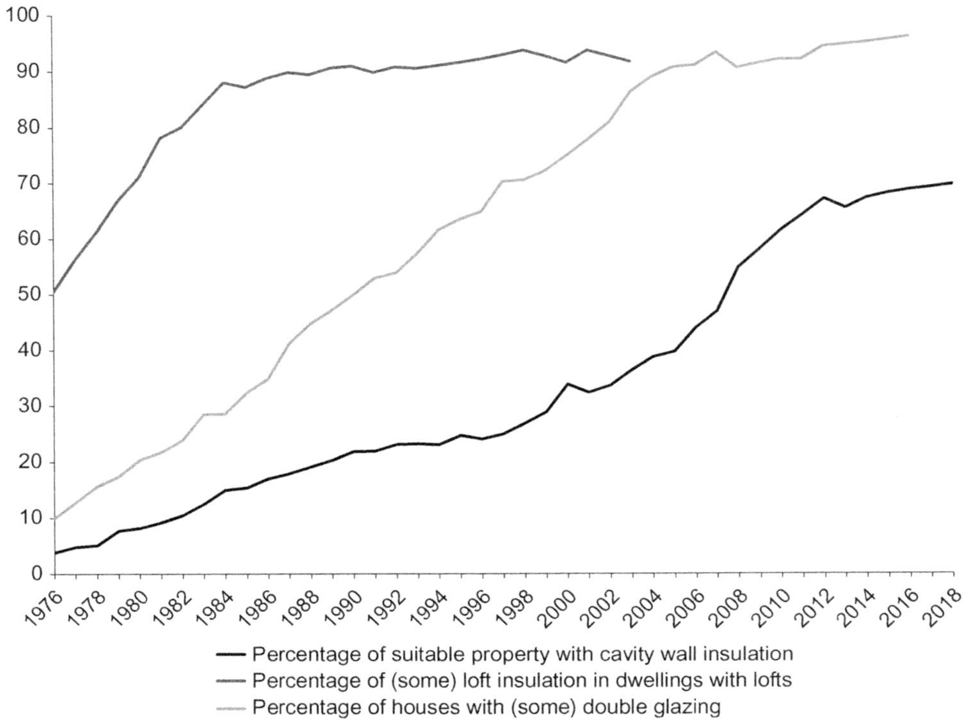

Figure 6.11 Diffusion of different home insulation measure in UK homes, 1976–2018 (constructed using data from DUKES; Energy Consumption in the UK; Supplementary tables S9, S10, S11)

relatively easy and inexpensive measures (so-called low-hanging fruits). This means there is still large potential for further thermal insulation improvements as the government also noted in 2013: 'There are hardly any homes with no insulation, but more than two thirds of the stock still has insufficient insulation by modern standards' (DECC, 2013b: 51). Nevertheless, recent policy changes, which are further discussed in Section 6.3.3, have made this problem more acute because they led to a collapse in delivery rates of key insulation measures (Figure 6.14).

Since the late 1960s, the housebuilding sector has also experienced substantial decreases in the number of new homes that have been built (Figure 6.15). This is particularly due to major declines in social housing construction by local authorities and a slump in private enterprise construction after the 2008 financial-economic crisis (Figure 6.15), which only recently has begun to recover. These trends helped to create serious housing shortages and limited the role of the public sector in driving low-carbon innovation and renovation. The houses that were built in the past decade do not meet high insulation standards, which not only suggests

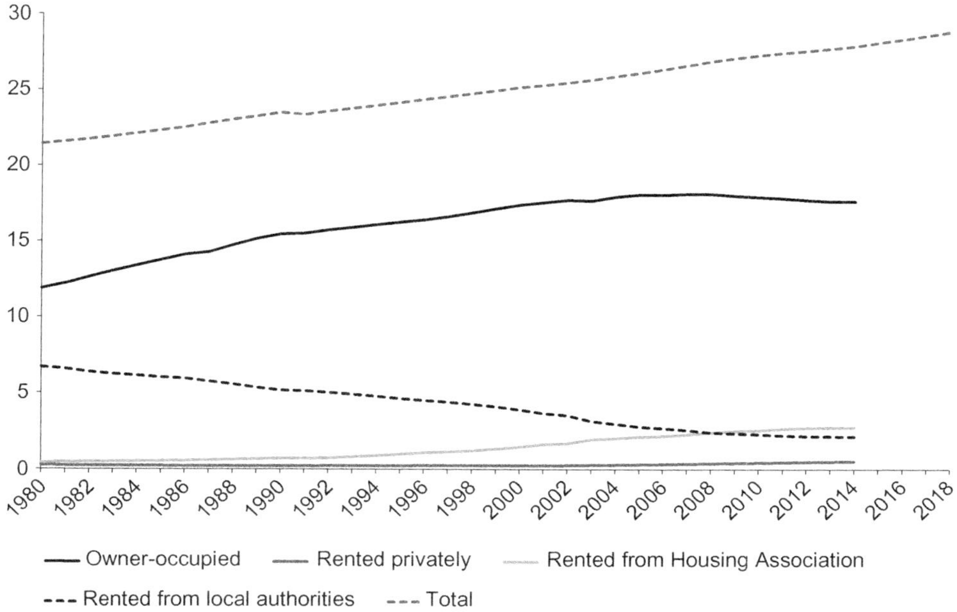

Figure 6.12 Total number of UK dwellings (in millions) and per tenure category, 1980–2018 (constructed using data from Statistics at Ministry of Housing, Communities & Local Government (MHCLG); Live tables on dwelling stock; Table 101,[6] Office of National Statistics Dwelling Stock by Tenure, UK)

that the housing crisis has been a more pertinent issue than climate change but also that future adjustments may be needed: 'Since the Climate Change Act was passed, nearly two million homes have been built that are likely to require expensive zero-carbon retrofits and have missed out on lower energy bills' (CCC, 2020: 19).

6.3.2 Actors

Housebuilding Sector: The UK housebuilding sector is diverse and includes contractors, developers, architects, builders and engineers, and many suppliers of specialised materials (e.g., doors, windows, insulation, construction materials). The UK house building sector is economically substantial (Table 6.3).

The sector is dominated by volume housebuilders, who minimise risk and maximise profit margins through the use of Standardised Design and Production Templates and the contracting out of physical construction to local builders (Barlow, 1999; Lees and Sexton, 2014). In 1960 the top 10 housebuilders

[6] This table was discontinued in 2014, because Northern Ireland data were no longer updated.

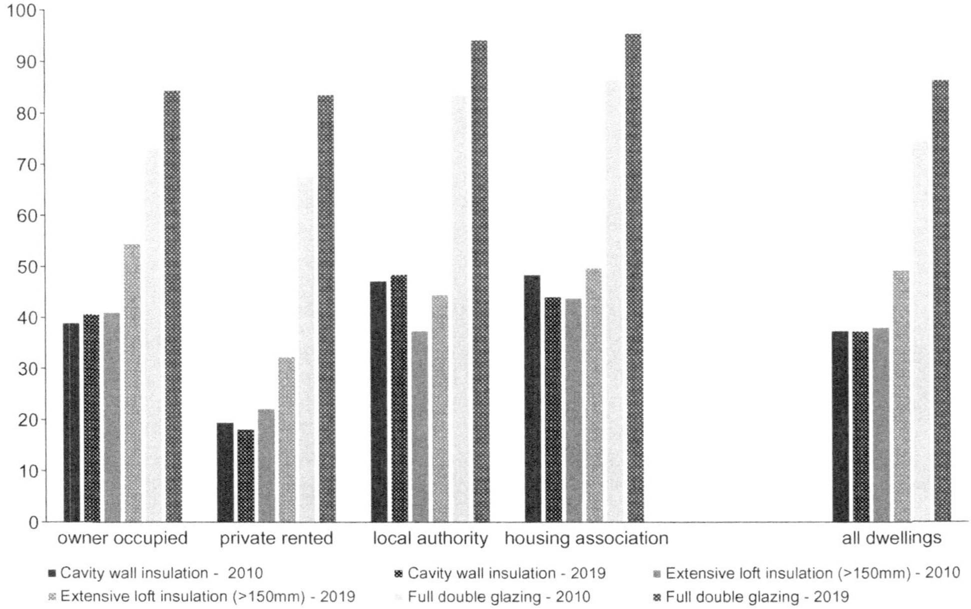

Figure 6.13 Insulation measures in English houses by tenure, 2010 and 2019 (constructed using data from English Housing Survey Tables, Table DA6201 (SST6.4): Insulation – dwellings)

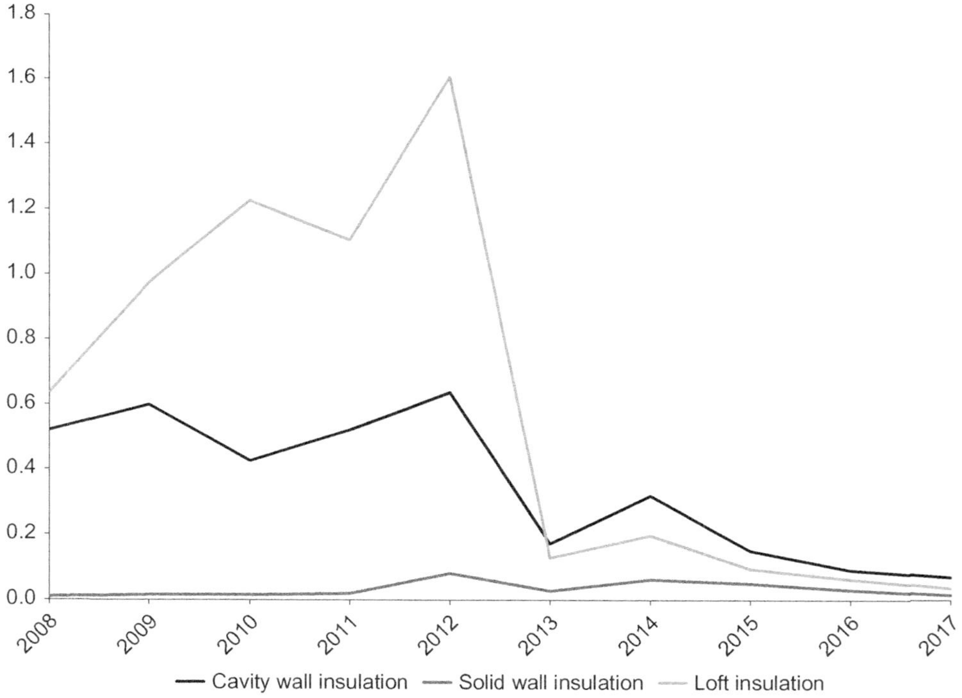

Figure 6.14 Delivery rates of key insulation measures in UK houses, in millions of installations (constructed using data from Committee on Climate Change (2018a))

Table 6.3. *House building sector in the UK, 2017 (HBF, 2018)*

	Total
Economic output (£ billion), including builders, contractors, suppliers, excluding induced economic output	38
Employed in on-site building	239,000
Indirect employment supported	119,500–186,420
Induced employment supported	174,470–272,270

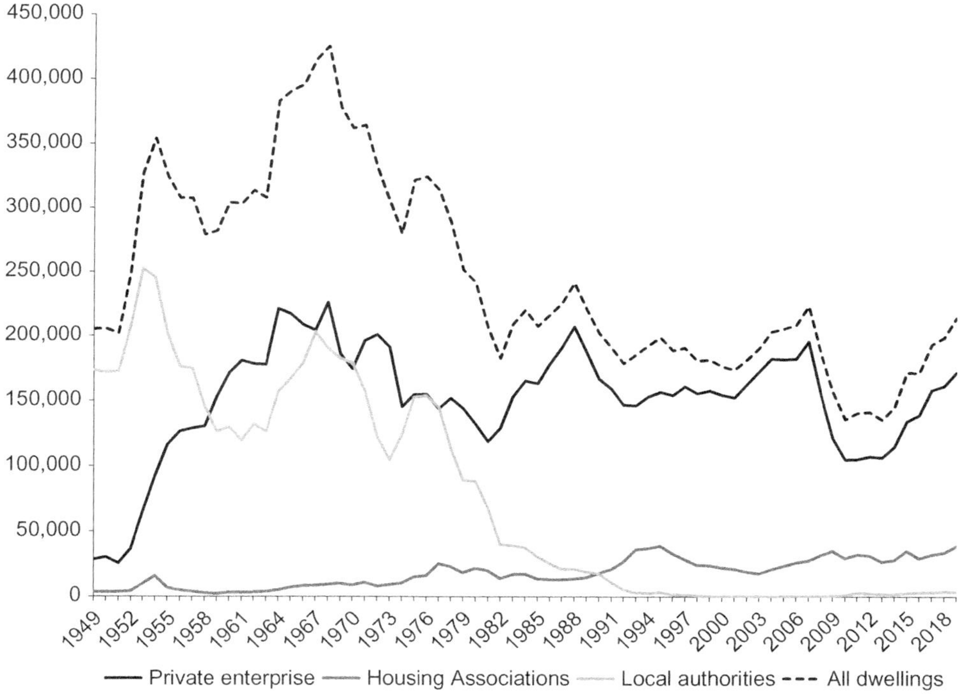

Figure 6.15 Number of permanent dwellings completed, by sector, United Kingdom, 1949–2019 (constructed using data from the Office for National Statistics; UK Housebuilding dataset; historical calendar year series, Table 3A)

accounted for about 9% of all new housing production. By 2004 this had increased to 46% (Archer and Cole, 2014). By 2015 large housebuilders (building over 2,000 units per year) had increased their market share to 59% (DCLG, 2017). This increasing concentration in the sector has enhanced the economic and political power of large housebuilders, who also make large profits through 'land banking', which is the practice of buying land and getting planning permission but then delaying actual construction to benefit from increasing land prices (Ryan-Collins et al., 2017). The structural power of large private housebuilders also increased because of changing market structures, particularly the virtual collapse of local authority commissioned buildings since the 1970s (Figure 6.15), which increased the influence of the private sector.

When building regulations started to address energy saving in the mid-1990s (Raman and Shove, 2000), most volume housebuilders took a compliance-only approach, focused on incremental changes that stayed within the Standardised Design and Production Templates (Lees and Sexton, 2014). Only a few incumbents (e.g., George Wimpey Ltd., Crest Nicholson) experimented with low-energy innovations in the early 2000s, but the experiences did not lead to strategic reorientation or commitment. Measures to reduce energy demand continued to be viewed as an add-on to existing practices rather than encouraging a change in building design (Smith, 2007).

When, in 2006, the Labour government introduced the Zero Carbon Homes (ZCH) target (which would see all new homes zero-carbon by 2016), housebuilders responded in two ways. On the one hand, they more seriously started to explore radical technical house designs that would be needed to meet this target. Several incumbent firms investigated the possibility of building to the Passivhaus standard, while others (e.g., Barratt, Stewart Milne) worked with the Building Research Establishment on test houses that met the highest level of the Code for Sustainable Homes (CSH) (Lynch, 2014).[7] The industry also funded the Zero Carbon Hub (created in 2008), which engaged in research and technical demonstration, provided advice, hosted events to create forums for discussion, and aimed to strengthen coalitions between industry and government. These learning processes and coalition building efforts were relatively weak, however, and did not generate widespread commitment in the housebuilding sector (Heffernan et al., 2015).

On the other hand, the industry mounted a sustained counter-lobby, which by 2011 succeeded in the watering down of the definition of 'zero-carbon', and by 2015 succeeded in the removal of the ZCH target altogether (Gibbs and O'Neill, 2015). One reason the industry's lobbying activities succeeded was that the Conservative-Liberal Democrat coalition (2010–2015) and the Conservative (post-2015) government felt less committed to the Labour government's target. Another reason is that the 2008 financial-economic crisis and subsequent recession and austerity politics changed the economic and political priorities, leading to more concern about growth, business profitability, and reducing regulatory constraints: 'While the government wanted to speed up house building amid concerns about housing affordability and economic recovery, the ZCH agenda was viewed as a

[7] The CSH, which was introduced a voluntary standard in 2006, rates the environmental performance of houses on a wider range of criteria, including energy, CO_2 emissions, water, materials, surface water run-off, waste, pollution, and health.

threat to productivity rather than an opportunity to place the UK at the forefront of green building and retrofitting' (O'Neill and Gibbs, 2020: 126). Aligning their lobbying strategies to these changed political concerns, 'industry actors like the Home Builders Federation, as well as some large-scale house-builders, took advantage of an opportunity to attack and unpick the ZCH agenda' (O'Neill and Gibbs, 2020: 126). A third reason is that the housing crisis increased the dependency of the government on housebuilders to build more homes. The building industry used this structural dependency to push for scrapping the ZCH (Edmondson et al., 2020).

Reluctance to engage with the low-carbon agenda characterises not only volume housebuilders but also smaller building firms, contractors, and suppliers who mostly lack the required competences, skills, and building templates. A low-carbon transition in housing would thus require extensive retraining, particularly in support of more energy efficient building and retrofitting (Martiskainen and Kivimaa, 2019), as well as a unified framework for standard setting, monitoring, and enforcement of quality improvements (e.g., insulation) (Bonfield, 2016).

There are a few organisations that promote the development of low-energy skills and buildings, but these operate mostly at the fringes of the industry. The Association of Environmentally Conscious Building (AECB), for instance, brings together contractors, trades people, architects, and builders to help develop, share, train, and promote sustainable building best practices. The UK Green Buildings Council also promotes greener approaches in the construction sector and lobbies the government to prioritise energy efficiency in buildings. The Energy Saving Trust provides information about low-emission retrofitting standards and practices. And the Zero Carbon Hub, which was disbanded in 2016, developed and disseminated knowledge and information regarding low and zero-carbon new homes.

Policymakers: Housing and housebuilding relate to different policy competences and responsibilities. The Ministry of Housing, Communities & Local Government (MHCLG), formerly the Department for Communities and Local Government (DCLG), has responsibilities for ensuring the safety and quality of buildings (e.g., through regulations and building standards) as well as for driving up housing supply and increasing home ownership. The UK has been characterised as a 'property-owning democracy' (Daunton, 1987): home ownership has been an explicit policy goal since the Thatcher government, with a range of policies to encourage user demand (e.g., Right-to-Buy, shared ownership).

Environmental, energy, and climate change issues in relation to buildings are responsibilities of other ministries, for example, DEFRA, DECC, BEIS.[8] BEIS

[8] Department for Environment, Food & Rural Affairs (DEFRA), Department of Energy and Climate Change (DECC, 2008–2016), Department for Business, Energy, and Industrial Strategy (BEIS, since 2016).

also aims to drive innovation and improvements in the construction business, notably by encouraging the long-term development of suitable supply chains and skills. Local authorities built many houses in the 1950s and 1960s but their active contribution to construction has shrunk greatly since then (Figure 6.15). One of their main current roles is to inspect new homes to ensure that they meet building regulations and standards.

These split responsibilities not only create coordination challenges but also mean that energy and climate change issues have been layered on top of the Housing Ministry's core remit. Building regulations and standards are an important policy instrument in the building sector, enabling the coordination of many dispersed actors. Building regulations historically focused on health, safety, and materials. Energy saving was added as an additional consideration in the mid-1990s (Raman and Shove, 2000), leading to gradual inclusion of incremental insulation measures in new buildings. Successive energy savings obligations on energy suppliers (further discussed in Section 6.3.3) also led to piecemeal retrofits and incremental insulation measures in existing buildings in the 1990s and 2000s. The 2013 Green Deal policy, which was meant to further accelerate the implementation of housing retrofits, was poorly designed and led to a collapse in the installation of key insulation measures (Figure 6.14), as will be further discussed in Section 6.3.3.

The 2006 Zero Carbon Homes (ZCH) target was a radical top-down policy that was meant to drive more radical low-carbon innovation in new homes. But this policy was first watered down in 2011 and then scrapped in 2015, because of industry counter-lobbying and changing political priorities (due to the financial-economic crisis and changes in government), as discussed previously.

These developments suggest that policymakers (have come to) perceive other issues as more important than climate change. These other issues include the limited supply of new buildings (which underpins the housing shortage crisis), limited availability of affordable housing (affecting younger generation), low quality of (many) new houses, and limited innovativeness in the construction sector. The weakening commitment of policymakers to climate change mitigation has led to less ambitious policies since the mid-2010s, which are presently not driving a low-carbon transition in housing: 'Policies to support low-carbon measures have been weakened or withdrawn, including Zero Carbon Homes and the Code for Sustainable Homes. This has led to many new homes being built only to minimum standards for water and energy efficiency' (CCC, 2019b: 11).

Users: Since the 1970s, people have gradually installed piecemeal insulation measures in their homes (Figure 6.11). For people with relatively low incomes, millions of insulation measures have been the result of government policies that required energy suppliers to make energy efficiency improvements in people's homes (further discussed in Section 6.3.3). Non-subsidised insulation and retrofit decisions have been motivated by household interests in improved thermal comfort, long-run energy cost savings, environmental benefits, addressing immediate problems (draughts, condensation, air quality, health), increased property value, and improved aesthetic appearance (for an excellent summary see Wilson et al. (2015)). But these motivations are not pertinent for all households, particularly not for low-income groups and non-homeowners.

Additionally, people may refrain from efficiency measures because of a range of barriers or concerns, including limited interest in energy or environmental issues, lack of money to pay upfront costs (or high interest rates on loans), uncertainty about benefits (e.g., cost savings, improvements in comfort, or health), concerns about contractor reliability and quality, fear of disruption caused by building works, information search costs, cognitive burden to process specialist information, or fear of time-consuming or frustrating dealings with builders or contractors (Wilson et al., 2015).

This multitude of barriers, which include but go beyond financial ones, helps explain why 'there are numerous, cost-effective measures that could be installed in many, if not most, houses, but the building owners are not putting them in' (Boardman, 2007: 41). The balance of motivations and barriers varies significantly across type of insulation measures, type of house, type of household, and type of occupancy. With regard to the latter, private tenancies (19% of the UK housing stock) offer the least motives for efficiency refurbishments due to the principal-agent problem (it is the tenant, not the landlord, who would reap the benefits of investment).

Wider Publics: Environmental movements and NGOs have long advocated for more attention to energy efficiency. National Energy Action, the Green Alliance, and WWF, for instance, have lobbied for more ambitious policy action, while the Association for Environment Conscious Building (AECB), which represents sustainable construction professionals in the UK, has lobbied for the adoption of more ambitious low-carbon home delivery targets and standards. It runs its own self-certification scheme for newbuilds, largely in line with the German PassivHaus standards.

Thermal improvement and comfort also relate to issues such as energy poverty, decent living conditions, and health benefits (Martiskainen and Kivimaa, 2019), which have been recognised as a particular form of social inequality and injustice. Low-energy housing can significantly reduce energy bills, improve thermal comfort, and improve indoor air quality if combined with appropriate ventilation (Chenari et al., 2016). Several NGOs have advocated energy poverty issues in the UK, such as National Energy Action, the National Right to Fuel Campaign, the End Fuel Poverty Coalition, and the Centre of Sustainable Energy. Other campaigns have been oriented towards information and emergency assistance, such as Beat the Cold, the Big Energy Saving Network, Citizens Advice, or the Energy Saving Trust.

Civil society organisations have also led energy poverty projects such as the Warm Homes for Health project that installed double glazing in social housing in Sunderland; the Warm Zones programme that provided advice and subsidies for household insulations in specific areas; or Green Doctors, who offer simple energy-efficiency measures (e.g., draught-proofing) and advice to energy poor households across London.

6.3.3 Policies and Governance

Formal Policies

Incremental building improvements have been stimulated by a range of policies since the mid-1990s. Part L of the Building Regulations (which relate to conservation of fuel and power) were tightened in 1995, requiring new buildings to meet higher insulation standards. This was complemented by the introduction in 1995 of the Standards Assessment Procedure, which allowed the energy performance of homes to be measured and compared (Mallaburn and Eyre, 2014). Three successive Energy Efficiency Standards of Performance programmes (1994–1998, 1998–2000, 2000–2002) further required energy suppliers to meet increasing energy-saving targets by making improvements in people's homes, which were mostly met through insulation measures.

The 2002 European Energy Performance of Buildings Directive (EPBD) increased regulatory pressures, setting minimum energy performance standards for new buildings and requiring owners or landlords to provide Energy Performance Certificates (EPC) when selling or renting existing buildings. UK policymakers introduced EPCs in 2007 (with Energy Efficiency Rating ranging from G to A+++) to provide information to users. The Energy Efficiency Commitments (EEC), the Carbon Emissions Reduction Target (CERT), and the Community Energy Saving Programme (CESP) were further programmes that placed energy savings obligations on energy suppliers, which between 2002 and 2012 led to the installation of millions of insulation measures in existing homes (Table 6.4), focusing particularly on disadvantaged customers.

Although these policies stimulated incremental, piecemeal improvements in new and existing buildings, they substantially contributed to reduced heat loss in buildings (Figure 6.10) and reduced greenhouse gas emissions in the 1990s and 2000s. The number of annual insulation improvements collapsed in the 2010s (Figure 6.14), because the Energy Company Obligations (ECO) and the Green

Table 6.4. *Number of insulation measures installed under EEC and CERT (data from UK Housing Energy Fact File 2012)*

	Cavity wall insulation	Loft insulation	Solid wall insulation
EEC1 (2002–2005)	792,000	439,000	24,000
EEC2 (2005–2008)	1,336,000	799,000	35,000
CERT (2008–2012)	2,103,000	4,549,000	47,000

Deal policies, both introduced in 2013, were weaker and poorly designed: 'Insulation rates fell very significantly after installation programmes (i.e., CERT and CESP) ended in 2012, with the replacement obligation (ECO) less ambitious than its predecessors and the Green Deal failing to deliver' (CCC, 2019a: 84).

The Green Deal, in particular, was a major failure, because it was introduced as the government's flagship policy to drive a mass rollout of low-energy housing retrofits. The Green Deal was a finance-based energy efficiency policy that deviated substantially from the previous regulatory approach (e.g., energy savings obligations, building regulations). It aimed to stimulate the use of private finance through a pay-as-you-save finance mechanism (in which households would pay back loans with money saved on energy bills). The Green Deal failed because the loan interest rate was too high, because the marketing campaign only emphasised financial savings (and ignored the multitude of other user barriers or motivations), and because policymakers failed to listen to critics and make adjustments (Rosenow and Eyre, 2016): 'The Green Deal too suffered from a lack of flexibility after initial poor design, which was heavily criticised at the time' (CCC, 2020: 99).

To stimulate *radical* innovations and drive a low-carbon transition, the government introduced the Zero Carbon Homes plan in 2006, which mandated that all new buildings from 2016 would be carbon-neutral or -negative (Kern et al., 2017). This was a radical top-down policy that was introduced rather suddenly, without much consultation. The ZCH-plan was accompanied by the 2006 launch of the Code for Sustainable Homes (CSH), which was a voluntary certification scheme that included performance measures for energy, CO_2 emissions, and other sustainability indicators, as part of a commitment towards 100% zero-carbon homes for newbuilds by 2016 (Heffernan et al., 2015). The Code had six levels, with Code 1 representing a 10% energy efficiency improvement over the 2006 building regulations, while Code 6 referred to zero carbon homes.

Policymakers hoped that the ZCH-target, increasingly stringent Building Regulations, and voluntary standards would encourage reorientation of incumbent housebuilders (O'Neill and Gibbs, 2020). To support this reorientation, the government also helped to create the Zero Carbon Hub (which was primarily funded by industry actors) to investigate, test, and demonstrate various zero-carbon options and create a platform for discussion and network building between industry and government (Edmondson et al., 2020). As discussed in Section 6.3.2, some housebuilders did indeed start exploring more radical low-carbon building options. But most industry efforts focused on political lobbying, which strategically aligned with changing political priorities in the early 2010s, leading to dilution of the ZCH-target in 2011 and its dismantling in 2015.

The suddenness of policy changes (which characterised both the introduction of the ZCH-policy and its removal) has been characterised as a major policy shortcoming (CCC, 2020). The limited effort to build stakeholder support has also been identified as a weakness:

> The creation of the Zero Carbon Hub Task Force may have been intended to create a coalition to support a green building transition, but governments failed to build strong alliances with those groups such as the Royal Institute of British Architects and the Association of Environmentally Conscious Building, which could have supported the policies. (O'Neill and Gibbs, 2020: 126)

The failures of the Green Deal, Energy Company Obligation, and Zero Carbon Homes policy have left UK housing policy without effective low-carbon policy for new and existing homes. The 2018 progress report from the Committee on Climate Change therefore rightly concluded that: 'Energy efficiency must urgently be improved across the building stock. Current policy is failing to drive uptake, including for highly cost-effective measures such as loft insulation. Policy needs to incentivise efficient long-term investments, rather than piecemeal incremental change' (CCC, 2018a: 85). And its 2019 progress report further assessed that: 'The progress made on buildings remains insufficient ... In order to go further to meet net-zero ambitions, bold and decisive action is urgently needed from Government' (CCC, 2019a: 67).

The 2018 Construction Sector Deal, which was developed together with the building sector, mentions energy-efficient homes in passing, but overwhelmingly aims to 'transform the sector's productivity through innovative technologies and a more highly skilled workforce' (p. 3). BEIS's heat-oriented clean growth plan (BEIS, 2018a) includes a Buildings Mission that aims to halve the energy use of new buildings by 2030. It also refers to £170m of public money to back the mission, which the government hopes will be matched by £250m of private sector investment (BEIS, 2018a: 110–111). Further operationalisation of this mission-oriented policy is still lacking, however. In its 2019 Spring statement, the government mentioned a Future Homes Standard by 2025 with a possible zero-carbon target for new homes. The technical specifications for the standards are expected to be ready for public consultation in 2023, legislation in 2024, and implementation in 2025. It remains to be seen, however, if this standard will indeed be adopted and what complementary policies will be advanced to reach it.

Governance Style

In terms of governance style, low-carbon building policy has largely been characterised by a satisficing approach to efficiency performance improvements, primarily driven by standards and sporadic demand-side incitation measures. This has delivered incremental but insufficient efficiency improvements, which have not substantially reconfigured the housebuilding sector. Policies supporting low-carbon building and retrofit measures in the UK are therefore not currently set to reach decarbonisation and net-zero objectives.

Until the early 2010s, the governance style largely rested on regulation (such as energy savings obligations and mandatory performance standards for buildings), which delivered incremental efficiency improvements through isolated insulation measures.

More recently, however, efforts to drive innovative housebuilding and retrofitting techniques have failed to deliver, for a range of reasons including lack of consistency over time (e.g., shifting priorities, policy termination), lack of consistency across interventions (e.g., fragmented or single instrument approach), and lack of coordination between policy competences (e.g., BEIS, Treasury, MHCLG), leading to conflicting objectives.

While a more hands-on approach may be needed, the sector is still characterised by a governance style oriented towards regulation and standards, which are presently not stringent enough to drive ambition and have major loopholes (particularly regarding the existing housing stock) due to significant industry influence and counter-lobbying. Low-carbon building and retrofitting has not entered the mainstream of the housebuilding and renovation sector; it lacks dedicated innovation, skilling, and large-scale market roll-out policy components.

6.4 Niche-Innovations

Radical niche-innovations have emerged and to some extent diffused within the heating and buildings systems. This section first provides a general discussion of developments that affect most niche-innovations. The subsequent sections discuss five niche-innovations for heating (heat pumps, biomass heating, solar thermal heating, heat networks, and gas grid repurposing to hydrogen or biomethane) and two for buildings (passive house designs, whole-house retrofits). For each niche-innovation, we first analyse techno-economic developments and then actors and institutions.

Techno-Economic Developments: Although renewable and low-carbon heat technologies have developed quite steadily in a number of European countries (notably Sweden and Denmark), their diffusion in the UK has remained limited.

Table 6.5. *Average capital cost data for domestic low-carbon heating systems and conventional systems (adapted from Rosenow et al., 2018)*

Building type	Individual low carbon heating system	Installed cost per dwelling (£)	Lifetime (years)
Existing building	Air source heat pump (ASHP)	7,000	15
	Ground source heat pump (GSHP)	14,000	20
	Biomass boilers	5,500	20
	Gas boiler	1,500	15
	Wood stoves (with chimney liner)	2,000	15
New building	GSHP	10,500	17.5

Although many technologies can be considered "mature" globally, the UK market in low-carbon heat technologies is only just emerging. This is due to the dominance of the gas boiler market driven by the wider availability of natural gas. Low carbon heat technologies remain niche, either because the target market is small or due to immature supply chains and low customer awareness. (Chaudry et al., 2015: 626)

There are significant differences in the installation and equipment costs of various domestic low-carbon heating systems (Table 6.5). The upfront capital costs of heat pumps and biomass boilers are significantly higher than gas boilers, which cost between £500 and £2,500, depending on size, brand, and type. Table 6.5 suggests that 1) heat pumps and advanced biomass boilers involve substantial upfront capital costs, 2) within heat pumps, GSHPs are considerably more expensive than ASHPs, 3) installing GSHPs in new buildings presents important cost savings (over installations in existing buildings).

The various niche-innovations also have different technical characteristics, which means that a low-carbon transition in heat and buildings may well lead to a more diverse overall system.

- Heat pumps (Section 6.4.1) require more space than gas boilers and may thus be especially suitable for large, sub-urban houses and off-grid locations.
- Heat networks (Section 6.4.5) are more suitable in locations with concentrated heat use (i.e., dense urban areas, tower-blocks, industrial/commercial applications). While they enable significant efficiency gains due to scale advantages accruing from collectivisation, their low-carbon nature is not automatic as it depends on the fuel source (which is presently mostly gas).
- Biomass heating (Section 6.4.2) is mostly used as add-on technology in the living room, where people like to enjoy a cosy fire when they relax. Exhaust of particulate matter is, however, contributing to air pollution problems in dense residential areas, leading to increased policy concerns. They are different from

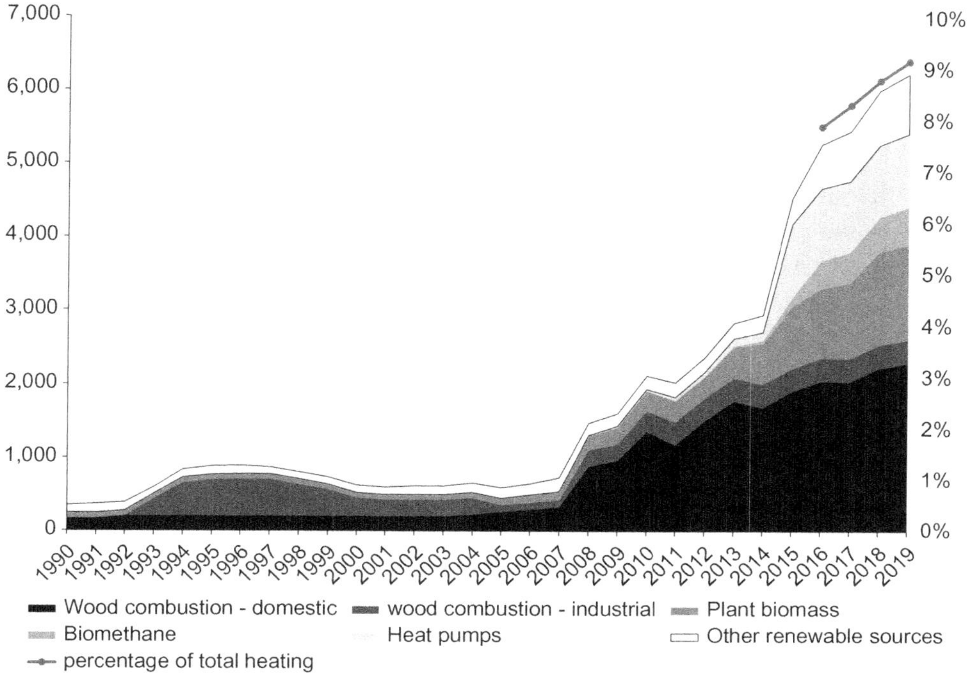

Figure 6.16 Renewable sources to generate heat (all sectors) in kilotons oil equivalent, 1990–2019 (constructed using data from DUKES; Renewable sources; Table 6.1.1 and U1)

advanced biomass boilers, which can be used as primary heat source with high efficiency ratios but are also more expensive.

- Gas grid repurposing towards low-carbon energy carriers (e.g., biomethane or hydrogen) makes use of the existing infrastructure (Section 6.4.4).
- Low- to zero-carbon housing relates to new building and renovation techniques focussed on radically improved insulation, ventilation, heat exchange innovations, as well as general design principles. While newbuilt zero-carbon housing (Section 6.4.6) enables the applications of such principles to design and delivery, whole-house retrofits (Section 6.4.7) are important because they apply to the existing building stock, which is the vast majority of UK housing.

Since 2007, the uptake of renewable heat technologies has increased to a non-negligible portion of heating in combined sectors (residential, public administration, commercial) (Figure 6.16). Domestic wood combustion is the largest category, but heat pumps and biomass have also started to diffuse more substantially since 2014. Most heat pump deployment, however, has been in the commercial rather than the domestic sector. Regardless of end use and sector, renewable heat[9] made up just over 9% of overall heat use in 2019 (BEIS, 2020a: 106; Figure 6.16). Despite this rapid deployment, GHG emissions have still grown between 2014–2019 (Figure 6.2), because of increased heat demand (Figure 6.6) consistent with a growing number of heated dwellings (Figure 6.12) and because these renewable sources are not all low-carbon.

Policies: Since the 2008 Climate Change Act, radical niche-innovations have been central in strategic policy visions on long-term transition pathways. The content of these visions changed several times, however, which created deep uncertainties (Winskel, 2016). Initial views, articulated in the *Low Carbon Transition Plan* (DECC, 2009), *2050 Pathway Analysis* (DECC, 2010), *The Fourth Carbon Budget* analysis (CCC, 2010), and *Carbon Plan* (DECC, 2011b), primarily emphasised a transition towards electric heat pumps, as part of the wider 'all-electric society' vision (in which first the electricity system would be decarbonised and then heat and transport would be electrified through heat pumps and electric vehicles). Winskel (2016) suggests that: 'The "all electric" vision for UK energy transition that emerged from these early scenarios was a rather simple blueprint, based on limited techno-economic research and modelling capacities which neglected many social, institutional and behavioural issues'.

Further analyses of the barriers for electric heat pumps (e.g., lack of installer skills, lack of space in dense urban areas, limited user confidence) and better understanding of potentials of other options led to revised visions a few years later. *The Future of Heating: A Strategic Framework for Low Carbon Heat in the UK* (DECC, 2012d) and *The Future of Heating: Meeting the Challenge* (DECC, 2013b) envisaged a smaller role for heat pumps than before and a more prominent role for heat networks. While both documents also identified an important transitional role for hybrid heat pumps (using gas and electricity), they envisaged that natural gas boilers would be phased out by 2050. Although some of the modelling still suggested a dominant role for heat pumps by 2050, the overall strategic vision of both documents involved a more diversified range of heat technologies than before.

A few years later, the strategic vision changed again. Both the *Clean Growth Strategy* (BEIS, 2017b) and its dedicated heat application (BEIS, 2018a) identified not only heat pumps and heat networks as possible low-carbon options but also the decarbonisation of gas grids by substituting natural gas with hydrogen or biomethane. This addition was partly the result of lobbying from gas industry actors (Lowes et al., 2020), who worried that a low-carbon transition to heat pumps and heat networks might threaten the gas grid. The changing and diversifying

[9] 'Renewable heat' includes options that are not considered low-carbon, such as wood combustion and unabated biomass more generally.

visions reflected and created deep uncertainties, leading policymakers to conclude that 'at present it is not certain which approaches or combination of them will work best at scale and offers the most cost-effective long-term answer. Decarbonising heat is our most difficult policy and technology challenge to meet our carbon targets' (BEIS, 2017a: 75). The uncertainties and weak policies also hindered industrial and market development in the low-carbon heat sector, jeopardising decarbonisation ambitions (CCC, 2018a; Li and Pye, 2018; Rosenow and Eyre, 2016).

To stimulate radical niche-innovations, the government introduced the Renewable Heat Incentive (RHI) in 2011, which provided subsidies for the installation of renewable heat technologies in non-domestic buildings. In 2014, the RHI was extended to domestic buildings, which stimulated demand for heat pumps, solar thermal, and biomass systems (Figure 6.17). The spike in 2014 and 2015 was caused by a large volume of 'legacy applications', which are RHI applications for systems that were installed between 2009 and 2014. RHI qualification criteria were adjusted after the 2015 Spending Review, leading to the introduction of a spending cap, new sustainability criteria that had to be met (especially for biomass), and a more restricted focus on strategically important technologies such as heat pumps and biomethane. Annual RHI-funded heating systems remained relatively stable after the initial legacy-induced peak, leading to steady growth of cumulative installations. Annual RHI-funded installations markedly increased from 2018, especially of pumps, which reached 21,500/year in 2019 (Figure 6.17). Subsidy levels for the RHI have fluctuated over time, according to policy priorities (Figure 6.18).

According to recent government proposal for the future of clean heat the policy priority is to 'provide support for biomethane injection into the gas grid through the Green Gas Support Scheme and provide support for buildings technologies (heat pumps and in limited circumstances, biomass) through the Clean Heat Grant' (BEIS, 2020b). The main instrument currently considered to support the deployment of low-carbon heating technologies in buildings is an upfront grant (as an alternative to a tariff-based mechanism) at a 'technology-neutral, flat-rate grant of £4,000 for all technologies eligible under the Clean Heat Grant' (BEIS, 2020b: 29) for all sizes of installation up to 45kW capacity, to be rolled out as a successor scheme to the domestic RHI from 2022.

In its consultation on 'Future support for low carbon heat', BEIS (2020b) spelled out some of the preferred technological options going forward, representing increased enthusiasm for heat pumps and biomethane, unchanged views of heat networks, and downgraded expectation for biomass and solar thermal:

- Heat pumps 'offer the greatest heat decarbonisation potential for the majority of buildings off the gas grid' and 'could enable us to almost completely decarbonise heat alongside the decarbonisation of electricity generation' towards 2050 (BEIS, 2020b: 26–27)

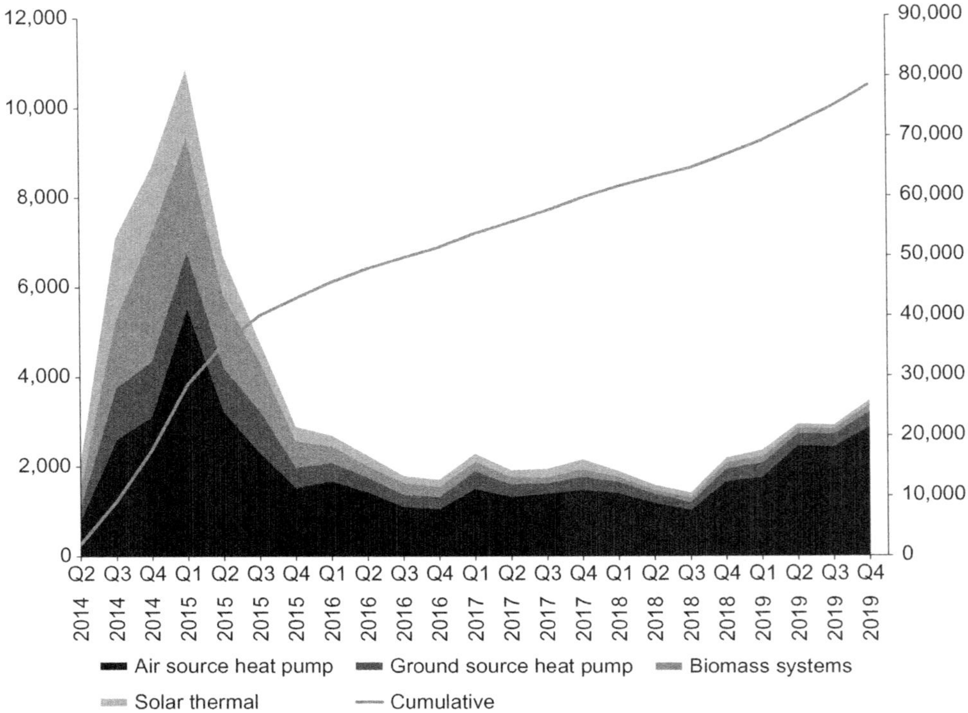

Figure 6.17 Quarterly (left-hand Y-axis) and cumulative (right-hand Y-axis) accreditations (new and legacy) of low-carbon heating installations under domestic RHI (number of installations). (RHI monthly official statistics, Domestic RHI deployment data)

- Biomethane injection into the gas grid 'accelerates the decarbonisation of gas supplies, by increasing the proportion of green gas in the grid. This transition is a necessary step towards meeting our carbon reduction targets' (BEIS, 2020b: 11)
- Heat networks 'are expected to play a crucial role in the decarbonisation of heat' (BEIS, 2020b: 42).
- Biomass 'although [it] has a wider strategic role to play in overall UK decarbonisation, its use in heating buildings should be limited' (BEIS, 2020b: 27)
- Solar thermal is no longer a priority, as 'given current cost data and recent deployment trends, we do not have any strong evidence to suggest that supporting solar thermal water heating through this scheme would prove to be an effective measure for preparing supply chains for the future phase-out of high carbon fossil fuel heating' (BEIS, 2020b: 40)

Most recently, the Government issued an Energy White Paper (HM Government, 2020a) and a Ten point Plan for a Green Industrial Revolution (HM Government, 2020b). The stated ambitions for domestic heating and buildings in these policy

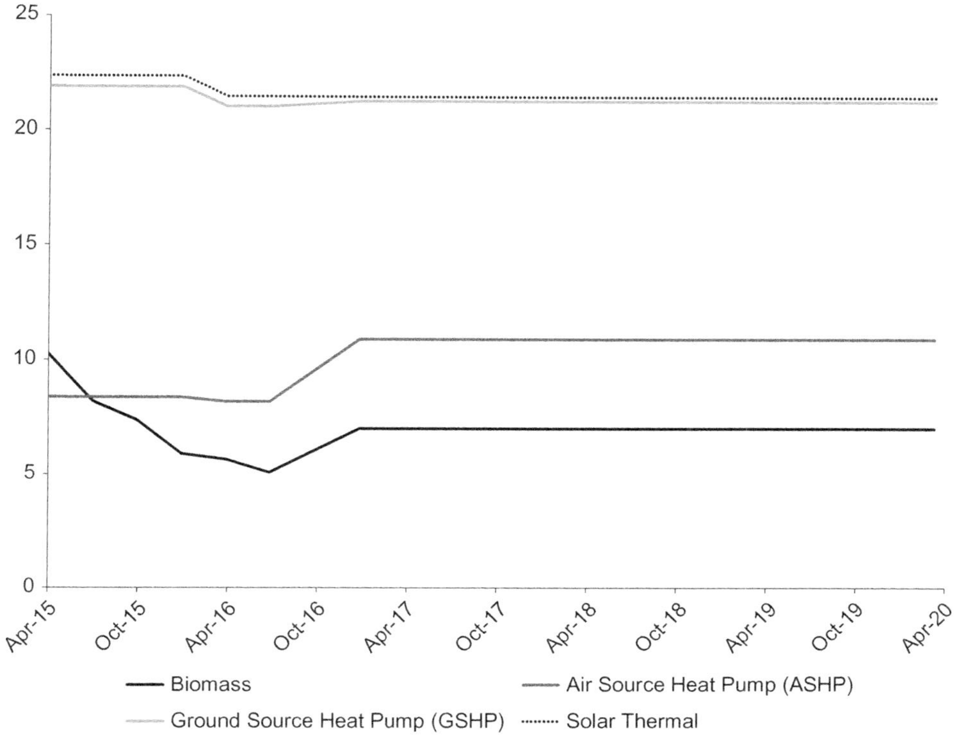

Figure 6.18 Variations in domestic RHI subsidy levels between 2015 and 2020 (p/kWh), adjusted by the Retail Price Index (RPI) before 1 April 2016 and by the Consumer Price Index (CPI) thereafter (constructed using data from OFGEM (2020))

documents are to roll out 600,000 heat pumps per year by 2028, support the delivery of biomethane to 230,000 homes by 2030, initiate demonstration trials for hydrogen heating in selected locations from 2023, and work with Local Authorities to enable the designation of new heat network zones by 2025. The implementation means and policy specifications to deliver these objectives are expected to be articulated in the delayed Heat and Buildings Strategy.

In September 2020, the government launched the £2bn Green Homes Grant scheme as part of the 2020 COVID-19 stimulus package, which would release vouchers for up to £5–10,000 (depending on income) for home insulation and low-carbon heating installations in homes already fitted with at least one primary insulation measure. It included primary insulation measures (solid wall, under floor, cavity wall, loft, flat roof, room in roof, insulating a park home), primary low-carbon heat measures (air or ground source heat pump, solar thermal, biomass boilers), and a range of secondary measures (e.g., draught proofing, double/triple glazing, heating controls).

Subsequent implementation problems with administering and paying out vouchers led to concerns that the scheme would not deliver (Laville, 2021). On 22 March 2021, the Environmental Audit Committee (2021: 27) provided a very critical evaluation of the implementation of the Green Homes Grant and its ability to deliver on retrofitting existing homes to higher efficiency levels:

The Green Homes Grant has been rushed in conception and poorly implemented. In its haste to create a scheme to deliver economic stimulus, the Government failed to consult industry adequately on its delivery, set a timescale which was overly short term and has presided over scheme administration which appears **nothing short of disastrous**. If the ambition for the scheme to retrofit 600,000 homes envisaged completion of the work by the end of the current financial year, then the Government has been **wildly optimistic** in its scheme planning and industry engagement. The impact of its **botched implementation** has had **devastating consequences on many of the builders and installers that can do the work**, who have been left in limbo as a result of the orders cancelled and time taken to approve applications. (Our emphasis)

A few days later, the Green Homes Grant was cancelled, owing to conflicting objectives between BEIS and the Treasury, which is indicative of a lack of policy coherence and coordination that further weakens supply chains and industry confidence (Institute for Government, 2021), and is ultimately detrimental to driving long-term heat decarbonisation.

Because of repeated problems and U-turns, there is substantial uncertainty about the upcoming Heat and Buildings strategy and the political desire and ability to develop visions and policy instruments that can deliver on decarbonisation and low-carbon heating deployment objectives.

6.4.1 Heat Pumps

Techno-Economic Developments

Heat pumps (HPs) are electrical or gas-powered devices that extract low-temperature heat from the ground, the ambient air, or even water (e.g., pond) to the desired heat sink (Greening and Azapagic, 2012). HPs are a mature technology that requires more installation and operational space than gas boilers. They are therefore presently most suited for newbuilt homes and/or off-grid housing. HPs are relatively new to the UK context, which means that many socio-technical dimensions (e.g., supply chains, installation skills, user confidence, standards) are under-developed:

Heat pumps are an established solution in many other countries, but not yet in the UK. Establishing them as a mass-market solution will take some time, with strong progress required during the 2020s. There are particular opportunities in new-build properties, homes off the gas grid, non-residential buildings and for hybrid heat pump systems retrofitted around existing gas boilers. (CCC, 2019a: 69)

Table 6.6. *Comparison of UK and major European heat pump markets (Data: EHPA (2018: 62–64))*

	UK	France	Italy	Spain	Sweden	Germany
Total heat pump installations (2018)	179k	2,028k	1,717k	581k	1,702k	947k
Annual heat pump sales (2017)	19k	242k	171k	106k	104k	92k

Heat pumps are not only bulkier than gas boilers but also more expensive to buy and install. Total costs are between £6,000 and £11,500 for ASHPs and between £9,000 and £20,000 for GSHPs, with significant variation depending on size and complexity of installation (BEIS, 2018a). For this reason, HPs require significant support mechanisms and lower costs to be able to compete with gas.

The UK market for heat pumps is small by European standards (see Table 6.6) but has somewhat increased since 2018 (Figure 6.17. Although annual sales have remained relatively small, the cumulative number of domestic HPs has gradually increased. A mass roll-out of heat pumps faces significant barriers related to cost (and incentive mechanisms), the adequacy of supply chains (i.e., scaling up production), skilled installer base (i.e., number of skilled installers, level of consistency, and adequate dimensioning), and user trust and demand.

Actors and Policies

Since 2009, government heat decarbonisation scenarios have envisaged important roles for heat pumps. Policymakers therefore tried to stimulate HP uptake with the Renewable Heat Incentive (RHI), which was introduced in 2011 for non-domestic buildings and extended to domestic buildings in 2014. The RHI subsidy for heat pumps varies according to type and size of heat pump, ranging from 18.8–21.16 p/kWh for GSHPs to 7.3–10.85 p/kWh for ASHPs (Figure 6.18). The actual eligible amount also varies on based on heat demand (itself depending on building size and insulation) and will range between £3,000–10,000 for ASHPs and £6,500–33,000 for GSHPs in different size houses over the eligible seven-year period, covering initial investment costs in most cases.

Under the currently proposed strategy for clean heat, the deployment scenario of the Clean Heat Grant aims for 12,500 new heat pump installations per year by 2024 (BEIS, 2020b: 18), which compares palely to the current 1.7 million yearly gas boiler installations (Rosenow and Thomas, 2020). According to the Energy White Paper, however, this objective has been significantly raised, with intentions to deliver 600,000 heat pumps per year by 2028, and an expected associated 20,000 new jobs (HM Government, 2020a). This objective is, however, unlikely to be met with current trends and policies, as it requires substantial investment as well as the rapid expansion of skills and supply chains.

On the industry side, the Heat Pump Association (www.heatpumps.org.uk) is supporting the tightening of building regulations for newbuilt homes towards greater efficiency requirements and the elimination of current loopholes. It is also keen to see the materialisation of the Future Homes Standard in 2025, the gradual tightening of emissions standards for heat in existing buildings, and the implementation of a successor scheme for the RHI, which is scheduled to end in 2021 but has been extended to March 2022 when it will be replaced by the Clean Heat Grant. If these policies materialise, the industry association says it is ready to drive a mass roll-out of heat pumps, first in newbuilt homes and off-grid retrofits from 2020 to 2025, and then in on-grid retrofits from 2025, to reach 1 million heat pump installations per year from 2030 (Heat Pump Association, 2019), which is more than the objectives set out in the 2020 Energy White Paper. This ambitious deployment trajectory would, however, require a rapid increase in the number of heat pump installers from the current 916 installers to roughly 40,000 by 2030 (Heat Pump Association, 2019), which is a challenge that requires dedicated skills and training programmes.

Additional challenges concern user engagement and acceptance, since HPs require greater user engagement: ‘Given the relative complexity of heat pumps compared to conventional heating systems, good technical support and advice for users is especially important’ (Caird et al., 2012: 292). Another user obstacle is that HPs require additional space, notably for GSHP that require outdoor underground pipes and are hence more suitable for detached housing (Hannon, 2015).

On the positive side, heat pumps, if properly installed and operated, can improve levels of warmth and comfort in the home. Some, though not all, users may also enjoy a sense of empowerment from greater control over their own heat service provision:

> Control and operation of a heat pump positions the user as participating in the provision of their own energy services and redefines their consumer role from “captive consumer” associated with a previous universal mode of service in multiple ways [. . .], creating new possibilities for users not only to unwittingly collaborate in the reproduction of energy systems but to act as “co-providers” of energy services. Consumers turned “co-providers” are able to generate some of their own technological and institutional services. (Judson et al., 2015: 34)

6.4.2 Biomass Heating

Techno-Economic Developments

Biomass heating includes different types of fuel input (e.g., wood logs, pellets, or chips) and different kinds of appliances ranging from conventional wood stoves and room heaters (which are often used in combination with another heat source) to advanced biomass boilers, which are larger and costlier (around £5,500) and quite rare in the UK. Indeed, most UK domestic biomass heating uses conventional technology: ‘Currently around half of wood grown and used for heating homes in the UK is burnt on open fires (based on the 2014 domestic wood fuel survey), with most of the remainder consumed in wood-burning stoves' (CCC, 2018b: 19).

Domestic wood combustion is the largest UK source of renewable heat (see Figure 6.16). While domestic wood combustion replaces some coal or oil-fuelled systems in rural, off-grid areas (Jeswani et al., 2019), it is mostly used as *additional* heat source for reasons of cosiness and ambience:

> Today, comparatively few households burn solid fuel as a sole or primary heating source, yet 2.5 million households use open fires, enclosed stoves and range-ovens to heat their home to some degree [...] Now relegated to a supplementary role, open fires and other "traditional" technologies are predominantly used for heating a single room in both on- and off-gas properties. (Roberts, 2020: 3)

Traditional forms of biomass combustion are rather inefficient and in dense areas contribute significantly to air quality problems. More efficient domestic biomass heating applications involve the combination of technologies, such as in combination with hybrid heat pumps or in local district heating systems (CCC, 2018b).

The total number of domestic biomass stoves has been estimated at around 1 to 1.5 million, with annual sales under 200,000, while the market for biomass stoves with back boilers is estimated at under 20,000 yearly (AEA, 2012). About 13,000 homes use modern biomass boilers (CCC, 2018b). The diffusion potential in urban setting is limited because of air quality regulations and fuel storage limitations.

Actors and Policies

While most woodburning in the UK is currently for power generation and commercial applications, supply chains and actors partly overlap. The Wood Heat Association represents the modern wood heating trade in the UK, bringing together wood suppliers (e.g., wood logs, wood chips, wood pellets, straw, miscanthus), biomass boiler and stove installers, and so on.

Domestic wood burning is popular because it has significant cultural and aesthetic value in the British context. Indeed, cosiness, comfort, and aesthetic value are important explanations for the continued desirability and appeal of traditional biomass heating practices:

> Often associated with a welcoming atmosphere, warmth, relaxation, comfort and cosiness; fireplaces, ranges and stoves are deeply valued in Britain and comparable Western contexts, precisely because they are symbols of homeliness [...] Within this romanticised understanding of wood heating as part of home-making, the aesthetic and sensory qualities of the fire have been shown to be highly valued [...]. As desirable home amenities, the aesthetic qualities of traditional wood-burning technologies – open fireplaces in particular – have been found to override considerations of energy efficiency when thermally retrofitting homes, as greater value is ascribed to the historical significance and appearance of the surrounds and mantelpieces. (Roberts, 2020: 3)

Compared to other forms of domestic heating, wood burning requires significant user engagement including wood sourcing (or chopping if using on-site supply), storage, regular refuelling and stoking, stove cleaning (ashes and residues), chimney sweeping, and so on. This does not seem to dissuade people from biomass heating, which suggests that cultural appeal trumps practical considerations.

In recent years, biomass heating has come under significant public pressure, owing to concerns about local air pollution, carbon emissions (which are positive without reforestation efforts), biodiversity impacts from forest clearings, and competition with other land uses such as food production. In response to these concerns, policymakers reduced RHI subsidy levels for domestic biomass heating after 2015 (Figure 6.18). More generally, UK policymakers aim to limit domestic biomass heating unless it is the only possible alternative:

> Although biomass has a wider strategic role to play in overall UK decarbonisation, its use in heating buildings should be limited, as the Committee on Climate Change (CCC) says, to maximise the overall carbon abatement that is possible from sustainable biomass. As far as it is proportionate to do so, we propose to introduce eligibility criteria so that biomass is not installed in individual buildings that would be suitable for a heat pump. We propose that support for biomass will not be permitted in urban areas. (BEIS, 2020c: 10)

Accordingly, only 700 domestic biomass heat installations are foreseen to be supported under the Clean Heat Grant scheme 2022–2024. These policy changes imply that the future potential for conventional residential biomass heating is limited and restricted to off-grid locations. Alternative biomass heating, such as with advanced biomass boilers or in the form of biofuels in suitable boilers, does present a legitimate option looking forward, but it is not a policy priority.

6.4.3 Solar Thermal

Techno-Economic Developments

Solar thermal systems come in different forms, but in essence absorb solar radiation, often in rooftop solar thermal collectors, and transfer the heat via linkages with conventional water or space heating systems (Greening and Azapagic, 2014). Owing to uneven daily and seasonal solar radiation, such systems are usually combined with other heat sources. Water heating applications are most common, and the most vibrant markets are in Southern and Central European countries, although applications in higher latitudes are also technically possible.

Although solar thermal technology is a mature microgeneration technology, the UK market has remained small compared to other countries, with only 8.5 kWth per 1,000 inhabitants, while the EU28+ average is 69.2 kWth per 1,000 inhabitants (Solar Heat Europe, 2018). Solar thermal capacity has gradually expanded since the early 2000s, but annual installations and sales have declined significantly since 2011 (Figure 6.19), indicating that this has become a stagnant niche-innovation.

Figure 6.19 Diffusion of solar thermal in the UK, annual installed capacity, in m2 (Constructed using data from EurObserv'ER online database, RES Capacity and Generation, Statistics time series, Solar Thermal Annual installed capacity, www .eurobserv-er.org/online-database/)

Actors and Policies

The post-2011 decline of solar thermal systems was largely due to the introduction of Feed-in-Tariffs (2010), which included attractive payments for domestic solar-PV electricity generation. Households subsequently chose solar-PV panels over solar thermal systems, because of much shorter paybacks on the former (Fiorentini, 2013). The revision of the RHI scheme at the end of 2015 generated further uncertainty, as it opened up the eligibility of solar thermal for discussion. Although RHI subsidies for solar thermal were maintained, and actually slightly increased (see Figure 6.18), the UK market remained very small, compared to other renewable heat technologies (see Figure 6.17) and compared to significantly more vibrant markets in European countries such as Austria or Greece (Solar Heat Europe, 2018).

Shrinking post-2011 markets negatively affected the installer base (McVeigh, 2017). Nevertheless, UK manufacturing capacity remained strong, with notable global actors such as AES Solar (flat plate collectors), Kingspan (evacuated tube collectors), or Viridian Solar (integrated rooftop solar panels). The Solar Trade

Association[10] promotes the development and deployment of solar energy in the UK (thermal and PV) and abroad. It successfully lobbied government for continued inclusion of solar thermal systems in the domestic RHI scheme (McVeigh, 2017), and later under the Green Homes Grant scheme (2020), although these were not included in the prior consultation (BEIS, 2020c) and ex ante policy impact assessment (BEIS, 2020b).

6.4.4 Greening the Grid

Techno-Economic Developments

Recently, the repurposing of gas networks for biomethane or hydrogen injection has also been considered as a heat decarbonisation option. This is not only due to increasing recognition of technical, market, and user barriers for heat pumps in the UK context (CCC, 2016; Speirs et al., 2018) but also relates to significant lobbying efforts by incumbent gas supply actors who aim to protect their sunk investments in the existing gas infrastructure (Lowes et al., 2020). Decarbonised hydrogen and biomethane could both replace (fractions of) upstream natural gas supply but maintain the gas infrastructure, perhaps with some modification.

Decarbonised hydrogen can be produced from fossil fuels or biomass by steam reforming or gasification with carbon capture and storage (which is not yet commercially viable), or through water electrolysis using renewable electricity (which currently accounts for a small portion of UK hydrogen production) (Speirs et al., 2018). The distribution of hydrogen, which has physical and chemical properties that differ from natural gas, would require modifications of gas pipelines, especially at higher concentrations. Unmodified gas pipelines do appear to be able to distribute hydrogen blends of 20% (Murray, 2021). At the point of use, hydrogen-based heating would also require modification or replacement of heat appliances such as boilers, especially at higher concentrations.

Decarbonised hydrogen could become part of a hybrid heat decarbonisation strategy (e.g., as back-up for heat pumps in winter months) from 2030, but this would require a dedicated strategy (CCC, 2018c). At present, the hydrogen-for-heating niche does not exist commercially, but does have some presence in visions and through demonstration projects, which have multiplied in Europe over the past 15 years, with Germany leading the way and the UK playing a smaller role (Wulf et al., 2018). The UK-based H21 programme, launched in 2016, brings together a number of projects exploring the conversion of existing gas pipelines to carry 100% hydrogen. The H21 Leeds City gate project proposes to convert the local gas

[10] The STA adopted a new name, Solar Energy UK, in January 2021, to reflect its expansion in the realm of energy storage technology.

distribution grid to hydrogen by 2028, with the aim of serving 3.7 million properties in the wider surroundings by 2038 (www.northerngasnetworks.co.uk). In 2017, as part of H21, BEIS invested £25m in the four-year Hydrogen for Heat project (Hy4Heat), led by Arup and involving major industry players to explore the potential and feasibility of hydrogen gas for heating in the UK. This project includes an exploration of possible quality standards, certification, the development and testing of heating appliances, and demonstration facilities.

The injection of biomethane in existing gas grids does not require significant infrastructure modifications, and only minor repurposing of boilers, given that its physical properties are largely equivalent to natural gas. However, biomethane production is costly, potentially displaces land use and available biomass, and its net carbon emissions vary significantly according to conversion technology and biomass source. Biomethane can be produced from different feedstocks and processes (e.g., landfill gas, sewage sludge digestion, anaerobic digestion), and it can serve different kinds of uses (e.g., power, heat, transport, or gas grid injection), which means that intersectoral competition is likely, especially if supply is limited. Currently in the UK, the rapidly growing volume of energy produced from anaerobic digestion (biogas and biomethane) primarily supplies the power sector, while less than 10% is used for heat (DEFRA, 2020).

Biomethane grid injection was demonstrated in 2010 with biogas production from a sewage treatment plant in Didcot. Subsequent commercial scale application started in Dorset in 2012. Since then, biomethane use for heating has gradually increased. By 2017, 81 biomethane plants were accredited under the non-domestic RHI (HoP, 2017), 93 by 2019, and an additional 30 are expected by 2021 (Mieke Decorte et al., 2020). Biomethane from anaerobic digestion has grown considerably in recent years, from a mere 0.2% of renewable heat generation in 2010 to 9.1% by 2019 (Figure 6.16). Landfill gas and sewage sludge digestion respectively accounted for 0.2% and 2.3% in 2019 (Figure 6.16). Total volumes, however, remain low compared to natural gas use and are unevenly distributed. In 2019, biomethane blending in the gas grid represented only 0.4% of total gas distributed, but can be as high as 12% locally in some parts of South-West England, notably North Gloucester, where significant production is concentrated (Regen, 2020). CCC (2018a) suggests that raising the proportion of biomethane injection (up to 5%) pertains to cost-effective low-regrets options that should be pursued today. Higher levels of blending are currently limited by supply-side constraints, particularly the limited supply of wet feedstocks for anaerobic digestion and competition for other uses in the case of Bio-Synthetic Natural Gas (HoP, 2017).[11]

[11] Indeed, BioSNG is not eligible under RHI payments and hence more likely to be used as transport fuel, where it can benefit from the RTFO (Richards and Al Zaili, 2020).

Actors and Policies

The development of the UK greening-the-grid niche has been supported by multiple actors. The social networks are larger and denser for biomethane, which substantially contributed to low-carbon heating, than for hydrogen, which mostly exists through visions and demonstration projects.

Incumbent gas suppliers and distributors (including National Grid and regional operators) have strongly advocated and supported greening-the-grid options in recent years to protect their vested interests in the existing gas infrastructure. They have been successful in persuading policymakers to include these options in heat decarbonisation strategies (Lowes et al., 2020). To defend the gas infrastructure from the threat of becoming a stranded asset in all-electric scenarios, these gas supply actors claim that the greening-of-the-grid is cheaper than a transition to heat pumps or heat networks. The evidence for these claims is spurious, given the significant technical, industrial, and cost uncertainties for both biomethane and hydrogen grid repurposing and alternative options (Speirs et al., 2018).

Gas distributors are trying to develop the green gas market by offering 'green gas' tariffs to their customers, who would pay a premium price for gas that contains a percentage of green gas, mostly biomethane. The green gas tariffs, which are led by smaller energy suppliers, show significant variation in terms of proportions of green gas blend and price premiums (see Figure 6.20). To support this market construction strategy (and user confidence in it), the Renewable Energy Association (REA) introduced a Green Gas Certification Scheme that aims to guarantee the origin and quality of biomethane injected in the grid by distribution companies.[12] For future diffusion, the Energy Networks Association is advocating a cluster-based approach that prioritises biomethane grid injection or hydrogen delivery in regions with strong local supply chains ('biomethane zones' and 'hydrogen zones') (ENA, 2020a).

Upstream producers of biomethane and hydrogen also support the greening-the-grid niche because of economic and industrial opportunities. The development and expansion of biomethane markets in the last decade has attracted actors from the agricultural and waste sectors (e.g., farmers, landfill site and sewage plant operators) who produce and process biomass resources, often from local supply chains. A rapid expansion of biomethane production projects (2014–2016) was followed by a period of market consolidation and a focus on scale economies and cost reduction (Mieke Decorte et al., 2020). Scale increases were enabled by specialised investors such as Privilege Finance, which invested over £500m in the anaerobic digestion and biogas sector (ADBA, 2020). The UK presently has over

[12] 'Each kWh of green gas is labelled electronically with a unique identifier known as a Renewable Gas Guarantee of Origin (RGGO). This identifier contains, for each kWh of gas, information about where, when and how it was produced' (from www.greengas.org.uk/).

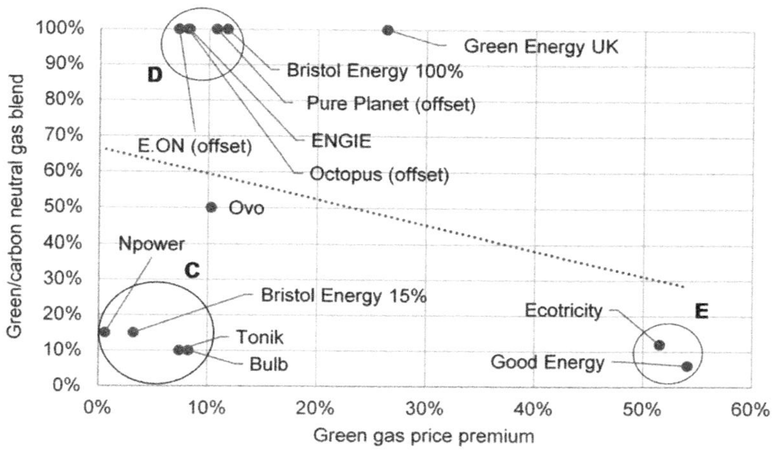

Figure 6.20 Green gas price premium versus blend percentage (Richards and Al Zaili, 2020: 52) (Note: Bristol Energy appears twice due to having two green gas tariffs: one with pure biomethane, and one with a 15% blend with natural gas. 'Offset' denotes 'carbon neutral' gas tariffs)

100 biomethane plants that supply into gas grids (ADBA, 2020), operated by about 65 biomethane supply organisations (GGCS, 2020).

The UK biogas industry is represented by two national associations (the Biogas Group and the Anaerobic Digestion and Bioresources Association ADBA), which lobby policymakers, organise knowledge circulation and transfer activities, and facilitate intra-industry dialogue. In 2017, ADBA developed an Anaerobic Digestion Certification Scheme (ADCS) to raise and harmonise industry practices, notably around operational, environmental, and health and safety performance.

The Energy Networks Association, which is 'the voice of the [energy] networks', has taken up an important advocacy role for a green gas future, relying on a combination of biomethane and hydrogen in on-grid areas. It has recently set out its vision and strategy to deliver 'a 100% hydrogen network' (ENA, 2020b). To support this vision, the current plan is to focus on large-scale 100% hydrogen pilots and up to 20% blending in parts of the network by 2030; scale up hydrogen pipelines between industrial clusters and roll out 100% hydrogen conversion for use in homes during the 2030s; and have a national hydrogen network in place in the 2040s (ENA, 2020b).

The government is more cautious about long-term hydrogen plans, highlighting that 'exact mix of different end uses for clean hydrogen in 2050 will depend on a variety of factors including cost, availability and technical application'. Nevertheless, it aims for '5GW of clean hydrogen production capacity in 2030, equating to 42TWh, and supporting up to 8,000 jobs by 2030 across our industrial heartlands and beyond' (HM Government, 2020a: 128). Collaboration with industry is an integral part of these plans.

Policymakers also support green gas as part of wider visions for the bioeconomy and the hydrogen economy. The European Commission's bioeconomy strategy, issued in 2012 and updated in 2018, seeks to further develop bio-based sectors and deploy local bioeconomies across the whole of Europe by 2030. The UK has developed its own bioeconomy strategy (HM Government, 2018) to support the development of bio-based energy and materials, including biomethane supply. The European Hydrogen strategy has gradually emerged in recent years, but accelerated with COVID-related green recovery packages that led several European countries (Germany, France, Spain, UK) to announce billions of euros of investments in hydrogen production and use (Vivid Economics, 2021). While most of these hydrogen investments target transport and heavy industries, the UK strategy also aims to develop hydrogen for domestic heating purposes.

Policy support for green gas has changed over time. The 2012 Heat Strategy did not consider it a priority, stating that 'Large-scale biomethane injection into the grid at a national level is not a realistic option, especially when efficiency losses are taken into account' (DECC, 2012d: 53). Since 2014, however, biomethane production has qualified for support under the non-domestic RHI, leading to a relatively unexpected increase in production, which by 2018 accounted for 22% of renewable heat delivered by the RHI (Lowes et al., 2019). The increasing political profile of and policy support for green gas is partly due to lobbying by energy suppliers and requests for more certainty concerning the long-term policy support mechanism (Richards and Al Zaili, 2020). In 2020, the government announced plans for a £150 million Green Gas Support scheme, which aims to further encourage biomethane injection into the grid, to be financed via a levy on gas suppliers.

Users have, so far, been limitedly involved. Most users do not notice that small amounts of biomethane are blended in their gas supplies, and active demand (through the uptake of green gas tariffs) is still small. Hydrogen is not yet commercially used in heating, but there is the potential for social acceptance problems over safety concerns (e.g., explosions, odourless leakage). Users may also be reluctant to replace appliances such as combustion boilers in order to use hydrogen for heating.

6.4.5 Heat Networks

Techno-Economic Developments

There are two types of heat networks: communal heat networks that serve multiple customers in one building (e.g., a flat) from a central gas boiler or heat pump in the basement, and district heating (DH) systems that use more extensive heat distribution infrastructures to serve multiple buildings (e.g., houses, schools, hospitals, office blocks), often in areas with dense heat demand such as city centres. While heat networks cover substantial parts of heat demand in many European countries (e.g., Austria 22%, Germany 14%, Denmark 70%, Finland 49%, Sweden 50%), they only cover 2% in the UK (BEIS, 2018a; DECC, 2012d).

UK heat networks have a long history and saw rapid growth in the 1960s and 1970s, owing to four kinds of deployment: a) in private sector tower blocks (mostly small communal schemes), b) single-owner hospital and university campus estates (medium-size DH), c) social housing blocks (medium-size DH), and d) several larger local authority-led schemes in cities and towns such as Woking, Sheffield, Southampton, and Aberdeen (Karvonen and Guy, 2018). Heat network deployment slowed down in the 1980s and 1990s due to energy market liberalisation and the penetration of domestic gas-based central heating. Deployment increased somewhat in the 2000s and 2010s but remained constrained by various barriers (DECC, 2013b).

In 2015, there were 11,908 communal UK heat networks and 2,087 district heating systems. The former provided 7,074 MW heating capacity and the latter 12,288 MW. The majority (90%) of all these systems used natural gas feedstocks (data from BEIS Heat Network Statistics, 2018, tables 1, 4, 5). Other fuel sources are electricity (5%), bioenergy and waste (2%), oil (1%), and coal and unknown (2%).

Heat networks have three main cost components related to: a) installing or adjusting heat interface units in the building, including heat meters, b) creating a heat pipe infrastructure, and c) installing and operating heat generation technology. Building the pipe-infrastructure is the largest cost component (Leveque and Robertson, 2014), which can be very substantial for district heating systems. The Greater London Authority (GLA, 2014) estimates that large-scale DH-systems, which may involve several tens of kilometres of heat pipe supplying 100,000 customers, can have infrastructure capital costs of £100 million or more. Medium-sized DH schemes (e.g., the Olympic Park and Stratford City project), which can support up to 20,000 homes, public buildings, and commercial users, can cost between £10 and £100 million.

Costs per dwelling can vary by a factor of four, depending on housing density and heat network configuration (Table 6.7). Heat networks for high-rise apartment blocks, which have high housing densities, are the most cost-effective. Costs per dwelling are highest for (semi)detached houses because these require longer pipe infrastructure.

At an aggregate level, DECC (2013a) estimates that heat networks can be cost-effectively applied in areas with heat demand density greater than 3 MW/km^2. DECC (2013a) further estimates that about 20% of UK heat demand (in the most

Table 6.7. *Cost estimates for different heat network configurations (TCPA/CHPA, 2008: 44)*

Building type	Form	Housing density (dwellings per ha)	Pipe length per dwelling (m)	Cost per dwelling (£)
High-rise apartment block	Corridor access, 10 to 15 stories	240	6.75	2,500
Medium-rise apartment block	Corridor access, 5 to 6 stories	120	8.0	2,800
Perimeter block of flats and townhouses	Stairwell or street-level access, 3 to 4 stories	80	11	4,100
Terraced street of row houses	Street-level access, 2 to 3 stories	80	13	5,300
Detached/semi-detached houses	Street-level access, compact street layout	40	19 to 24	7,700 to 9,550

populous towns and cities) has this density or more, so there is plenty of growth potential. Although heat networks are more efficient than individual gas boilers, payback periods are long (several decades for larger schemes). The cost-effectiveness of investments in heat networks therefore also depends on the discount rate (or level of return) that financiers require. With a discount rate of 10%, only 0.3% of heat demand can be cost-effectively met by heat networks. With a discount rate of 3.5%, networks may be economic for 6–14% of heat demand (DECC, 2013b).

Actors and Policies

Lead actors for historical heat network construction were local authorities (for application in social housing blocks or area-wide neighbourhoods) and private actors (for tower blocks, offices, campus estates). The construction role of local authorities has declined since the 1980s, as discussed in Section 6.3.1, which has eroded the capabilities for building medium- and large-scale DH-schemes. Limited construction in the 2000s focused mainly on communal heat networks, which 'remained small-scale, fragmented and hence technically sub-optimal' (Hawkey and Webb, 2014: 1229). Because of the small market size, the 'number of companies and individuals that specialise specifically in heat networks in the UK is small' (DECC, 2013a: 44). Engineering consultants, construction workers, plumbers, and energy companies have *general* skills that are relevant for heat network construction, but they miss more *specific* skills and knowledge. Consequently, there is a 'lack of common technical standards ... for design, installation, operation, and maintenance of [heat network] schemes' (DECC, 2013a: 50–51). Some components (like highly insulated district heat pipework) are not domestically produced, and the need to import them may increase material costs by 50% (Leveque and Robertson, 2014: 56).

Additional barriers for UK heat network deployment are the following:

- Limited knowledge and internal resources to instigate larger district heating systems, where local authorities tend to play a substantial role: 'UK local government is constrained by statutory duties prescribed by central governments, and is principally dependent on central government grant funding rather than local taxation' (Hawkey and Webb, 2014: 1232).
- Lack of generally accepted contract mechanisms: 'There is a lack of standardisation of the commercial arrangements for heat network construction and operation (including models for risk sharing)' (DECC, 2013a: 49).
- Limited technical and economic expertise for initial feasibility and design work, including detailed mapping of heat demand, technical planning, and estimating costs and benefits (Webb and Hawkey, 2017).
- Obtaining capital funding for heat network construction: This issue is not only challenging because of the large sums involved but also because many uncertainties increase investment risks and the cost of capital: 'With limited recent experience of constructing heat networks in the UK and little UK-based manufacturing, the costs of developing heat networks are particularly uncertain and there are significant commercial barriers which inflate costs' (Leveque and Robertson, 2014: 54). These uncertainties also include low-carbon heat generation options (e.g., bioenergy, large-scale heat pumps, waste heat from power stations), which heat networks are expected to use in future to lower greenhouse gas emissions.
- User acceptance: It may be challenging to persuade homeowners to switch from their current gas boilers to heat networks, but 'there may be less resistance to heat networks in new build' (DECC, 2013a: 49). Since households may also have concerns about the monopoly of heat network providers, it is concerning that 'as of December 2018, there is no regulator for heating networks ... meaning consumers have less security' about the quality of heat and pricing (Millar et al., 2019: 14).

Policy support for heat networks was piecemeal in the 2000s and fragmented across multiple schemes with 'none specifically designed for promoting the development of heat networks' (DECC, 2013a). Policy attention for heat networks increased substantially since the 2008 Climate Change Act. Although the *Low Carbon Transition Plan* (DECC, 2009) and *Carbon Plan* (DECC, 2011b) envisaged heat pumps as the main low-carbon heat provision technology, the two successive *Future of Heating* documents (DECC, 2013b, 2012d) saw a more prominent role for heat networks, potentially providing 20% of UK heat demand.

This new vision triggered further analyses of heat network potential and barriers (DECC, 2013d, 2013e), which was followed by policies to address the barriers. The Heat Networks Delivery Unit (HNDU), for instance, was created in 2013 to support local authorities in heat mapping, technical planning, and feasibility studies. HNDU also aimed to develop technical and commercial standards and templates and 'encourage knowledge sharing between local authorities who are developing heat networks' (DECC, 2013d: 5). In 2016, the government launched the £320 million Heat Networks Investment Project (HNIP), which aimed to alleviate the capital funding problem by providing substantial grants to district heating schemes. Following a two-year trial period, HNIP has since 2018 awarded £40m funding to seven projects, including £5.9m for the Gateshead District Energy Scheme, £6.6m for the Cardiff Heat Network, and £14.8m for the Meridian Water Heat Network. By reducing commercial risks, HNIP hopes to leverage £1bn in private funding for heat network investments.

To improve user security, BEIS (2020d) also launched a consultation about giving Ofgem powers to regulate domestic heat networks. The consultation document repeated the vision from the 2017 Clean Growth Strategy that 18% of UK heat could come from heat networks by 2050, which would require up to £16 billion of capital investment. Meeting those goals would require 'a step-change in the pace of rollout and adoption of heat networks with lower-carbon heat sources to meet our carbon reduction targets' (BEIS, 2020d: 10). While it is not yet clear how this step-change will be achieved, the policy and implementation momentum of heat networks has substantially increased in recent years and looks set to accelerate further in the near future.

6.4.6 Passive Housing

Techno-Economic Developments

Passive house (or 'passivhaus') is a radical whole-house design approach that reduces space-heating demand by more than 90% (Mlecnik, 2013). It combines multiple interdependent low-energy innovations such as super-insulated roofs, walls, and floors, triple glazing, mechanical ventilation with heat recovery (MVHR), passive solar design,[13] and technical solutions that enhance air tightness and reduce thermal bridges.[14] Focusing on thermal properties of the building shell, passive houses (PH) meet very high performance standards (Pitts, 2017): high thermal insulation values (e.g., U values between 0.6 and 0.15 W/m^{2}K), high air tightness standards (air flow lower than 0.6 air changes per hour at 50 Pc pressure), and very low annual space heating demand (lower than 15 kWh/m^2 of net living space). Because of these performance characteristics, PH-designs are seen as one

[13] Passive solar design uses windows, walls, and floors to collect, store, reflect, and distribute solar energy.

[14] A thermal bridge is an area or component of an object that has higher thermal conductivity than the surrounding materials, creating a path of least resistance for heat transfer. They most commonly occur at junctions between two or more building elements.

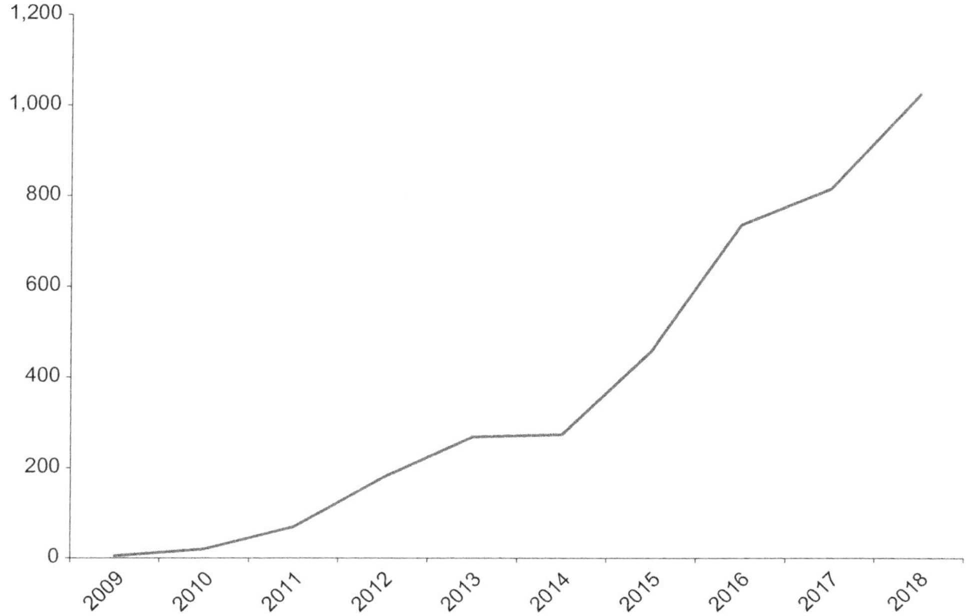

Figure 6.21 Cumulative number of passive house units in the UK (based on estimated data from www.passivhaustrust.org.uk/news/detail/?nId=787#:~:text=The%20Passivhaus%20Trust%20is%20delighted,many%20again%20in%20the%20pipeline)

of three possible routes (known as 'extreme fabric') to achieve zero-carbon homes (Zero Carbon Hub, 2013).[15]

PH-designs originated in Germany in the 1980s. Since then, about 65,000 PH-buildings have been constructed worldwide (www.passivhaustrust.org.uk/). Since 2009, the cumulative number of certified PH-buildings in the UK has increased steadily, to more than 1,000 in 2018 (Figure 6.21). Compared to mainstream new built homes (of which two million were constructed between 2008 and 2020), PH-buildings are still a very small niche.

PH-buildings are more costly to build than mainstream homes. For a standard-size Belgium home, PH-designs in the mid-2000s cost about €39.000 (or 16%) more than normal designs (Audenaert et al., 2008). Early PH-buildings in the UK had even higher additional costs because many of the materials and much of the installation expertise had to be imported from Germany, Switzerland, and Austria (Lynch, 2014). However, as domestic supply chains developed over time, costs of

[15] The other two routes are 'extreme low carbon technologies' (in which homes produce and store most of their own energy using heat pumps, solar-PV, biomass, or batteries) and 'balanced' pathway (using a mix of insulation, low-carbon 'on site' technologies, and low-carbon energy from district heating or community energy).

PH-components such as MVHR and triple glazing decreased. In 2015, the UK Passivhaus Trust (Passivhaus Trust, 2015) estimated that PH-buildings cost between 15 and 20% more than mainstream houses, which is between £33,000 and £44,000. Estimates in 2019 suggest that best practice additional costs were 9%, which is about £21,000 (Passivhaus Trust, 2019).

Actors and Policies

Low-energy house designs were pioneered from the 1970s in the sustainable building movement, which in the UK consisted of new entrants such as pioneering architects, developers, specialised suppliers and consultancies, and highly skilled self-builders (Lovell, 2008; Smith, 2007). Many of these actors were motivated by ecological, social, and cultural concerns, critiques of modern architectural practice, and an interest in using natural and local materials. Early designs were developed through small-scale, private projects using personal resources and academic research funding. Niche-actors experimented with construction principles, combining traditional and state-of-the-art materials (Lovell, 2008; Smith, 2007).

While the early sustainable building movement addressed a range of issues (e.g., water, waste, materials, energy, climate), a more narrowly focused low-energy and low-carbon niche emerged by the early 2000s (Lovell, 2008). This niche was carried by a heterogeneous mix of demonstration projects of competing designs, which varied in scale, actors, and motivations. The award-winning 2000–2002 BedZED project (Beddington Zero Energy Development) in London, for instance, constructed 82 units using both insulation materials and low-carbon technologies (e.g., solar panels, district heating). Focusing specifically on the building shell, PH-designs emerged within this niche in the late 2000s, with the first UK PH-building being completed in 2009 as a self-build project.

The two main application domains for subsequent PH-houses have been self-build by motivated individuals and social housing by local authorities, registered social landlords, and housing associations. Actors in both application domains were willing to accept the higher costs of PH-designs. Since 2009, PH-designs gradually scaled up, moving from single dwelling projects by self-builders to schemes of 30 to 90 homes by social landlords and housing associations (Passivhaus Trust, 2019). These larger schemes comprised low-rise developments of 2–3 bed apartments, and terraced and semi-detached houses, often for a mix of tenures such as ‘social rent’, ‘affordable rent’, or shared ownership. Self-build and social housing are small UK niches, which limits the potential for PH diffusion.

PH-designs have only marginally entered mainstream commercial markets, because of relatively limited interest from homebuyers (who are often deterred by higher PH prices) and volume housebuilders (Lynch, 2014; Pitts, 2017). The introduction of the 2006 Zero Carbon Homes (ZCH) target initially stimulated some industry hedging, because it was complemented in 2007 by the announcement of government plans to build five eco-towns (with 5,000–20,000 homes built to zero-carbon standards), which formed an attractive commercial opportunity (Edmondson et al., 2020). Incumbent building firms (e.g., Barratt, Stewart Milne) started to explore PH-designs and also funded the Zero Carbon Hub (created in 2008) to engage in relevant research and technical demonstration (Lynch, 2014). These hedging activities stagnated, however, when the ZCH-regulations were weakened in 2011 and scrapped in 2015. The post-2009 austerity politics also removed funding for the eco-town plans, which weakened the building industry's belief in the commercial prospects of PH-designs. Mainstream UK homebuilders see higher capital costs, low market demand, and the lack of supply chain skills as major barriers for passive house development, leading to limited engaged with the design concept (Heffernan et al., 2015).

Intermediary organisations such as the Passivhaus Trust, the Passive House Centre, and the Association of Environmental Conscious Builders have continued to develop, share, and disseminate PH-knowledge through workshops, courses, and national conferences, aimed at educating the general public and housing associations (Martiskainen and Kivimaa, 2018). But in the absence of supporting policies and wider industry interest, PH-diffusion is likely to remain relatively slow.

6.4.7 Whole House Retrofits

Techno-Economic Developments

Given the composition of the UK housing stock, which is predominantly made up of old buildings (see Figure 6.9) with around 80% of the current building stock projected to be still in use in 2050 (Dowson et al., 2012), the potential for an approach to building efficiency based entirely on new builds with Passive Housing standards is limited. Despite significant improvements with the implementation of individual energy efficiency measures over the last decade (see Section 6.3.1), most British houses have relatively low energy performance (see Figure 6.10). For the existing housing stock, whole-house retrofits are therefore crucial to meet carbon reduction targets.

Whole-house retrofitting of existing homes to significantly improve their energy efficiency (i.e., by 50% or even 80%) consists of a systemic approach to housing insulation that mobilises different measures, with particular emphasis on the building fabric and the combined interactions of measures. Whole-house retrofitting is not a technology in the conventional sense, but rather a systemic improvement strategy involving several techniques, insulation technologies, and site-specific considerations.

Retrofitting a home is more difficult than building an energy efficient home from scratch. It requires assessing opportunities for improvement, adapting existing design and building features, negotiating space and systemic constraints, as well as juggling building and conservation regulations. The main strategies seek improved insulation, air tightness, and ventilation. Depending on building characteristics, conventional technical measures may include solid wall insulation,[16] cavity wall insulation, loft insulation, floor insulation, glazing upgrades (double or triple), draught proofing, and may also be combined with heating measures (boiler improvement, heating controls, hot water tank insulation). Further measures include the replacement of heating and ventilation systems, and renewable energy technologies. However, there are technical barriers to the diffusion of retrofitting: an estimated 40% of the existing housing stock is considered 'hard-to-treat' as they 'possess solid walls, no loft space to insulate, no connection to the gas network or are high-rise' (Dowson et al., 2012: 296), and can therefore not easily be retrofitted using conventional techniques. Furthermore, the appropriate construction skills required for high performance retrofits are rare and the supply of high efficiency standard components is underdeveloped in the UK.

Whole-house retrofits are also costly. The Energy Technologies Institute, based on recent experiences, estimates 25% and 50% thermal improvements of a typical semi-detached house to cost an average of £12,000 and £19,000, respectively, and that these costs may be expected to drop to £7,500 and £13,750 in the period 2030–2050, with advances in materials and techniques (ETI, 2016). Looking at deeper retrofits achieving net-zero energy results as put forward by Energiesprong, current costs range between £60–90,000 per unit, with a possible medium-term horizon to bring these costs down to £40,000 (Brown et al., 2019).

Given the complexity of whole-house retrofits, and the eventuality of discovering building defects and surprises in individual projects, these costs provide only a rough estimate, which could be higher in practice for individual houses but also considerably lower per dwelling in tower blocks (IET, 2020). Regardless, for whole-house retrofits to become a credible option in the UK context, these large sums need to be brought down through supply chain improvements but will also require public subsidies as well as innovative ways of accounting for and valuing the added benefits (e.g., comfort, long-run savings). Spontaneous customer demand for energy efficient retrofits in the UK is low and its actual diffusion is hard to estimate:

> There are no centralised records of how many sustainability-related renovations projects are carried out, but there is consensus among both practitioners and academics that the rate of such renovations is stubbornly low. (Bobrova, 2020: 17)

Barriers to the adoption of whole-house retrofits include financial considerations (e.g., costly measures with long payback periods), technical considerations and

[16] Opportunities for external insulation are restricted in the UK because they alter historical façades.

challenges (e.g., 'hard to treat' buildings, uncertainty concerning energy saving calculations), supply-chain considerations (e.g., limited availability of skills and competences, difficult collaborations, and logistics involving multiple contractors), regulatory constraints (e.g., planning restrictions in conservation areas, lack of standards, low level of enforcement), and user considerations (e.g., lack of trust, lack of interest, lack of information, preference for historic features, disruptions) (Martiskainen and Kivimaa, 2019; Yeatts et al., 2017). Yet, as suggested by a number of commentators, whole-house retrofits promise not only to deliver long-term energy savings and GHG emissions reductions, but they could also underpin significant industry and quality jobs: 'expertise in highly energy efficient buildings also represents an industrial opportunity for the UK' (CCC, 2019c: 59)

Actors and Policies

Incremental energy retrofits have been an important part of emissions reductions in the UK housing system, notably through the installation of more efficient heating appliances and targeted insulation measures. The period 2004–2012 has been particularly effective, owing to the combination of 1) relatively easy and cost-effective technical options, and 2) an effective dedicated policy framework (Eyre and Baruah, 2015). Comparatively, whole-house retrofits stretch current cost-benefit calculations and technical capabilities, and lack a dedicated policy framework (Martiskainen and Kivimaa, 2019). The deployment of whole-house retrofits on a large scale will require innovation in business models, finance mechanisms, and logistics (Brown, 2018), as well as a coherent policy mix that addresses regulatory barriers, provides financial incentives for owner and users, and engages contractors in the development of new practices and supply chains.

Advocates of whole-house retrofits include environmental NGOs, architects, and designers seeking to pioneer sustainable building practices, but also increasingly local authorities and councils. However, many non-profit associations, such as the Association for the Conservation of Energy (ACE), the Energy Saving Trust (EST), or the UK Green Building Council (UKGBC), have experienced declining resources in recent years (Kivimaa and Martiskainen, 2018). The UK Green Building Council brings together industry organisations with the mission to 'radically improve the sustainability of the built environment' (ukgbc.org), notably by campaigning for higher standards, sharing experience, and developing technical guidance. Concerning the scaling of retrofits in the UK, it is particularly concerned with enabling local authorities to leverage resources and stakeholders to deliver solutions best suited to local needs (UKGBC, 2021). For instance, local authorities have a key role to play with respect to the maintenance and improvement of social housing. Local authority and housing association homes provide a unique setting for demonstrating deep energy retrofits, because they account for 17% of the housing stock, offer opportunities for the development of solutions on a large scale (large individual estates, standardised building design), have an explicit social mission, and because energy retrofits could be combined with required renovation programmes (notably to improve safety standards) (IET, 2020).

In terms of policy, despite ambitious long-term goals and decarbonisation targets for the housing sector, current policy instruments are fragmented and do not add up to a coherent policy mix that can drive innovation, supply chain development, and market uptake of whole house retrofits at the required scale:

> In many areas current policy is failing to drive uptake, including for highly cost-effective measures such as loft insulation. Policies have yet to be set out to deliver the stated ambition on home retrofits (EPC band C by 2035), including for those households deemed "able-to-pay", and a delivery mechanism for social housing minimum standards. Policy needs to incentivise efficient long-term investments, rather than piecemeal incremental change. (CCC, 2019c: 61)

Indeed, current policies primarily cover minimum performance standards and regulations (which tend to water down ambitions), incentives for individual measures (which do not really encourage integrated whole-house approaches), and seed-funding for demonstrators.

Standards provide systematic metrics and benchmarks for building performance. The Energy Performance Certificates (EPC), introduced in 2007, provide information on energy performance to potential tenants or buyers, without introducing any specific requirements. The UK's Buildings Regulations are largely concerned with setting minimal acceptable standards and regulations for new buildings and refurbishments. However, they primarily encourage 'reasonable' provisions for energy efficiency and fuel savings within technical, functional, and economic feasibility limits, and so do not actively influence more radical interventions such as whole-house retrofits. The Government launched the Zero Carbon Homes target in 2006 and the Code for Sustainable Homes in 2007, as a voluntary initiative to push the boundaries beyond what is required by regulations, with expectations that it would develop into a hard target for new builds that could also be applied to refurbishment. However, subsequent policy reversals on the Code for Sustainable Homes and the Zero Carbon Homes standard indicate policy reluctance to encourage more ambitious industry norms and standards. More recently, the Clean Growth Strategy set a target to upgrade homes to EPC band C by 2035, but the phrasing remained vague and uncommitting:

> We want all fuel poor homes to be upgraded to Energy Performance Certificate (EPC) Band C by 2030 and our aspiration is for as many homes as possible to be EPC Band C by 2035 where practical, cost-effective and affordable. (BEIS, 2017b: 13)

A number of voluntary standards and certification schemes have been developed, with and without public policy involvement. The Bonfield Review (Bonfield, 2016) led to a revision of the Publicly Available Specifications (PAS 2030:2017) for 'the installation of energy efficiency measures in existing buildings' (Laganakou, 2019), notably to take account of the need to adopt a 'whole dwelling' view on building efficiency by attending to the interaction of measures (e.g., insulation and ventilation). It also paved the way for the higher specifications in PAS 2035:2019 for 'retrofitting dwellings for improved energy efficiency', which covers 'how to access dwellings for retrofit, identify improvement options, design and specify Energy Efficiency Measures (EEM) and monitor retrofit projects'. Voluntary standards and assessment methodologies seek to further push measurable levels of building thermal performance, including:

- BREEAM (Building Research Establishment Environmental Assessment Method) is a housing assessment methodology developed by the BRE for new and refurbished buildings
- The Passivhaus refurbishment standard (EnerPHit) is set on achieving a heating demand of 25 kWh/m^2/year
- Energiesprong has pioneered a new build and whole-house retrofit standard based on actual measured energy consumption rather than modelling based on implemented measures. In terms of space heating, it aims for 30 kWh/m^2/year for demonstrators but accepts performances of 40 kWh/m^2/year for pilots on a case-by-case basis (Energiesprong UK, 2018). Initially launched in the Netherlands, backed by Dutch Government funding, it is being adopted in demonstrations in the UK, including the Nottingham City Council and London Borough of Sutton demonstration programmes. The Energiesprong model, though still reliant on public subsidies to be viable, rests on innovative value and supply-chain propositions: it is specifically oriented towards whole-house retrofits, offers a guarantee of net-zero energy consumption, emphasises the co-benefits of home improvement (e.g., aesthetics and comfort), and seeks to build up an integrated supply chain to address the complication of dealing with multiple suppliers and installers (Brown et al., 2019).

Concerning incentives, funding mechanisms have been introduced in relation to targeted energy efficiency improvements in homes, but few instruments have targeted whole-house retrofits as an integrated and rather radical form of efficiency improvement. The Green Deal, initiated in 2013, was aimed at a mass rollout of energy efficient retrofits by addressing the main economic barrier: upfront costs. Its finance mechanism, making £120m available for private energy efficiency improvement, displaced upfront costs by spreading them over time. However, the Green Deal was ineffective and terminated (see Section 6.3.3). It has, however, been criticised as targeting the low-hanging fruits made up by the easiest retrofits (Rosenow and Eyre, 2016), and it is not clear how less cost-efficient retrofits will be funded in order to meet the targets. The termination of the Green Deal has been detrimental to the development of a homegrown retrofitting industry.

Concerning investments in R&D and demonstration, there are interesting projects being funded, but these remain small, relatively experimental, and involve very little funding and commitment:

- the £17m Retrofit for the Future programme (2009–2011), which aimed to demonstrate the feasibility of ambitious (80% reduction) retrofits in UK social housing stock and 'kick-start' an industry (Laganakou, 2019)
- Retrofitting British houses to passive house standards remains a challenge for which it is important to multiply demonstration projects and proper evaluation and data collection. In an effort in that direction, the Technology Strategy Board launched the 'Retrofit for the Future' competition in 2009, funding over 100 high energy standard retrofit demonstrations in the UK, encouraging collaborations between housing providers, designers, contractors, and researchers. It has helped stimulate new business opportunities in the UK retrofit market.[17]
- Recently, the low impact buildings innovation platform secured £60m to support innovation in buildings over 2014–2019.
- In 2020, the Government awarded £7.7m to support whole-house retrofit demonstration projects in different settings (London Borough of Sutton, Nottingham City Council, Cornwall Council), each supported by public–private partnerships.

Concerning societal objectives, the main motivations behind retrofits are energy performance improvements to reduce energy use and carbon emissions. Retrofits can also reduce individual energy bills, increase energy independence and so contribute to addressing fuel poverty and excess winter deaths, particularly if deployed on social housing (IET, 2020), and more generally improve comfort and safety. Given the high costs of deep retrofits, advocates have argued for new forms of investment assessments to take into account the wider benefits that they entail.

From the perspective of users, there are major barriers to energy efficiency refurbishment (costs, disruption, uncertainties about long-term return on investment in terms of reduced bills, etc.), which reduce the appeal of whole-house retrofits relative to other options with more immediate benefits and shorter-term investments. Furthermore, the immaturity of the supply chain may lead to unnecessary delays, inexperience, mistakes, and cost increases, which are

[17] See retrofit.innovateuk.org

additional sources of uncertainty discouraging an already small customer base. Indeed, 'many households find a whole-house approach impractical, and are likely to be more attracted to retrofit measures that take place over time, spreading the cost and disruption' (Kerr and Winskel, 2020: 109778).

6.5 Low-Carbon Transition through Whole System Reconfiguration

Synthesising the analyses of sub-systems and niche-innovations, this section first assesses low-carbon whole system reconfiguration through the three dimensions (techno-economic, actors, policies) and then addresses speed, scope, and depth of change.

6.5.1 Low-Carbon Innovations Driving GHG Emission Reductions

Focussing on the deployment of technical innovations, we conclude that the notable reduction in GHG emissions from residential heating between 2001 and 2014, despite an increasing building stock, resulted primarily from incremental efficiency improvements in existing building and heating systems, as well as some diffusion of renewable heat, which continued after 2014.

The incremental changes included efficiency improvements in existing gas boilers, which were stimulated by gradually tightening standards and energy savings obligations. They also included the diffusion of incremental housing insulation measures, although these remained piecemeal and uneven across the housing stock. These incremental component improvements and piecemeal additions to existing systems only required localised adjustments in the heating and housing systems without affecting their structure, organisation, or reliance on gas infrastructures.

Concerning low-carbon niche-innovations, there are notable differences between the heating and buildings systems. In the heating system, renewable heating sources increased substantially to over 9% of water and space heating (all sectors) in 2019, supplied by a variety of renewable sources (e.g., biomass heating, heat pumps) (Figure 6.16). Domestic wood combustion, which has some negative sustainability effects such as air pollution, diffused farthest. Wood combustion is only carbon-neutral when abated, and since this was often not the case, the diffusion of renewable heat did not translate into equivalent GHG emissions reductions. Heat pumps, biomethane, and plant biomass (for industrial use) also started diffusing after the introduction of the non-domestic RHI in 2011 and domestic RHI in 2014. But markets and supply-chains for these more recent niche-innovations are still uncertain and fragile.

Table 6.8. *Mapping system reconfiguration opportunities in the UK heat domain*

		Core elements			
Linkages (coupling) between system components		**Reinforced**	**Substituted**		
	Unchanged	***Modular incrementalism*** – Efficiency innovation in appliances (boilers) – Efficiency innovation in housing (piecemeal insulation)	***Modular substitution***		
			System–system switching	***Niche-system add-on and hybridisation*** – Solar thermal as additional heat source – Biomass heat as additional heat source	***Niche-system replacement*** – Biomass boilers for gas boilers – Heat pumps – Greening the grid with biomethane
	Changed	***Architectural stretching*** – *Whole-house retrofits*	***Architectural reshaping*** – Passive house designs – Heat networks – Greening the grid with hydrogen injection – Electrification of heat with 100% RETs		

In the housing system, niche-innovations such as passive house and whole-house retrofits have remained very small, owing to the detrimental effect of policy reversals (e.g., Zero Carbon Homes, Green Deal) on adoption and market development, creating further difficulties for fragile supply-chains.

6.5.2 Techno-Economic Reconfiguration

To further interpret the pattern of heat and building system reconfiguration, we use summary Table 6.8, which positions the various low-carbon innovations in the techno-economic mapping framework that we developed in Section 2.2.1. Combining this table with the dynamic analyses in the previous sections, we identify the following pattern.

Modular incrementalism (in the form of incremental boiler or piecemeal insulation improvements) has been the dominant type of change for much of the studied period. Although these 'low-hanging fruits' are relatively easy and cheap to implement, they are not sufficient to deliver deep decarbonisation. In fact, GHG emissions from heat and buildings have slightly increased since 2014 (Figure 6.2).

In recent years, political debates and future visions have started to focus more on modular substitutions that may enable deep decarbonisation but are largely compatible with existing infrastructures and only require add-on or substitution in isolated components of heat systems. Some of these options, such as conversion to biomass boilers, heat pumps, or greening the grid with biomethane involve niche-system replacement and substitution of heat generation devices and energy carrier. Other options, such as add-on heating options (e.g., solar thermal or domestic wood combustion) that are used besides gas boilers, involve niche-system hybridisation since they offer reconfiguration pathways whereby existing components are not abandoned but complemented. By 2019, niche-system add-on and hybridisation options were larger than niche-system replacement options:

- domestic wood combustion, which represented 36.9% of renewable heat, generated 5% of total heat in 2019 (Figure 6.16)
- grid-injected biomethane, which represented 9.1% of renewable heat, generated 1.1% of total heat in 2019 (Figure 6.16)
- heat pumps, which represented 17.4% of renewable heat, generated 1.6% of total heat in 2019 (Figure 6.16).

Whole-house retrofits can be considered an architectural stretching option because they keep the main element (existing buildings) unchanged but involve a systemic modification of linkages between building envelope, ventilation, and energy use to maximise thermal efficiency and comfort, through the integrated introduction of refurbishment measures. While whole-house retrofits are a promising option with significant potential, they have not diffused significantly to date due to high costs, a poorly understood value proposition, and weak supply-chains.

There are also several niche-innovations with architectural reshaping potential, which require changes in both components and system architectures.

- Passive-house designs involve changes in roofs, walls, floors, windows, ventilation, and overall design parameters to minimise heat losses and optimise solar heating.
- Heat networks involve both infrastructural changes and a shift towards a heat service business model. Architectural changes would be even larger if heat networks also use renewable inputs (which would be needed to achieve deep decarbonisation).
- Greening the grid with hydrogen involves changes in upstream inputs and technical production processes, new or adjusted pipeline infrastructures, and changes in end-use appliances (because hydrogen has different chemical and physical properties than natural gas). Delivery visions and strategies seek to minimise associated disruptions by advocating a cluster-based approach.

All of these architectural niche-innovations have significant decarbonisation potential, but are currently very small, because they face significant hurdles. Current barriers include infrastructural and cost challenges (e.g., large proportion of old buildings unlikely to be retired from the market, significant costs of heat networks and hydrogen transmission infrastructure), technical challenges (particularly for hydrogen), regulatory challenges (e.g., building codes and standards for passive housing, long-term multi-partite contracts for heat networks, safety standards for hydrogen production, storage, delivery, and appliances), and capabilities challenges (e.g., shortage of skilled installers for passive housing). While hydrogen injection is a long-term option, the gradual scaling of heat networks and passive housing is possible today (as evidenced by successes in other countries), provided significant financial incentives are provided (to cover start-up and scale-up costs) and new policy mixes reduce barriers and uncertainties.

6.5.3 Actor Reconfiguration

For *existing* heating and housing actors, we conclude that there have been significant developments towards greater acceptance of climate change and energy efficiency as guiding principles, but that these mostly remain layered on top of other considerations (e.g., keeping costs down, reducing regulatory burdens) and are framed as competing with other issues (e.g., fuel poverty, access to affordable energy and housing, energy security, reliability). Accordingly, most change has been incremental, aimed at preserving existing market positions, infrastructure, business models, and skillsets. This has hampered more radical reconfiguration pathways and innovations.

However, we also observe the emergence of *new* actor coalitions, articulating more radical visions around new technologies, developing new skills and competence bases. These coalitions have remained rather small in the housing system, where passive house designs and low-carbon retrofits remain on the margins of a powerful construction sector dominated by volume housebuilders.

In the heating system, however, we see a more dynamic picture, with more vibrant but still small new coalitions forming to develop and deploy renewable heating technologies (e.g., heat networks, heat pumps, biomethane), stimulate user awareness, and lobby for favourable policies. There is, however, only limited user interest in alternative heating options, and a reluctance to pay more. Under current circumstances (higher costs relative to conventional technology, even with RHI subsidies), diffusion has therefore remained limited to adoption by green consumers. Furthermore, current policy uncertainties and conflicting signals tend to leave strategic planning decisions up to market competition and the influence of industry lobby groups, which has exacerbated existing power dynamics and inertial tendencies as well as increased entrepreneurial risks for more radical alternatives. Policymakers have provided some support for alternative heat options through incentives (notably the RHI), which stimulated the emergence of niche-innovations but are not enough to drive large-scale diffusion or the development of strong supply chains.

Tables 6.9 and 6.10 systematically summarise and interpret actor reconfigurations in the heating and building systems. For the heating system, Table 6.9 shows that the actor reconfigurations that *support* low-carbon transitions have been 'low' to 'medium', while the lock-in mechanisms and competing issues that *constrain* actors' engagements with low-carbon transitions have been 'medium' or 'large'. This helps explain why low-carbon reconfiguration in heating has been mostly incremental, with some recent shift towards modular substitutions, but hardly involves architectural reshaping. For the building system, Table 6.10 shows that the actor reconfigurations that *support* low-carbon transitions have been mostly 'low', while the lock-in mechanisms and competing issues that *constrain* actors' engagements with low-carbon transitions have been 'large' or 'medium'. This helps explain why radical low-carbon reconfiguration options have been very slow. Even incremental building improvements, which used to be important, have slowed to a trickle in recent years due to weak and fragmented policies.

The overall picture is that existing heating and housing systems are still relatively stable due to the dominance of large incumbents, limited user interest, infrastructural inertia (gas grid, housing stock), and lack of coherent or stable policy frameworks, which generates uncertainty for reconfiguration. At the fringes of the heating system, several niche-innovations have emerged and secured a foothold in small (subsidised) markets. These innovations are small, but gradually growing. Radical niche-innovations in the housing system have remained very small and marginal, because incumbents have successfully lobbied to reverse support measures such as the Zero Carbon Homes policy. These niche-innovations are struggling to develop a toehold and have unclear future scaling pathways.

Concerning low-carbon heating options, two main visions co-exist, supported by different actor groups: 1) emerging actor coalitions support the deployment of low-carbon substitutes, and 2) established coalitions favour preserving pathways (e.g., efficiency improvements, appliance replacements, and greening of the gas grid). Given important uncertainties and competition between these visions, we currently observe the multiplication of niche dynamics (low-carbon innovations developing in spaces presenting favourable conditions) with unclear scaling pathways ahead.

There are two manifestations of this niche phenomenon in low-carbon heating. One manifestation is that low-carbon heating options are deployed by particular social and demographic groups such as households with significant disposable

Table 6.9. *Changes and lock-ins for actors in the heating system*

	Actor changes supporting low-carbon transition	Actor lock-ins and competing issues constraining low-carbon transition
Energy supply/ distribution appliance, installers	LOW-MEDIUM – Incumbents acknowledge climate change mitigation as important goal but take limited significant steps – Limited number of certified low-carbon heat installers – Limited incumbent reorientation towards deep decarbonisation options – But increasing activity from new suppliers (of heat pumps, biomethane, heat networks, biomass stoves)	MEDIUM-LARGE – Support solutions that protect sunk investments in existing grids: efficiency improvements, greening the gas grid. – Maintain large-scale operations and business model (in energy supply) – Resistance from appliance installer trade favours conventional gas heating – Shrinking gas supply market share is more important concern for incumbents than climate mitigation
Policymakers	LOW-MEDIUM – Heat decarbonisation on policy agenda since 2009, but no stable or comprehensive policy strategy – Domestic RHI provided some incentives for low-carbon heat, which supported emergence of niche-innovations – Lack of complementary measures to develop supply chains, skills, etc.	MEDIUM-LARGE – Changing policy visions, frameworks, and instrumentation generated uncertainty and favoured business-as-usual – Other issues (affordability, energy poverty, energy security) appear to be more salient than climate mitigation
Users	LOW-MEDIUM – Little interest and engagement with heating from mainstream users – But early adopters switch to heat pumps or biomass stoves (for reasons of cosiness)	LARGE – High user satisfaction with existing heating appliances and practices – User routines hamper heating systems that require more user engagement – Lack of trust in installers hampers adoption of low-carbon appliances – Concerns over rising energy costs hamper low-carbon heat diffusion
Civil society organisations, public debate	MEDIUM – NGOs have contributed to putting low-carbon heating on the policy map	MEDIUM – Fuel poverty is perceived as a more acute societal problem than climate change – Trade organisations have significantly engaged in public debates to protect vested interests

Table 6.10. *Changes and lock-ins for actors in the housing system*

	Actor changes supporting low-carbon transition	Actor lock-ins and competing issues constraining low-carbon transition
Housebuilding actors	LOW – Recognition of climate mitigation as additional issue, but pursued mainly through incremental and piecemeal insulation measures – Some experimentation with Zero Carbon Homes in late 2000s, but this did not lead to strategic reorientation – New entrants advance radical innovations (e.g., passive house, whole-house retrofit) in small niches	LARGE – The sector is concentrated and dominated by volume housebuilders seeking profit maximisation and risk minimisation – Smaller building firms equally reluctant to raise industry standards – Important lobbying against Zero Carbon Homes targets – Lack of skills, training, and learning for low-carbon building
Policymakers	LOW – Building regulations and energy saving obligations stimulated incremental building improvements from 1990s to about 2012 – Since then policies were poorly designed, watered down, and reversed (e.g., Green Deal, Zero Carbon Homes)	MEDIUM – Split policy responsibilities (environment and energy, building and housing) are a challenge for a coordinated low-carbon strategy – Climate and energy performance are less of a priority than other issues (e.g., supply, affordability, quality, skills, and innovation)
Users	LOW – Climate change not a major motivator – Incremental thermal improvements motivated by other concerns (e.g., thermal comfort, long-term savings) – Limited interest in radical innovations for multiple reasons (e.g., high cost, inconvenience, daily life disruption, limited confidence in builders)	MEDIUM – Ownership structure provides conflicting incentives for insulation on cost grounds for non-owners and low-income groups – Climate mitigation less important than other issues (e.g., house value, aesthetics, convenience)
Civil society organisations, public debate	LOW-MEDIUM – Advocacy by NGOs placed the issue on the agenda – Co-benefits of thermal improvements are an important argument	LARGE – Housing shortage, decent housing, low building quality, increasing house prices, and (un) affordability are more important issues in public debate on housing than climate change.

income and green interests, social housing residents benefiting from dedicated (pioneering) programmes, or households in off-grid locations with few alternatives. The second manifestation involves localised niches around specific regional innovation ecosystems, such as with the development of biomethane production in the south-west, which is paving the way for 'cluster' approaches to grid injection, or as can be witnessed for large-scale district heating schemes around pioneering urban development areas. What we are seeing in these latter cases is significant experimentation but also struggles with standardising approaches (e.g., long-term heat contracts and fragile collectives around heat networks) and market uncertainties (e.g., unviable biomethane production followed by a consolidation phase).

Concerning low-carbon housing, we currently observe significant inertia (more than in the heating sub-system), despite some exploratory diversification from incumbent actors between 2006 and 2011 (after the ZCH policy), which receded after policy weakening and reversal. This inertia results from resistance by the concentrated housebuilding industry against tighter building standards and an effective lobby against a shift towards low-carbon building codes. The inertia also has an infrastructural dimension, because the UK has a relatively old housing stock which is particularly hard to thermally improve.

The role of users is also problematic for low-carbon transitions in the heating and building systems, where the adoption of many options needs to be done by and in households (district heating and gas grid repurposing are different in this respect), which makes it different from the electricity system where adoption is done upstream and costs then passed on (often unknowingly) to users. The adoption of low-carbon options has so far been a slow process because: a) few households think much about their heating system or associate it with climate change, b) households mainly contemplate replacement when the heating device breaks down, c) costs are moderately high (in the hundreds or thousands of pounds for boilers and heat pumps, and in the tens of thousands for comprehensive low-carbon housing measures), d) installation disrupts daily household life and is therefore perceived as a nuisance, and e) there is much uncertainty and doubt about the suitability of low-carbon options and the skills and reliability of installers. These challenges for significant household adoption may have stimulated the government to explore with significant vim and vigour the possibility of working with incumbents (in gas supply and distribution) on upstream decarbonisation options that largely maintain the existing infrastructure (greening the gas grid).

6.5.4 Policy Reconfiguration

Policymakers only gradually recognised heat as an important domain for decarbonisation, evidenced by the development of high-level policy strategies in two distinct stages (2012–2013, 2017–2021). We also observe increasing awareness of the considerable challenges of heat decarbonisation. Policy reconfiguration has so far mostly consisted of broad outline strategies and policy visions with a limited arsenal of coherent policy instruments, which has led to a pattern of over-promising and under-delivering.

Formal Policies

Heat-related policies in the 1990s and 2000s mobilised a range of instruments, including regulations and obligations, which led to incremental but substantial improvements in boiler efficiencies. Since the early 2010s, policy focussed on supporting the deployment of low-carbon heat sources, but the instruments are narrower than before and mostly focused on market-based interventions (the Renewable Heat Incentive) with little technological prescription or innovation support. This strong reliance on market modulation has not been very effective, because it is fragmented, lacks a systemic strategy and complementary instruments (notably to develop supply chains and key competences for the delivery of large-scale roll-out of alternative heat appliances), and therefore denotes a lack of policy consistency. Visions and objectives, though ambitious, have not been translated into concrete and effective measures, leading to a delivery gap marked by over-promising and under-delivering. Policy signals and instruments also changed over time, which created a lack of policy coherence and an uncertain context for the development of supply-chains, which ultimately limited on-the-ground delivery.

Concerning policies targeting the *buildings* system, current interventions to support low-carbon buildings and retrofit measures are less developed than for heating appliances, and are insufficient to reach decarbonisation objectives. Until the early 2010s, regulatory instruments (e.g., energy savings obligations and performance standards) stimulated incremental efficiency improvements through piecemeal insulation measures. More recently, building decarbonisation policy has focussed on driving innovation in building and retrofitting techniques (e.g., Green Deal and Zero Carbon Homes plan). However, policy terminations, policy reversals, and a fragmented approach relying on isolated policy instruments has created policy incoherence and inconsistency, which has led to disappointing results such as plummeting deployment of key insulation measures since 2013.

Lobbying from the construction industry succeeded in watering down regulations and standards. Significant loopholes (e.g., the existing housing stock is not the object of stringent objectives) and the postponements of objectives further weaken the signals for the significant changes required. Currently, low-carbon building and retrofitting are only pursued by a small fraction of pioneering industry actors (and committed clients) who bear the associated risks. What is needed to deliver a large-scale decarbonisation of buildings is a more coherent and consistent approach, including tighter standards and regulations, innovation and market deployment support, and an industry-wide training and skilling strategy.

We conclude that heat decarbonisation is a relatively new policy concern that has risen on the mainstream policy agenda from 2012 but is facing significant hurdles. We also observe policy tensions and changing orientations, which generate uncertainties for deep low-carbon heat reconfigurations. These uncertainties relate both to the formulation of transformation visions and strategies and to the design of concrete instruments and policies.

The formulation of heat decarbonisation pathways and related visions has changed significantly over time. From 2011, future scenarios initially relied strongly on the electrification of heat and continued efficiency improvement, which a few years later was complemented with heat network visions. More recently, we observe policy interest in exploring the repurposing of existing gas infrastructure (in addition to electrification and heat network pathways). This change in the portfolio of options and pathways is not necessarily problematic. Indeed, technological variety may increase the likelihood of future transformations, and it is also largely plausible that the decarbonisation of heating in the UK will rely on a wide range of options, adapted to different settings (which in itself represents a major break from the current quasi-universal gas-based system).

On the one hand, changes in transformative visions are thus to some extent inevitable and can be an important marker of policy learning. On the other hand, however, there are also significant costs associated with delaying decisions about the preferred pathway: financial costs of investing in multiple options, policy legitimacy and credibility costs, as well as over-exposure to the influence of vested interests who may interpret policy indecision as opportunities for advocacy. Although the recent interest in options that preserve existing infrastructures has short-term cost advantages (such as avoiding stranded assets), it may lead to future lock-ins that hamper more radical and effective decarbonisation options.

The design of concrete interventions and formal policies has thus far primarily relied on single instruments (market incentives such as the RHI or more regulatory approaches such as ZCH). This fragmented approach has not been effective in delivering robust and vibrant markets and supply chains, because it tends to target only one element of the system, is vulnerable to policy reversals, and fails to provide an adequate degree of policy coherence and consistency. Policy interventions to support low-carbon heat have so far lacked an integrated perspective enabling the joint development of user demand, supply-chains, and skills. Beyond single instruments such as the RHI, the government is yet to deliver a consistent policy framework for heat and buildings, leading the CCC to conclude that: ‘Over ten years after the Climate Change Act was passed, there is still no serious plan for decarbonising UK heating systems or improving the efficiency of the housing stock, while no large-scale trials have begun for either heat pumps or hydrogen. The low-carbon skills gap has yet to be addressed' (CCC, 2019a: 66). Delivering low-carbon heating and housing will require a more diverse set of interventions, articulated in a policy mix, as well as safeguards against policy swings over time.

Governance Style

The governance style in the past 10 years has primarily followed a *hands-off, neo-liberal intervention logic*, characterised by strong reliance on market mechanisms and incentives with limited technological prescription, a preference for voluntary standards (as opposed to regulatory prescriptions for industry), and a reluctance to interfere with user preferences through behaviour change policies. This deviates from the preceding period (mid-1990s to 2010), when *a regulatory style* (e.g., energy savings obligations and performance standards) was reasonably effective in driving the uptake of more efficient appliances and insulation measures. Attempts at a more interventionist style (e.g., Zero-Carbon Homes policy for buildings, Green Deal for demand-side responses involving users) were short-lived because they faced significant opposition and disappointment, which seemingly confirmed the preference for a hands-off approach.

While this style is deeply engrained in current policymaking, we also note that this is not inevitable and may change in the future, given that policymakers in the UK have had significant influence over heat system transitions in the past (Hanmer and Abram, 2017) and have new opportunities to do so, notably with raised urgency and public concerns around climate change, and greater acceptance of more hands-on interventionist governance since the COVID-19 crisis.

Owing to this hands-off style, but also because of the relative novelty of policy interest in heating and buildings as a target area for decarbonisation, we also observe a *deficit of policy expertise and coordination*. Indeed, the UK currently lacks a dedicated body with oversight over heating systems and fuels (Connor et al., 2015) (despite Ofgem playing a *de facto* role), and there is only limited policy coordination concerning low-carbon housing, which intersects multiple policy portfolios and responsibilities. The Committee on Climate Change therefore called for a change in governance style, providing more leadership and coordination: 'Approaches based on heat pumps, hydrogen and heat networks will only be realised with strong Government leadership at both local and national levels because all of these solutions will require coordination' (CCC, 2016: 10).

In parallel, we also observe a tendency for *working closely with incumbent actors* for decisions concerning heating (e.g., renewed interest for biomethane injection and grid repurposing) and housing (e.g., attention to building industry concerns about imposing regulatory constraints). This is a double-edged sword, however. On the one hand, working closely with incumbents may lead to policy pathways with more buy-in from key actors, who may be willing to mobilise their own resources (e.g., finance, expertise, infrastructure, research) towards implementation. On the other hand, this may lead to policy capture by dominant actors (especially when relatively weak sectorial public expertise does not offer the appropriate means to evaluate industry proposals), a lowering of overall ambitions towards what is deemed 'cost-effective' rather than in line with long-term policy objectives, as well as policy legitimacy problems if policymakers are seen as catering exclusively to 'big business' interests.

A notable governance style problem relates to *policy coherence over time*, which we see as symptomatically lacking in the UK heat policy landscape. Indeed, there have been multiple significant direction changes, policy reversals, and failures. These are linked to poor policy design but also to a tendency for grand strategising, reliance on abstract modelling, and comparatively little attention to detailed implementation or reliance on sector-level expertise. These difficulties in setting out a systemic and temporally consistent approach to heat policy have tended to exacerbate market uncertainties, creating a negative environment for supply-chain development. More generally, uncertainty with energy policy directions is emerging as a British syndrome (Keay, 2016; Kuzemko, 2016; Rosenow and Eyre, 2016).

Since 2017, we observe steps towards the formulation of an *industrial strategy* concerning heating (and a lack thereof for buildings), notably through funding of demonstrations and trials and a shift towards more technology-oriented policy (e.g., biomethane production, hydrogen production, electrification of heat). One-shot demonstration investments should not, however, detract from the need to develop a *coordinated* policy framework and long-term *coherence*. Indeed, considerable uncertainties remain regarding the delivery of such an approach, notably in the absence of a more integrated policy mix (e.g., targeting skills and supply-chain development) and long-term policy coherence, as well as in the absence of targeted deployment measures beyond trials. The upcoming Heat and Buildings Strategy may be an opportunity to put forward such an approach, particularly as it is oriented by ambitious objectives set out in the Ten Point Plan for a Green Industrial Revolution (HM Government, 2020b) and the Energy White Paper (HM Government, 2020a), and as it may benefit from recent cash injections from the COVID-related recovery package. Nevertheless, very serious concerns remain about the feasibility of implementation under current circumstances (Institute for Government, 2021).

6.5.5 Scope, Depth, and Speed of Reconfiguration

Low-carbon reconfigurations in the UK heating and building systems are rather limited to date. There are differences, however, in terms of the scope, depth, and speed of reconfigurations.

The scope of the unfolding reconfiguration in the heating system is moderate, because only a subset of low-carbon innovations is being supported and experiences some diffusion. These innovations represent modular incremental and modular substitution options, which require only limited adaptation of established actors. The more radical innovations (which require more substantial system and actor reconfigurations) are less developed and stuck in small niches.

The depth of reconfiguration in the heating system is limited, because market niches remain small and have not yet led to substantial reorientation of established actor activities and business strategies (within the energy supply and conventional appliance trades), despite some noticeable activity from new entrants (e.g., biomethane production, and heat pump and biomass stoves suppliers).

The speed of reconfiguration is limited to moderate because diffusion of most renewable heat technologies (e.g., wood combustion, plant biomass, biomethane, heat pumps) has remained limited. While, cumulatively, renewable heat reached about 9% of overall heat production in 2019 (across all sectors), up from negligible contributions before 2007, this is far from enough to support significant GHG emissions reduction given that many renewable heat options are not low-carbon if unabated and given that domestic energy consumption for heat has increased since 2014 (Figure 6.6). The most recent deployment targets for heat pumps seem over-optimistic under current policy conditions, and the deployment of biomethane remains at an experimental demonstration stage.

The scope of reconfiguration in the building system is limited, with few radical alternatives diffusing to deliver substantial low-carbon improvements (passive housing and whole-house retrofits), despite some more success with deploying incremental improvements with piecemeal insulation measures.

The depth of reconfiguration in the building system is limited, because radical alternatives are only supported by small new entrants with weak supply-chains, and because the established building sector remains committed to established practices and has successfully lobbied against more ambitious programmes.

The speed of reconfiguration in the building system is very limited, with radical innovations remaining at the demonstration stage with poorly developed market niches. The road ahead for deep decarbonisation of the building system remains uncertain and currently uncommitted, despite significant urgency to find workable pathways for a highly inertial system.

6.5.6 Future Outlook

The UK heating and housing systems are not on track to meet decarbonisation targets. Significant efficiency improvements have stalled over recent years, owing to a lack of consistent and coordinated policy interventions. Ambitious long-term transformative policy visions have been developed (e.g., mass roll-out of heat pumps or greening of the gas grid), but these have not (yet) been complemented with implementation policies and instruments. The unfolding low-carbon system reconfiguration thus has limited depth, scope, and speed, and does not qualify as a Great Reconfiguration.

Existing heating and building systems remain dominated by inertial forces (e.g., hard to retrofit building stock, gas grid as potential stranded asset), limited momentum of low-carbon alternatives (due to limited demand and user engagement, lack of skills, poorly developed supply chains), and strong influence of incumbent actors (e.g., energy suppliers, volume housebuilders) who are interested in maintaining exiting system configurations and thus mostly enact incremental changes.

To achieve deeper decarbonisation, future heat reconfiguration is likely to involve a mixture of options in housing, appliances, and supply/infrastructure. Continued reliance (above 10%) on the existing natural gas-based system beyond 2030 is incompatible with deep decarbonisation objectives, unless CCS is implemented on a massive scale (Mcglade et al., 2018). Because the various low-carbon heat and building options have different characteristics, multiple reconfiguration pathways should be pursued (Watson et al., 2019). We thus expect that future low-carbon reconfiguration will be from one dominant system (national gas distribution and individual gas boilers) towards a more diverse system with options that may vary according to locality, network access, and supply-chain location: heat networks in denser areas, heat pumps and biomass boilers in off-grid areas, and biomethane injection in regional networks related to production clusters.

7
Conclusions

This chapter consists of two parts. The first part (Section 7.1) answers the book's research questions by systematically applying the research protocol to the three focal systems and pulling together the empirical findings and conclusions from the previous chapters. Comparing, as it does, the speed, scope, and depth of unfolding reconfigurations across the techno-economic, actor, and policy dimensions of the three systems, this part is rather dense. The second part (Sections 7.2, 7.3, and 7.4) discusses broader issues such as cross-cutting findings across the three systems, policy recommendations, and future research suggestions.

7.1 Comparing Low-Carbon Transitions in Electricity, Heat, and Mobility Systems

The previous empirical chapters made comprehensive whole system analyses of unfolding low-carbon transitions in the electricity, heat, and mobility systems. Although these transitions-in-the-making vary in speed, scope, and depth, they all involve interactions between multiple niche-innovations and multiple existing (sub)systems. These analyses thus confirm the general usefulness and empirical validity of our system reconfiguration approach, which changes the transition imagery from singular 'bottom-up' disruption, with niche-innovations replacing existing systems, towards a more dispersed reconfiguration process that results from multiple change mechanisms including incremental improvement in existing elements, replacements of system elements, changes in the relative size of (sub) systems, and changes in how elements are linked together in system architectures.

The low-carbon transitions in the three focal systems are in different stages of a Great Reconfiguration. In terms of emission performance, the transition has progressed farthest in the UK electricity system, where GHG emissions decreased by 71% between 1990 and 2019. It is beginning to unfold in land-based passenger mobility systems, where GHG emissions decreased by 14% from their peak in

2007 to 2019, despite an 11% increase in passenger-kilometres in the same period. Transport GHG emissions declined by 29% in 2020 because of COVID-related lockdowns, which strongly affected passenger mobility systems. A low-carbon transition is not yet unfolding in residential heat. Although heat-related GHG emissions decreased by 24% from their peak in 2001 to 2019, a temporal breakdown shows that the low-carbon transition has stagnated: GHG decreased by 29% from the peak year 2001 to 2014 (despite a 9.5% increase in the number of dwellings), but increased by 6.4% between 2014 and 2019, partly due to a 4% housing stock increase over that period and weakened insulation activities.

This chapter aims to explain these salient differences between the three systems and answer the following four research questions from Chapter 1:

(1) Which innovations and system changes contributed directly to the varying GHG emission performances in the three systems?
(2) What are the underlying techno-economic, actor, and policy reconfigurations, and what do these changes imply for the scope and depth of socio-technical system reconfiguration?
(3) Are the unfolding low-carbon transitions moving in the direction of a Great Reconfiguration, characterised by high scope and depth of system changes?
(4) What explains the different speed between unfolding low-carbon system reconfigurations?

7.1.1 Low-Carbon Innovations Driving GHG Emission Reductions

To answer the first research question, this section summarises the findings from the previous chapters about the low-carbon innovations and system changes that directly contributed to the varying emission reduction performance in different systems. For the ***electricity system***, we found that the substantial emission reductions resulted from the following four main innovations and sub-system changes, which we characterise in bold in MLP terms:

- **niche-innovations:** the diffusion of renewable electricity technologies (e.g., onshore wind, offshore wind, bio-power, and solar-PV), which increased their contribution to power generation from 2% in 1990 to 39% in 2019, and mostly displaced coal;
- **within system substitution:** a substantial and rapid switch from coal to gas in the power generation sub-system in the 1990s, which reduced coal's contribution to power generation from 73% in 1990 to 29% in 1999 and increased gas's contribution from 0% to 41% in the same period;
- **niche-innovations:** the diffusion of CFLs and LEDs, which has started to replace incandescent light bulbs in the electricity consumption sub-system, leading to a 38% reduction in electricity use for lighting between 2007 and 2015;

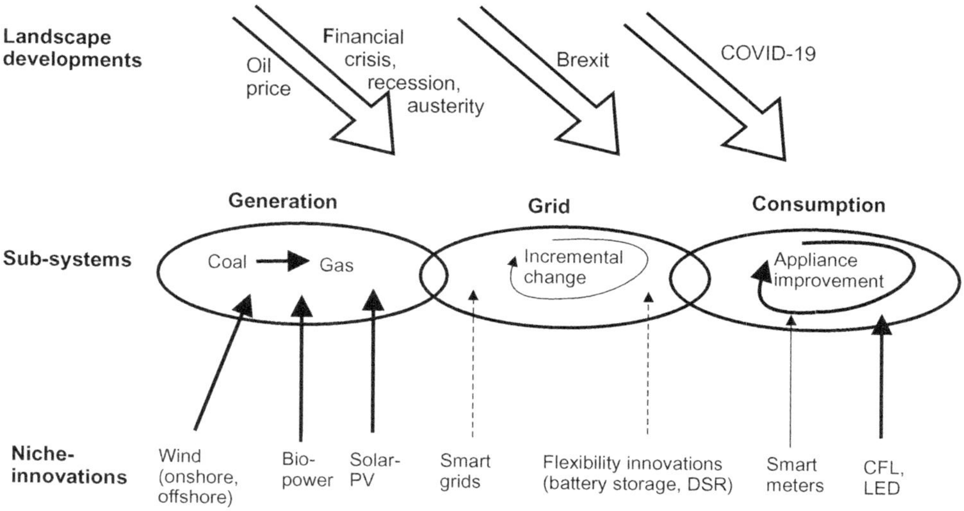

Figure 7.1 Schematic MLP-representation of electricity system reconfiguration with bold, normal, and dotted lines representing the relative contributions of different innovations to unfolding carbon reductions

- **incremental system improvement:** efficiency innovation in appliances, which (combined with lighting substitutions and industrial offshoring) reduced electricity demand by 15% between 2005 and 2019, despite substantial appliance proliferation (particularly consumer electronics and computers).

While not directly contributing to GHG emission reduction, incremental innovations in transmission grids (extensions, strengthening, new offshore grids, new interconnectors) complemented and enabled diffusion of renewable electricity technologies (**incremental system improvement**). Several **niche-innovations** with transformative potential (e.g., smart grids, battery storage, demand-side response) have not yet widely diffused, while smart meters did diffuse, but have been less transformative than anticipated.

Figure 7.1 provides a schematic MLP-representation of the unfolding UK electricity system reconfiguration, summarising the relative contributions of various innovations to GHG emission reductions.

In ***passenger mobility systems***, we found that emission reductions resulted mostly from the following innovations and system changes, which we characterise in bold in MLP terms:

- **incremental system improvement:** incremental engine innovations that improved the fuel efficiency of new petrol and diesel cars by respectively 25% and 27% in the 2007–2016 period (although the manipulation of emission test results by automakers cast doubt on the reliability of these numbers);

- **within system substitution:** a gradual shift since 2001 from petrol to diesel cars, which are more fuel-efficient (this trend reversed after the 2016 Dieselgate scandal);
- **between system substitution:** a doubling of rail travel since the mid-1990s, including some modal shift from cars to trains, especially for commuting in and out of London;
- **niche-innovation:** the diffusion of biofuels, which increased from 0.8% of fuel blends in 2007 to 4.3% in 2019;
- **niche-innovations:** the diffusion of electric vehicles (HEVs, BEVs, PHEVs), which increased from 0.3% of the passenger car fleet in 2007 to 3.3% in 2020; electric vehicles accounted for 20.8% of all passenger car sales in 2020;
- **niche-innovation:** tele-working at home, which increased from 2.7% of the working population in 2007 to 5% in 2019 and 8.5% in 2020 (although the influences on GHG emission reduction are uncertain).

Several **niche-innovations** with transformative potential (e.g., car sharing, inter-modality, self-driving cars) have not yet widely diffused, while ride-hailing did diffuse but is not very radical. Figure 7.2 provides a schematic MLP-representation of the unfolding UK reconfiguration in passenger mobility systems.

In the ***heat system***, we found that emission reductions resulted mostly from the following innovations and changes, which we characterise in bold in MLP terms:

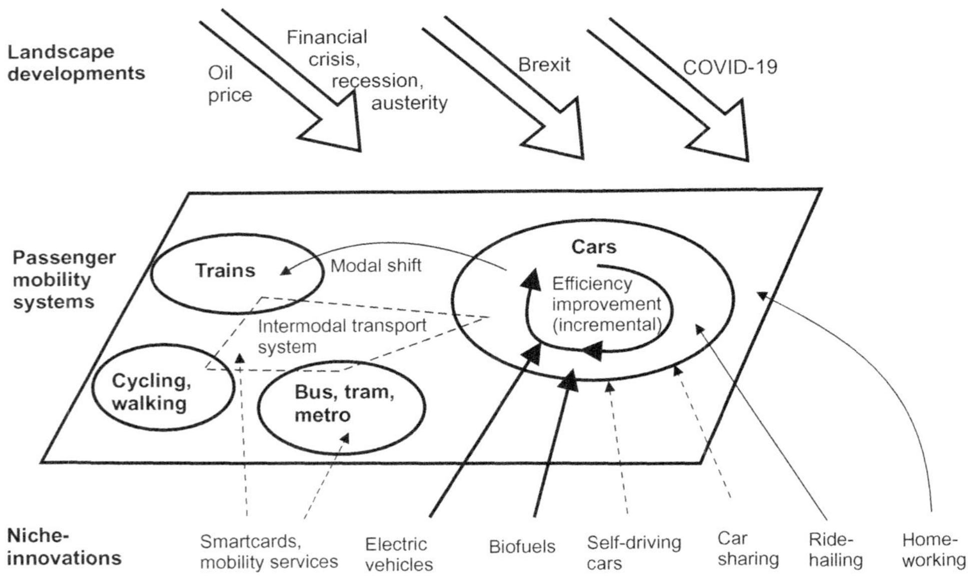

Figure 7.2 Schematic MLP-representation of passenger mobility systems reconfiguration with bold, normal, and dotted lines the relative contributions of different innovations to unfolding carbon reductions (adapted from Geels (2018))

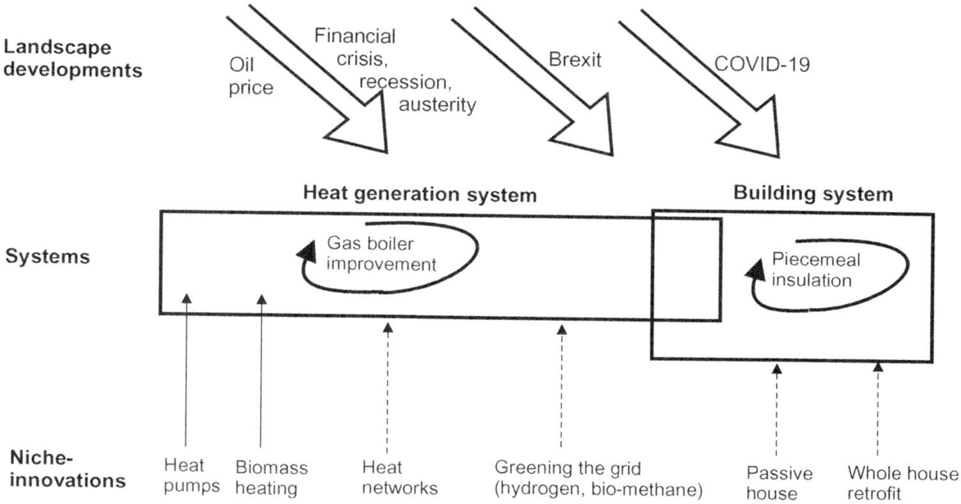

Figure 7.3 Schematic MLP-representation of heat system reconfiguration with bold, normal, and dotted lines the relative contributions of different innovations to unfolding carbon reductions

- **incremental system improvement:** efficiency improvements in gas boilers and piecemeal insulation measures (e.g., double glazing, cavity wall insulation, loft insulation), delivery of which has collapsed since 2013.
- **niche-innovations:** some diffusion of renewable heat technologies (such as domestic wood combustion, heat pumps, plant biomass, biomethane), which significantly increased their contribution to heat generation in all sectors (residential, commercial, public administration) to 9% in 2019. Not all renewable heat technologies are zero-carbon, however, so the GHG emission reduction effects are lower than the diffusion rates.[1]

Several **niche-innovations** with transformative potential (e.g., heat networks, greening the grid, passive house, whole house retrofit) have not yet widely diffused. Figure 7.3 provides a schematic MLP-representation of the unfolding UK heat system reconfiguration.

7.1.2 Depth and Scope in Techno-Economic Reconfigurations

To answer the techno-economic aspect of the second research question, this section summarises and evaluates the findings from the previous chapters with regard to types of change that represent different depth: ‘modular incrementalism’ (which

[1] Biomass and wood combustion, in particular, are not zero-carbon for various reasons (related to collection, transport, processing, and agro-forestry practices).

represents *limited depth*), 'architectural stretching' (which represents *limited depth*), 'modular substitution' (which represents *moderate-depth* if it involves radical niche-innovations and limited depth if it involves relative size changes in existing systems), and 'architectural reshaping' (which represents *substantial depth* involving radical innovations and changing system linkages). To evaluate *scope*, we assess the difference between the low-carbon innovations that have diffused (i.e., actual scope) and the low-carbon innovations that have emerged but remained stuck in small niches (i.e., potential scope).

For the *electricity system*, the identified innovations imply that the current techno-economic reconfiguration has mainly taken the form of 'modular incrementalism' and moderate-depth 'modular substitution' but is beginning to be complemented by 'architectural stretching' and deeper 'architectural reshaping'.

- 'Modular incrementalism' has been the dominant type of change in the electricity consumption sub-system, where continued incremental innovations in existing appliances have substantially improved energy efficiency, leading to electricity demand reduction. The *depth of change* has thus been limited, but the *scope* was substantial since many appliances have been improved.
- Various forms of 'modular substitution' have been the dominant type of change in the electricity generation sub-system: 1) coal-to-gas switching represents a partial substitution between existing system technologies, representing *limited depth* of change, 2) bio-power has mostly taken the form of converting existing coal plants to burn biomass, which represents a *moderate-depth* hybridisation of niche-innovation and existing system, and 3) onshore wind, offshore wind, and solar-PV are niche-technologies that replaced existing system technologies (mostly coal), representing *moderate-depth* change. The replacement of incandescent light bulbs by CFLs and LEDs also represents a *moderate-depth* niche-to-system modular substitution in the consumption sub-system.
- *Limited depth* 'architectural stretching', which builds on existing technical capabilities, occurred in transmission grids that were extended to connect remote wind and solar generation sites and strengthened to enable more electricity flows from Scotland to England. Stretching also took the form of building new offshore grids and new interconnectors to European countries, including France, Northern Ireland, Ireland, and the Netherlands.
- *Substantial depth* 'architectural reshaping' and modification of linkages among the generation, consumption, and grid sub-systems are only just emerging in response to increasing amounts of variable renewable generation (especially from wind power and solar-PV). Because electricity generation and consumption need to be closely aligned to avoid blackouts, the deployment of intermittent

renewables is creating challenges for grid management and load balancing. To address these problems, various niche-innovations are being introduced such as battery storage (which can rapidly provide additional capacity), demand-side-response (DSR, which enables peak shifting and temporary demand reductions), smart grids (which improve the management of bi-directional electricity flows from centralised power stations to users and from decentralised wind and solar assets into the grid), smart meters (which may enable new time-variable tariffs and remote control), and capacity markets (which pay providers of back-up capacity). Changes in the generation sub-system are thus having knock-on effects on the grid and consumption sub-systems, which are beginning to make system boundaries more porous, leading to new operational principles such as intelligent and flexible load management, peak shifting, and demand-follows-supply (instead of the present supply-follows-demand).

The types of changes that are reconfiguring UK electricity systems are mostly large-scale options that fit incumbent interests and have been systematically favoured by policymakers. These moderate-depth large-scale options have thus played more prominent roles than the various small-scale options that represent *deeper* forms of reconfiguration, including a shift from centralised to decentralised systems:

- large-scale wind farms (by energy companies, investors, project developers) are more prominent than smaller-scale community wind projects;
- large-scale solar-PV farms (operated by landowners, investors, project developers) are more prominent than small-scale roof-top solar-PV (by households);
- large-scale bio-power (e.g., co-firing and biomass conversion of coal-fired plants) is more prominent than smaller-scale dedicated biomass plants (by sawmills or poultry farms);
- large-scale battery storage (by new and incumbent companies) is more prominent than decentralised batteries by households with rooftop solar-PV;
- smart grids in relation to the flexibility agenda are more prominent than micro-grids in relation to decentralisation and energy independence;
- Capacity Markets policies favour conventional back-up capacity (by utilities) rather than domestic DSR.

Deeper 'architectural reshaping' based on small-scale decentralised renewables, local storage, and micro-grids has thus remained limited in favour of moderate-depth large-scale technical options and flexible grid management. **For the overall electricity system, the realised *reconfiguration scope* has thus remained more limited than the potential scope, because of choices by incumbent firms and policymakers.**

For the *passenger mobility system*, the identified innovations also imply that the unfolding techno-economic reconfiguration is mainly taking the form of 'modular incrementalism' and moderate-depth 'modular substitution'.

- *Limited depth* incremental innovations in system modules (e.g., car engine improvements, shift from petrol to diesel cars, railway extension) have been relevant since the early 2000s. Various forms of 'modular substitution', which replace system components but do not require deeper changes in the system architecture, have gained importance during the last decade: 1) some *limited depth* system-to-system switching occurred in the form of a small modal shift from automobility to railways, 2) *moderate-depth* hybridisation occurred within the automobility system in the form of biofuel blending and hybrid-electric vehicles, 3) battery-electric vehicles, plug-in hybrid-electric vehicles, ride-hailing, and tele-working represent *moderate-depth* niche-innovations that replaced existing technologies or business models (for ride-hailing), or reduced mobility demand (in the case of tele-working).
- *Substantial depth* 'architectural reshaping' has remained limited in passenger mobility systems. There are seeds for deeper architectural reshaping (such as car sharing, Mobility-as-a-Service (MaaS), driverless cars, and intermodal transport), but the associated innovations have remained relatively small and have uncertain growth paths. The only exception is London, which has succeeded in creating an effective intermodal transport system with frequent bus, tube, and rail services that are well-aligned through transfer hubs and an integrated electronic payment scheme. Using this system, more than 20% of London's commuting trips involve multiple transport modes.

For the overall passenger mobility system, however, the realised *reconfiguration scope* has remained much more limited than the potential scope, for lack of an integrated approach to mobility (demand reduction, multimodality), apart from London, and a focus on engine substitution in automobility.

For the *heat system*, the identified innovations imply that limited-depth 'modular incrementalism' is the most prominent type of change, complemented with some moderate-depth 'modular substitution' options.

- *Limited depth* incremental improvements in gas boilers and piecemeal home insulations have been the dominant type of change since the 1990s (with insulation measures experiencing marked slowdown since 2013). Various forms of 'modular substitution' have become somewhat more prominent since the mid-2010s: 1) domestic wood combustion has diffused somewhat as an add-on niche-innovation that is used besides gas boilers (mostly for reasons of cosiness), while small amounts of biomethane (from anaerobic digestion) are injected into gas grids; both options represent *limited depth* change, 2) heat pumps and plant biomass (in commercial settings) somewhat replaced gas boilers, representing *medium-depth* change.

- *Substantial depth* 'architectural reshaping' options (e.g., passive-house designs, heat networks, green hydrogen) and 'architectural stretching' options (e.g., whole-house retrofit) have remained very small, because they face significant hurdles. More generally, efforts to reduce heat demand by deeply reconfiguring the housing stock have stagnated, which means that 'modular substitution' efforts in heat technologies presently dominate. **The realised *reconfiguration scope* has thus remained substantially more limited than the potential scope, because of little or fragile policy support to increase the attractiveness of low-carbon heating and high-efficiency housing for mainstream users.**

This comparison shows that 'modular incrementalism' and 'modular substitution' are the dominant types of change in all three systems. Only the electricity system is beginning to experience some degree of *substantial depth* 'architectural reshaping' because supply-side modular substitutions are having knock-on effects in other sub-systems. For all three systems, however, the realised *reconfiguration scope* has been more limited than the potential scope, because more radical niche-innovations involving deeper architectural change have remained small. The rationale behind this finding is that incremental and modular changes are generally easier to implement and are less threatening to incumbent interests than changes that alter the entire system architecture.

Generalising from these findings, we can formulate the following propositions:

P1: 'Modular incrementalism' and 'modular substitution' are more common techno-economic reconfiguration pathways because they are easier to implement, do not stretch existing business models and system configurations, and are likely to face less resistance from incumbent interests (even though firms may need to acquire new capabilities).

P2: Deeper kinds of techno-economic reconfiguration options, which require radical innovations and some degree of architectural reshaping, often struggle to diffuse beyond initial niches, even when they hold substantial low-carbon promise, and thus require more policy support to gain momentum.

7.1.3 Depth and Scope in Actor Reconfigurations

To answer the actor-related part of the second research question, this section summarises and evaluates the findings from the previous chapters with regard to realised actor reconfigurations that represent different depths: small adjustments in behavioural templates, routines, or habits (*limited depth*); changes in innovation strategies, (technical) capabilities, and resource allocation (*moderate-depth*); and changes in cultural-cognitive beliefs, repertoires, or paradigms (*substantial depth*). To establish the *scope* of reconfiguration in different systems, we evaluate the degree of substantial change across the main actor groups (firms, consumers, policymakers, civil society).

In the *electricity system*, substantial actor reconfiguration was especially enacted by incumbent firms and policymakers, in the context of pressures from changing public debates and discourses.

The *scope* of actor reconfiguration in the electricity *generation sub-system* has been substantial because many actors (except consumers) changed substantially.

- Utilities, energy companies, and project developers enacted *substantial depth* reconfiguration, as they substantially changed their views of low-carbon transitions and their strategies towards renewable electricity technologies (RETs), moving from reluctant acknowledgement and engagement to strategic reorientation in the past 15 years. They also adjusted their investment strategies and (technical and operational) capabilities as they realised that substantial public subsidies and ongoing techno-economic developments made RET deployment financially attractive while allowing them to retain their large-scale business models.
- Policymakers also made *substantial depth* reorientations by adjusting capabilities, instruments, governance style, and policy goals (further discussed later in the chapter). Public discourses experienced *moderate-depth* change as they came to express positive views on renewables, negative views on coal, and increasing concerns about climate change in general.
- Consumer reconfiguration in the generation sub-system has remained of *limited depth*: rooftop solar-PV adoption by households has stagnated, while 'prosumers' (who use self-generated low-carbon electricity) remain a marginal sub-group. Consumers did, however, ultimately pay for low-carbon generation assets through their electricity bills and through general taxation (which underpinned government subsidies) – although this resulted from policy decisions rather than consumer choices. Because electricity bills are very opaque, utilities could pass extra costs on to consumers without the latter consenting to (or being aware of) the upstream low-carbon investments. This specific payment mechanism thus helped to create 'indirect' or 'involuntary' market demand for low-carbon electricity, which is an important difference with the mobility and heat systems where consumers need to make active decisions to purchase low-carbon technologies.

The *scope* of actor reorientation in the electricity *consumption sub-system* has remained moderate to limited, with only appliance manufacturers and policymakers enacting moderate to substantial change.

- European and UK policymakers in the *electricity consumption sub-system* increasingly accepted demand reduction and climate mitigation as important goals, which led to tightening energy efficiency standards for appliances (e.g., refrigerators, televisions, washing machines), representing *moderate to substantial depth* change. From the mid-1990s to 2012, UK policymakers also imposed increasing energy savings obligations on UK energy suppliers (to drive deployment and uptake of energy-efficient appliances).
- International appliance manufacturers initially contested the new energy efficiency regulations, but in the mid-1990s changed their strategic and technological orientations, representing *moderate-depth* change, which led to substantial energy efficiency improvements in their products through mostly incremental innovations. Appliance manufacturers did not alter their business model, which remained focused on rapid product lifecycles, increasing functionalities, persistent innovation, product differentiation, and market expansion.
- Public debates on energy efficiency remained relatively muted, representing *limited depth* change. Consumers did not substantially alter their social practices or cultural conventions, but they did purchase more energy-efficient appliances, which represents *limited depth* change.

Actor reorientation in the *electricity grid sub-system* remained limited in *scope* and *depth*, as few actors enacted substantial change.

- Transmission Network Operators (TNOs) build on existing capabilities to make expensive but incremental changes in transmission grids (e.g., extensions, strengthening, new offshore grids, new interconnectors).
- Distribution Network Operators (DNOs) engaged in some R&D and demonstration projects, but mostly remained locked in their traditional operational model (around passive distribution), capabilities, and strategy (low cost 'sweating the assets').

In the *passenger mobility system*, there has been deep reconfiguration by some actors as well as substantial lock-in and inertia for other actors.

In the ***automobility system*** there has been some substantial actor reorientation (by automakers and policymakers, and some segments of the user population) towards electric vehicles. For all other low-carbon innovations, however, actor reconfiguration has been moderate or limited depth, which suggests that the *overall scope* of actor reconfiguration is limited.

- Car manufacturers substantially changed their views of low-carbon transitions and their strategies for electric vehicles, moving from reluctant acknowledgement and engagement to strategic reorientation in the past 10 years. These changing views and strategies, which represent *substantial* reconfiguration, were accompanied by large and increasing investments in new technical capabilities and factories, and were largely conditioned by an accelerating innovation race that threatened to leave behind car manufacturers not involved with electric vehicles.

Automakers also started to invest in the development of driverless cars, often in collaboration with IT companies such as Microsoft to bring in new technical capabilities. But since this technology is in an early developmental stage, and because of recent deflations in the hype-cycle, we evaluate this as *moderate-depth* reconfiguration. Automakers also have some *limited depth* engagement in the car sharing segment to learn about its potential.

- Transport policymakers also *substantially* reoriented regarding electric vehicles, developing new goals, governance strategies, and policies in the past 10 years (including phase-out policies for diesel and petrol cars) and creating new organisations (such as the Office for Low Emission Vehicles (OLEV) and the Automotive Council) to facilitate coordination with industrial actors.

Transport policymakers have provided far less support to car sharing, mobility services, teleworking, and other transport systems (bus, rail, cycling). They also invest substantially in new road building (£27 billion in the 2020–2025 period), which helps to further entrench the automobility system. The 2021 Transport Decarbonisation Plan and the new bus and cycling strategies, introduced in 2020 and 2021, represent attempts at policy reconfiguration because they emphasise modal shifts to public and active transport and hope to change local transport systems in the next decade. Their success is uncertain at present, despite £3 billion and £2 billion new funding for bus and bicycle system improvements.

- The past 10 years also saw the entry of new organisations such as car sharing organisations and ride-hailing companies (e.g., Uber), whose mobility services have disrupted the taxi market. We evaluate both changes currently as *limited depth*, because ride-hailing is not particularly radical and car sharing has remained a small application niche. They do, however, have *future* potential for deeper actor reconfiguration as both options have started to question the taken-for-grantedness of car ownership while user experiences with mobility apps may prepare the ground for further changes (e.g., intermodality or MaaS).
- User reconfiguration in the automobility system shows diverging trends, which reflects geographical and age-related variation between user groups as well as the relevance of other considerations such as comfort, safety, convenience, or speed, which many consumers find more important than climate mitigation. On the one hand, there are *limited and moderate-depth* reorientations in low-carbon directions through increased electric vehicle purchase (mostly by middle-aged,

affluent urbanites with an interest in new technology and environmental issues), increased use of ride-hailing, and some uptake of car sharing (mostly by highly educated, tech-savvy, young people in big cities). Driver's licenses and car ownership have also become less prevalent among younger generations, especially in big cities, although the relative importance of cultural reasons (less favourable attitudes towards cars) and economic reasons (unaffordability) is unclear. In our evaluation, this represents *limited to moderate-depth* change.

On the other hand, there are deep user lock-ins to automobility, especially in rural areas and for families with children, and even some trends in high-carbon directions, which represent *limited depth* or even *climate-negative* reconfiguration. Passenger-kilometres by cars have remained around 85% of total passenger mobility between 2007 and 2019. The percentage of households with two or more cars has gradually increased since the early 2010s. And the percentage of heavy SUVs in passenger car sales increased from 6.6% in 2009 to 21.2% in 2018. So, although new mobility practices are beginning to emerge on the fringes, a major behavioural shift away from cars does not yet appear to be underway. This conclusion is reinforced by the response to the pandemic, which substantially reduced car travel during the lockdowns but saw rapid rebounds to pre-pandemic levels when restrictions were lifted. So, while the pandemic created an opportunity for modal shifts away from cars, which led to many optimistic speculations by the green commentariat, these did not materialise.

Despite substantial expansion of the ***railway system***, actor reconfiguration remained *limited in depth* and *scope* until the pandemic, since no actors enacted deep change.

- Climate change has been of low importance to Train Operating Companies, whose strategies mostly focused on financial gaming of the complex franchise and lease system by privatising gains and collectivising costs (such as rail infrastructure investments that are mostly paid by the government and taxpayer).
- Rail use has more than doubled since the mid-1990s, despite soaring rail fares that are among the highest in Europe. But this increased rail travel represents *limited depth* change because it required no new skills and did not result from new user preferences or climate concerns. Instead, increased rail travel resulted from other reasons such as high rents and house prices in London (which forced many people to live elsewhere and commute to London) as well as car access and parking restrictions in London. The importance of Greater London in rail travel is underlined by the fact that almost two-thirds of all rail journeys start or end there, which relates to its socio-economic importance and population size (about nine million people). Rail travellers are more concerned about rising costs, overcrowding, and punctuality than about climate mitigation. Rail travel decreased

by 95% during the first COVID-lockdown and has only partly rebounded since then, reaching 50% of pre-pandemic levels by July 2021. It is thus possible that health concerns about travelling with others in confined spaces have structurally reconfigured user preferences.
- Policymakers invested in new rail infrastructure but did not substantially change their visions or governance style before the pandemic. Since March 2020, however, policymakers have provided £12 billion emergency support and in May 2021 introduced substantial institutional reform that did reconfigure actors and incentives. This is further discussed in Section 7.1.4.

Although ***cycling*** has increased in the last 10 years, especially in cities that have built dedicated cycling infrastructures (like London), there was limited mainstream actor reorientation towards this transport mode before the pandemic.

- Cycling has remained a small mobility practice, accounting for less than 1% of travel distance in 2019. Most cyclists are young urban professionals, who cycle to increase fitness and save travel time and money. Most non-cyclists perceive cycling as dangerous and unsuitable for the British weather, which is why we evaluate the current *scope and depth* of actor reconfiguration as limited. Cycling increased by 46% in 2020, with even larger expansions during and immediately after the first COVID-lockdown. Subsequently, however, bicycle use returned to pre-pandemic levels, as car traffic resumed, and safety concerns resurfaced.
- The new 2020 cycling and walking strategy aims to boost active travel in cities and stimulate broader and deeper actor engagement in the future.

Actor reconfiguration also remained *limited depth* in the ***bus system***.

- Bus companies have been more concerned about surviving in shrinking markets than about climate mitigation. The great majority of buses in Great Britain still use diesel fuel, which substantially contributes to local air pollution. Although a few UK city regions (e.g., Nottinghamshire, North Yorkshire, Greater Manchester) adopted some electric and hybrid-electric buses, only London's bus companies used government subsidies to significantly adopt hybrid and electric buses (amounting to 40% of the London fleet in 2019), which represents *limited depth* change.
- Across Great Britain, travellers have steadily abandoned buses due to increasing fares and decreasing service quality. London, in contrast, saw bus travel increases, partly because the city continued to regulate prices and service quality, and partly because the 2003 London congestion charge restricted automobile use in the city centre. Bus travel plummeted by almost 90% during the first COVID-lockdown and had subsequently rebounded to only 60% of pre-pandemic levels by July 2021, suggesting that lingering health concerns continue to shape user practices.

- Policymakers did not change their visions, governance style, or strategies before the pandemic. Since then, however, they first provided £1.4 billion emergency support and in 2021 introduced a National Bus Strategy for England, which aims to improve bus services outside London and reverse the decades-long decline.

Actor reconfiguration has been *substantial depth* in London, which enabled the creation of an effective intermodal transport system that is widely used and led to a pronounced modal shift, resulting in a 35–40% decline in car travel between 2002–2018. The changes were supported by *substantial depth* reconfiguration of policymakers, which included the articulation of new transport goals and political leadership by successive mayors, investments in improved public transport and cycling systems, the introduction of the London congestion charge, and the creation of Transport for London (TfL) in 2000, which produced a local governance organisation with substantial budgetary and policy discretion and the ability to coordinate and align the different transport modes.

Substantial depth actor configuration also underpinned the rapid diffusion of tele-working, which jumped from 5% of the working population in 2019 to 8.5% in 2020, reaching as high as 46.6% in April 2020. Although the tele-working niche had grown gradually since the 1990s, this jump was due to the COVID-lockdowns, which forced people to stay at home, and was enabled by technological progress in computers, internet, and videoconferencing. The external shock forced workers and organisations to make deep changes in office procedures, cultural norms, work practices, and physical arrangements (e.g., creating dedicated home office spaces). These changes were easier to make for highly skilled occupational categories, for whom they are likely to become structural (to some degree), than for workers in manufacturing, hospitality, leisure, or retail.

Actor reconfiguration has been *limited in depth* and *scope* in the heat system, owing to the dominance (and resistance) of powerful incumbent industries, limited consumer interest, and risk-aversion and weak governance by policymakers.

- Gas suppliers, boiler manufacturers, and appliance installers have mainly focused on *limited depth* incremental efficiency improvements in gas boilers, and limitedly diversified to low-carbon technologies and skills. Only recently have gas companies started to advance new visions and engaged in some demonstration projects with hydrogen and biomethane injection in gas grids, but we interpret this, at least partly, as a delay tactic intended to obfuscate public debates, which started to focus more on heat pumps and heat networks as low-carbon options.
- Volume housebuilders, which dominate the UK building system, engaged in some exploratory low-carbon diversification after the introduction of the

2006 Zero Carbon Homes (ZCH) policy, but they abandoned this strategy in the early 2010s when their political lobbying activities first succeeded in watering down and then removing the ZCH policy in 2015. Their low-carbon reconfiguration has thus remained *limited depth*.

- Policymakers engaged in *limited depth* reconfiguration, because low-carbon heat hardly figured on the policy agenda until the early 2010s, because it is overshadowed by other priorities (e.g., energy security, fuel poverty, supporting the construction industry), and because policy action has fluctuated and changed direction rather than consistently ratcheting up. Furthermore, close relationships with incumbent industry interests have contributed to watering down policies, which have been oriented towards incremental improvements and maintaining existing assets rather than shaking up industry practices.
- At the fringes of the heating system, new actor coalitions pioneered low-carbon niche-innovations (e.g., domestic wood combustion, heat pumps). But since the subsidised market niches remained small, we evaluate mainstream actor reconfiguration as *limited depth*. New actor coalitions also pioneered passive house designs and whole-house retrofits, but these niche-innovations remained even smaller, owing to limited policy support and disinterest from volume housebuilders.
- Niche-innovations in both systems also remained small because of limited interest from mainstream consumers, who did not reconfigure themselves, except for some green consumers purchasing low-carbon heat innovations. Most users remained disengaged from low-carbon heating, as they do not think much about their heating system, except when it breaks down. Low-carbon options are also more expensive and therefore less attractive than gas boilers (heat pumps cost thousands of pounds more, while comprehensive low-carbon housing measures cost tens of thousands of pounds). Consumer reorientation also remained *limited depth* because consumers struggled to navigate uncertainties about low-carbon options, including performance, cost, and the reliability and skills of installers.

In all three systems, the unfolding low-carbon system transitions were primarily enacted by existing mainstream actors (incumbents) rather than by new entrants whose contributions remained relatively small, except perhaps in the automobility system (where some new entrants introduced disruptive innovations). Although the academic literature sometimes speculates about consumer-led or civil society-led low-carbon transitions, we find little evidence for this in the UK cases, which are currently mostly incumbent-led, involving close interactions between existing industries and policymakers. Civil society actors have been important in shaping public debates but have not been much involved in decision-making and were not successful in driving the diffusion of local or grassroots innovations. Substantial and moderate-depth reconfiguration has only been enacted by incumbent firms and policymakers in the electricity generation sub-system, the electricity consumption sub-system, and the automobility system. Some user segments in the automobility system also enacted moderate-depth change. Other mainstream actors thus remained relatively locked-in or were more concerned about other issues than climate mitigation (especially in the electricity grid sub-system, railway system, bus system, cycling system, and heat system).

Generalising from these findings, we can formulate the following propositions:

P3: Incumbent actors are not inert and can reorient to drive low-carbon transitions, in which case they are likely to use their market power and political influence to reduce the potential scope and depth of reconfigurations.

P4: Because incumbent actors have pre-existing commitments to established goals, interpretations, and ways of doing things, deep reconfiguration to address new issues (such as climate change) is challenging and requires dedicated efforts and favourable conditions.

7.1.4 Depth and Scope in Policy Reconfigurations

To answer the policy-related part of the second research question, this section summarises and evaluates the findings from the previous chapters with regard to realised policy reconfigurations. New instruments and a change in governance style (including new goals) respectively represent moderate and substantial depth, while the number of new policy instruments determines scope. Policy reconfigurations varied significantly between the three systems, which is another major explanation for the differences in the low-carbon transitions in the three systems.

In the *electricity system*, policy reconfiguration was *substantial in scope and depth* in the *generation sub-system*, which experienced substantial changes in both governance style and policy instruments, *substantial in depth* and *moderate in scope* in the *consumption sub-system*, and *limited in depth and scope* in the *grid sub-system*.

In the late 2000s, the governance style in *electricity generation* experienced *substantial depth* change from a hands-off, technology-neutral approach towards a more interventionist and technology-specific style that shaped markets and supported the deployment of low-carbon technologies. Around the same time, the 2008 Climate Change Act and the 2009 UK Low Carbon Transition Plan introduced long-term targets (80% GHG reduction by 2050 and 30% renewable electricity by 2020), which created a sense of direction and acknowledged climate mitigation as an equally important policy goal as affordability and energy security (which was recognised by the introduction of the 'energy trilemma' concept). A suite of generic policy strategies,[2] technology-specific plans,[3] and new policy instruments[4] increased the *scope, consistency, and comprehensiveness* of the policy mix, while the creation in 2009 of a new Ministry (the Department for Energy and Climate Change[5]) provided a dedicated organisation with policy and budgetary responsibilities to drive implementation. The policy instruments not only provided attractive financial subsidies for the deployment of low-carbon technologies but also had a political dimension because they systematically favoured incumbent actors and large-scale options, as noted in Section 7.1.2. The ending of feed-in tariffs for small-scale renewables in 2019 further skewed the policy mix towards large-scale renewables and incumbents.

The *electricity consumption sub-system* also experienced *substantial depth* policy reconfiguration. The governance style has gradually become more interventionist as energy efficiency standards for appliances were steadily tightened from the mid-2000s, shaping markets and innovation strategies. The *scope* of policy reconfiguration remained more limited than in the electricity generation sub-system because the policy instrument mix was less comprehensive. Nevertheless, it contained several instruments that stimulated both the gradual development and adoption of more energy-efficient appliances such as performance standards, energy labels, energy savings obligations on electricity suppliers (which were removed in 2013), and a ban on incandescent light bulbs, which strongly shaped markets.

Policy reconfiguration had *limited depth and scope* in the *electricity grid sub-system*. The independent regulator Ofgem, which is dominated by mainstream economic thinking, reluctantly accommodated climate mitigation as a policy goal, but this remained an add-on to traditional goals (promoting competition and efficiency to lower costs). Because efficiency-oriented price control regulations reduced innovation activities by TNOs and DNOs, Ofgem introduced some new policy instruments[6] and a new policy framework, RIIO,[7] for the post-2015 period, to drive innovation. Although these add-on instruments stimulated R&D and

[2] These included the UK Low Carbon Transition Plan (2009); the amended Renewables Obligation (2009); the UK Renewable Energy Strategy (2009); the Carbon Plan (2011); the Energy Bill (2012); and the Electricity Market Reform (2013).

[3] These included the White Paper on Nuclear Energy (2008), UK Bioenergy Strategy (2012), UK Solar PV Strategy (2013; 2014), Offshore Wind Sector Deal (2020).

[4] These included Feed-in-Tariffs, Renewables Obligation, Contracts for Difference, and the Carbon Floor Price.

[5] In 2016, DECC morphed into the Department for Business, Energy & Industrial Strategy (BEIS), which signalled stronger alignment between climate change and industrial policy agendas.

[6] These included the Innovation Funding Incentive, Registered Power Zones scheme, and Low Carbon Network Fund.

[7] RIIO stands for Revenue = Incentives + Innovation + Outputs.

demonstration projects, they did not drive broad deployment of new technologies in distribution networks. Ofgem's negotiated investment model (in which infrastructure investments must be legitimated with regard to demonstrated needs) has been somewhat more successful in supporting incremental changes in transmission grids. Nevertheless, policies have remained too weak and limited to drive deeper and more comprehensive infrastructure change.

In the *passenger mobility systems*, policy reconfiguration before the pandemic was *limited depth and scope* in the railways, bus, and cycling systems. In the automobility system, policy reconfiguration was *substantial depth* for electric vehicles (EVs) but *limited and moderate-depth* for other innovations. The scope of automobility policy *goals* thus remained *limited* and focused on EVs, although recent policy strategies in 2020 and 2021 expanded the scope somewhat. But the scope of EV policy *instruments* was *substantial*, using multiple instrument types.

Overall, passenger mobility policy reconfigurations before the pandemic remained *limited in scope*, as policymakers were mostly committed to a 'greening of cars' strategy rather than a more multi-modal transition approach. This translated into continued investments in road infrastructures and significant support to alternative propulsion, notably EVs.

Looking at the level of automobility innovations, policymakers *reconfigured substantially* towards EVs, switching to an interventionist governance style that actively shaped markets and stimulated EV development through a *broadening scope* of policy instruments that steadily increased the consistency, comprehensiveness, and strength of the policy mix. These instruments included financial support for R&D, the build-up of a battery recharging infrastructure, and EV purchase subsidies. They also included tightening European CO_2 emission regulations (which since 2019 comprised stiff financial penalties for non-compliance) and a UK commitment to phase out petrol and diesel cars by 2030. Policymakers also created a dedicated organisation (the Office for Low Emission Vehicles), which since 2009 has strengthened the EV innovation system by stimulating interactions between automakers, policymakers, and research organisations. The EV-oriented policy reconfiguration was underpinned by an increasing alignment of climate mitigation and industrial policy goals, which aim to reposition the UK car industry in the global EV innovation race.

Policymakers also supported driverless cars (with substantial R&D subsidies) and biofuels (through R&D subsidies and RFTO-targets,[8] which have substantially increased since 2017). Although these policy reconfigurations represent *limited to moderate-depth and scope*, they indicate that UK policymakers expect cars to

[8] RFTO stands for Renewable Transport Fuel Obligation.

remain central in future low-carbon transitions. Policymakers have given much less support to low-carbon options with a stronger behavioural component such as car sharing, mobility services, or teleworking (which might reduce automobility demand), which means that general automobility policy has *limited scope*.

Policy reconfiguration before the pandemic remained *limited depth and scope* in the railways, bus, and cycling systems. Although these systems continued to receive support, the assistance was piecemeal, fragmented, and time-limited, often characterised by repeated stop-start dynamics that hampered long-term planning. **Railways** have been helped with operational support for Network Rail and through large-scale London-centric rail infrastructure investments (e.g., Crossrail, Thameslink, HS2) rather than through whole-system improvements (including in signalling and railway electrification). In response to the pandemic, policymakers provided substantial financial support (around £12 billion) and in May 2021 introduced an institutional shake-up which creates a new organisation from 2023 (Great British Railways) in charge of rail infrastructure, fares, timetables, and contracting. The reforms also include a shift from franchises to contracts for railway companies, which will incentivise them for improved coordination, better consumer services, efficiency improvements, and reduced public subsidies. Large investments in rail infrastructure (including £100 billion for new high-speed rail) also aim to grow rail networks and travel.

The **bus system** has been supported with bus service operator grants, some hybrid bus adoption subsidies, and bus travel passes for the elderly, but none of these support measures before the pandemic were transformative. Policymakers provided £1.4 billion COVID emergency support to bus operators and in March 2021 introduced the new National Bus Strategy, which represents an attempt at deeper and broader policy reconfiguration. In terms of high-level policy goals, the strategy aims to elevate the importance of buses in local transport systems and reverse their decades-long decline outside London. More specifically, it aims to improve bus systems by stimulating innovations such as bus priority lanes, 4,000 electric buses, contactless intermodal payment systems, and improved digital information, as well reducing fares and extending services. In terms of policy instruments, it provides £3 billion funding over five years and introduces institutional reforms, including Enhanced Partnerships that give local policymakers more influence. The future success of this top-down strategy is not guaranteed and depends on local uptake, renewed user interest in bus travel, and reductions in COVID-related concerns.

Policy support for **cycling** was limited before the pandemic, with fluctuating national funding for local cycle lanes resulting in ad-hoc improvements rather than interlinked cycling infrastructures (except for London). The new cycling and walking strategy, introduced in July 2020, represents an attempt at deeper and broader policy reconfiguration. In terms of goals, it elevates the importance of active transport, aiming for 'half of all journeys in towns and cities being cycled or walked by 2030' (DfT, 2020b: 12). It also broadens the scope of instruments, providing £2 billion funding for local initiatives, offering technical design advice for safe and high-quality cycling infrastructure, issuing statutory guidance advising local authorities on reallocating road space to cycling and walking, and creating a new body (*Active Travel England*) to oversee and inspect local schemes. In response to this strategy and in the pandemic context, many cities implemented local initiatives, including pop-up segregated cycle lanes and low-traffic neighbourhoods. Future success is uncertain, however, because numerous cities have subsequently abandoned or downscaled initiatives in response to local protests, and because the strategy does not include 'sticks' such as congestion charging to incentivise a shift away from cars.

While the new policy strategies are aspirational but inconclusive, the only realised and effective example of *substantial depth* and *substantial scope* policy reconfiguration in the last two decades is London, which created a dedicated local governance organisation in 2000 (Transport for London) with significant budgetary and policy discretion. Additionally, successive mayors (Livingstone, Johnson, Khan) provided political leadership and allocated financial resources to the improvement of public transport and cycling systems, which resulted in an effective intermodal transport system. The 2003 London congestion charge also dis-incentivised automobility, galvanising a substantial modal shift.

Policy reconfiguration has remained *limited in depth and scope* for the heat system, which helps explain stagnation in the low-carbon transition in this domain. Heat decarbonisation did not rise on the policy agenda until 2012, when it was layered on top of other policy goals such as energy poverty, affordability, and energy security (in relation to gas supplies), which have remained more important than climate mitigation. Energy saving was included in the building regulations in the mid-1990s, followed by climate mitigation in the mid-2000s, but both goals have remained add-ons to existing goals of the Housing Ministry, which traditionally focused on increased housebuilding (to address persistent shortages) and the safety and quality of buildings (through building regulations).

Because of the limited coherence of policy goals, policymakers have not been able (or willing) to develop an integrated governance approach to low-carbon transitions in the heat system. In the 1990s and 2000s, policymakers succeeded in stimulating incremental improvements in gas boilers and piecemeal building insulation measures with gradually tightening regulations[9] and successive energy savings obligations on energy suppliers, which required them to assist in the

[9] These included the 1992 European Boiler Efficiency Directive, the 2000 Building Regulations revision, the 2005 Building Regulations, and the 2018 Boiler Plus Standard.

deployment of efficient gas boilers as well as other measures (e.g., insulation, lighting, appliances).[10] Policymakers have been far less successful, however, in driving more radical and transformative low-carbon innovations.

One attempt was the 2006 Zero Carbon Homes (ZCH) policy, which mandated that all new buildings from 2016 would be carbon-neutral or carbon-negative. But this policy failed due to resistance and lobbying from volume housebuilders, which led to its watering down in the early 2010s and its removal in 2015. Another attempt was the 2013 Green Deal, which aimed to drive a mass rollout of energy efficient retrofits through a novel finance mechanism. But this policy also failed because design failures (such as a high loan interest rate) led to low consumer uptake. Both policy failures left the building system without effective low-carbon governance or policy instruments, which represents *limited policy depth and scope*. This problem has remained unaddressed since the mid-2010s, despite various government promises, and has contributed to low confidence in policy action and uncertainties for low-carbon heat interests.

To drive low-carbon innovation in the domestic heating system, policymakers introduced the Renewable Heat Incentive (RHI) in 2014, which provided subsidies for the installation of renewable heat technologies. Although the RHI stimulated some household adoption, the reliance on a single demand-oriented instrument means that the policy mix has *limited scope* and lacks complementary instruments to develop technology, supply chains, and installation skills needed for the delivery of a large-scale roll-out of alternative heat appliances.

Policy visions of low-carbon heating have changed substantially in the last decade, from an almost exclusive reliance on electric heat pumps in the early 2010s to a mix of heat pumps (for sub-urban and rural houses) and district heating (for dense city centres) in the mid-2010s. By the late 2010s, the strategic vision changed again to also include the injection of hydrogen and biomethane in gas grids as possible low-carbon options. This inclusion partly resulted from lobbying by gas industry actors, who became worried about the threat of heat pumps and heat networks to their sunk investments in the gas grid. On the one hand, this increasing diversity may prevent premature lock-ins and enable learning about the suitability of various options in different application contexts. On the other hand, this diversity creates deep uncertainties about future transition pathways, which hinder industrial and market developments because actors delay making commitments and investing resources.

[10] These obligations included the third Energy Efficiency Standards of Performance programme (2000–2002), the Energy Efficiency Commitments (2002–2008), the Carbon Emissions Reduction Target (2008–2012), the Community Energy Saving Programme (2009–2012), and the Energy Company Obligations (from 2013 onwards).

Low-carbon transitions in the heat system are thus hampered by the absence of clear long-term strategic frameworks, the limited translation into a consistent and comprehensive policy instrument mix, and a governance style of attending closely to incumbent interests, which explains preference for interventions aimed at preserving existing assets and infrastructures and reluctance concerning regulatory constraints. Effective governance is also impeded by a deficit of policy expertise and coordination, and the lack of a dedicated body with oversight and responsibility for low-carbon heating and buildings. Policy reconfiguration has not just remained *limited in depth and scope* but actually moved backwards over the past decade because policy terminations, policy reversals, and a fragmented 'hands-off' approach relying on isolated market-based policy instruments have created uncertainties, stakeholder frustrations, and disappointing results (except perhaps for some progress in low-carbon heat technologies).

The upcoming Heat and Buildings Strategy, supported by a new string of objectives set out in the Ten Point Plan for a Green Industrial Revolution (HM Government, 2020b) and the Energy White Paper (HM Government, 2020a), is an opportunity to deliver a more coordinated and coherent policy framework for low-carbon heat and to put an end to years of detrimental policy signals. It remains to be seen if this opportunity will be used, but repeated delays and concerns about funding and implementation do not bode well.

Generalising from these findings, we can formulate the following propositions:

P5: Where policy goals, instruments, and governance styles have been deeply reconfigured, they have substantially contributed to low-carbon transition pathways in selected (sub)systems by advancing particular innovations.

P6: Policy reconfiguration is challenging, because existing policy goals, instruments, and governance styles are locked-in and often linked to vested interests. Policy reconfiguration is therefore contested and liable to weakening or reversals, particularly in early stages when momentum is limited, and vested interests may effectively counter-mobilise.

7.1.5 Moving towards a Great Reconfiguration?

Table 7.1 summarises our conclusions with regard to the depth and scope of low-carbon system reconfigurations in the three focal systems and sub-systems. Answering the third research question, this table shows that none of the low-carbon transitions yet qualifies as a 'Great Reconfiguration', which Chapter 1 defined as a particular kind of reconfigurational change with *high scope* and *high depth*.

Table 7.1. *Summary evaluations*[11] *of the scope and depth of unfolding low-carbon reconfigurations in the UK electricity, passenger mobility, and heating systems (shading indicates degree of reconfiguration)*

	Depth	Scope
ELECTRICITY		
Electricity generation sub-system	*Substantial* (modular substitution; deep reorientation of firms and policymakers; change in governance style and policy mix)	*Moderate* (transformative small-scale options marginalised; limited consumer reorientation; broad policy mix)
Grid infrastructure sub-system	*Limited* (incremental technical change, restricted actor and policy change)	*Limited* (radical innovations remain small; narrow actor engagement; some new policy instruments)
Electricity consumption sub-system	*Moderate* (incremental technical change; deep reorientation of firms and policymakers; change in governance style and tighter instruments)	*Moderate* (many appliances; few actors involved; moderate instrument scope)
MOBILITY		
Automobility system	*Substantial* for EVs (modular substitution; deep reorientation of firms and policymakers; change in governance style and policy mix)	*Limited* (other low-carbon innovations remain small; narrow actor engagement; narrowly focused transport policy
Railway system	*Limited* (incremental change; limited low-carbon reorientation; limited transformative policy implementation	*Limited* (no substantial low-carbon change)
Bus system	*Limited* (incremental change; limited low-carbon reorientation; limited transformative policy implementation)	*Limited* (no substantial low-carbon change)
Cycling system	*Limited* (small system with limited mainstream actor reorientation)	*Limited* (only traction in a few cities; practiced by small user segment)
HEAT		
Heat supply system	*Limited* (incremental change and some modular substitution; limited mainstream actor reorientation; single policy instrument)	*Limited* (transformative low-carbon innovations remain small; narrow actor engagement; fragmented and narrow policy mix)
Buildings system	*Limited* (stagnated innovation; incumbent actor resistance; weak governance framework)	*Limited* (no substantial low-carbon change; policy failures and reversals)

[11] The summary evaluations aggregate the evaluations from the techno-economic, actor, and policy reconfiguration Sections 7.1.2, 7.1.3, and 7.1.4 and average the results (so, if two evaluations were 'substantial' and one 'moderate', then the summary evaluation is 'substantial').

The low-carbon electricity transition, however, has substantially progressed in the direction of a Great Reconfiguration as it achieved *substantial depth* and *moderate scope* in electricity generation and *moderate depth* and *moderate scope* in electricity consumption. With *limited depth and scope* of change, the grid infrastructure sub-system is increasingly becoming a bottleneck (due to actor lock-ins and ineffective governance). However, the emergence of radical innovations (such as smart grids, battery storage, capacity markets, demand-side response) does offer potential for deeper system reconfiguration in the coming years. Future grid transformations will become even more necessary when the likely diffusion of electric vehicles and the potential diffusion of electric heat pumps will increase the pressure on local distribution grids. The boundaries among electricity generation, consumption, and grid sub-systems are also likely to become more porous in the future, because the increase in intermittent renewables (especially wind and solar-PV) requires adjustments in the other two sub-systems, which will likely also entail architectural changes in the entire system and thus also in actor roles and capabilities.

In the passenger mobility systems, only electric vehicles (EVs) and tele-working presently represent *substantial depth* change. But since EVs are modular substitutions, they represent *limited scope* change, which means that EVs on their own do not qualify as a Great Reconfiguration. Tele-working can substantially shape mobility demand, but its *scope* has, so far, been limited to highly skilled professions, so on its own it does not (yet) qualify as a Great Reconfiguration. Other innovations have appeared, but they are not very radical (ride-hailing), in early developmental stages (driverless cars), or very small (car sharing, intermodal transport). The only exception, which has achieved *substantial depth* and *substantial scope* change, is London, which has created an effective intermodal transport system with well-aligned bus, tube, and rail services as well as a dedicated cycling infrastructure. Emulation of this change by other UK cities, which would require major institutional change and large investments, would qualify as a Great Reconfiguration for *local* transport. Recent bus and cycling strategies and the 2021 Transport Decarbonisation Plan aim to stimulate future changes in that direction, but their effectiveness remains to be seen. Even a major local transport reconfiguration, however, would not provide a solution for long-distance high-speed travel on major roads, which accounted for 62% of motor vehicle traffic in 2019. So, even if local transport systems were deeply reconfigured, EVs would still have to be part of a low-carbon transition to reduce the GHG emissions of the remaining passenger car travel (which is likely to remain substantial in the next one or two decades).

The heat system, where change is *limited in both depth and scope*, is not yet moving in the direction of a Great Reconfiguration. Even piecemeal improvements have slowed to a trickle in the building system, owing to weak and fragmented policies, incumbent actor opposition, and limited consumer interest. In the heating system, there have been some positive developments in recent years, notably the adoption of some low-carbon heat technologies. But there is limited actor commitment or policy drive to roll-out these modular technologies on a wider scale, or to implement more systemic or architectural changes.

7.1.6 The Speed of Low-Carbon System Reconfiguration

Since existing technologies, mainstream actors, and policies are stabilised by various lock-in mechanisms, the speed (as well as the depth and scope) of low-carbon system reconfiguration initially tends to be low. Our conceptual framework suggests, and our empirical findings confirm, that an increase in the speed of low-carbon system reconfiguration results from a strengthening momentum of emerging innovations and favourable opportunity structures resulting from weakening system lock-ins, both of which involve interactions between techno-economic, actor, and policy change mechanisms.

For increasing momentum of emerging innovations, these mechanisms include:

- *techno-economic*: a) technological performance improvements, b) increasing availability of complementary innovations, c) large-scale infrastructure investments, and d) decreasing costs, which can result from multiple mechanisms (e.g., scale economies in production, learning-by-doing improvements in deployment, lower financing costs, increased competition, overproduction, or price dumping);
- *actors*: a) an increase in the number of enrolled actors (through expanding social networks, alliances, coalitions), which enhances available skills, resources, connections, lobbying power; b) changes in actor orientations towards particular low-carbon innovations (e.g., in interpretations, views, expectations, preferences, strategies, commitments) and activities (e.g., investments, purchase, behaviour, deployment, learning); c) feedbacks between actor groups: for example, increasing consumer adoption creates larger markets, which attract more firms, which invest more in technologies, which then decrease in costs, which boosts further adoption; another feedback loop is that more positive public discourses stimulate demand and legitimate stronger policy support for firms or consumers, which stimulates demand or investments, which advance the technology and further improve public discourses;
- *policy:* a) changing strategies and governance frameworks to reach new goals, b) changes in supportive policy instruments (e.g., purchase subsidies, investment grants, direct infrastructure investments, regulations, information campaigns), c) the creation of dedicated agencies or ministries with strategic and implementation responsibilities, budgets, and political significance.

For weakening system lock-ins and opportunity structures, these mechanisms include:

- *techno-economic*: a) unaddressed technical performance challenges and 'reverse salients' (i.e., problems that hold back particular parts of a system), b) technical accidents, failures, or system breakdowns, c) increasing costs and decreasing competitiveness of existing technologies;
- *actors*: a) legitimacy pressures on established technologies or ways of doing, b) defection or disbanding of critical actors from incumbent alliances and networks;
- *policy*: a) dealignment between the existing system and changing political agendas and goals, which may result from increasing system problems, niche-innovation opportunities, or pressures from interest groups, the media, and wider publics, b) significant swing in electoral politics that shifts policy priorities away from the existing system.

The kinds and strengths of these interacting mechanisms are likely to vary depending on the types of innovations, and sectoral and socio-political contexts. The general answer to the fourth research question thus is that more of these mechanisms were activated in the electricity system than in the passenger mobility systems, while very few were activated in the heating and building systems.

More specifically, the empirical analyses of electricity and mobility systems included several instances of accelerated change, including the diffusion of renewable electricity technologies (RETs), the diffusion of electric vehicles, the diffusion of LEDs and CFLs, the diffusion of tele-working, rapidly increasing cycling in London, and London's modal shift from cars to other transport modes. We briefly summarise these acceleration instances with the aim of indicating the main interacting mechanisms and articulating more general lessons.

The rapid diffusion of RETs (e.g., onshore wind, offshore wind, bio-power, solar-PV) since 2010 was driven by the following momentum-enhancing mechanisms: a) rapidly falling costs,[12] which made many RETs more cost-competitive with coal and gas-fired power plants, b) technical improvements that enhanced capacity, load factors, and efficiency, c) a new Ministry (DECC, later BEIS) and a more interventionist governance style, which actively shaped markets and technologies through a range of financial support policies that made RET-investment more attractive and less risky, d) a change in the views and strategies of utilities and energy companies, who came to see RETs as attractive economic opportunities and invested accordingly, e) a positive public discourse about RETs, which legitimated increasing policy support.

[12] Between 2010 and 2020 cost decreased by 85% for utility-scale solar-PV, 56% for onshore wind, and 48% for offshore wind.

The following 'lock-in weakening mechanisms' also created opportunity structures for RET diffusion: a) decreasing competitiveness of coal compared to RETs, partly due to the Carbon Price Floor instrument, which increased the relative price of coal, b) civil society campaigns and negative public discourses about unabated coal-fired power plants, which eroded its legitimacy, c) the 2015 government decision to phase-out unabated coal by 2025.

Two structural characteristics of the electricity system also enabled rapid RET-diffusion. First, the separation of supply and demand by the electricity grid means that electricity companies could reorient towards RETs without directly affecting or involving consumers. Second, the specific payment mechanism of electricity bills means that consumers pay for RETs without being aware or agreeing to this. Earlier, we characterised this as 'indirect' or 'involuntary' market demand, which differs from the other two systems where consumers need to make active purchase or behaviour change decisions.

The accelerating diffusion of electric vehicles since the early 2010s was driven by the following momentum-enhancing mechanisms: a) rapidly falling battery costs, which decreased by almost 90% between 2010 and 2020, and the expectation that battery electric vehicles will be cheaper to buy than conventional cars by the mid-2020s, b) technical improvements resulting in longer ranges and shorter charging times, c) a more interventionist governance style, including the creation of a dedicated policy unit (OLEV), which actively shaped markets and technologies through purchase subsidies, R&D subsidies, and recharging infrastructure investment, d) increasing consumer interest (despite higher EVs prices, even with subsidies), e) positive public discourses about EVs, f) changing views and strategies of automakers, some of which started to substantially reorient towards EVs, leading to an innovation race involving the entire industry, g) the alignment of EVs with multiple policy goals, including climate mitigation, air pollution reduction, and supporting the UK car industry in the global EV race.

The following 'lock-in weakening mechanisms' also created opportunity structures for EVs: a) the 2015 Dieselgate scandal, which revealed widespread emission test cheating, damaged the legitimacy of automakers (particularly Volkswagen), and caused major sales declines of diesel cars, b) public debates about local air pollution problems, which eroded the legitimacy of petrol and diesel cars, c) stronger European CO_2 regulations (with stiff financial penalties) and a regulatory ban on the sale of diesel and petrol cars from 2030.

The rapid diffusion of LEDs and CFLs since the late 2000s was driven by the following momentum-enhancing mechanisms: a) technical changes that made CFLs 3–5 times and LEDs 7–10 times more energy-efficient than incandescent light bulbs (ILBs), b) rapidly falling prices, especially for LEDs, which decreased 96% in price between 2008 and 2015, c) consumer uptake because of decreasing prices and new functionalities (such as better light control to influence ambience).

The following 'lock-in weakening mechanisms' also created opportunity structures for LEDs and CFLs: a) civil society pressure and negative public debates since the 1990s about inefficient ILBs, b) increasing regulatory pressure as policymakers shifted their views and policies from voluntary measures to an ILB phase-out ban (2007 in the UK, 2009 in the EU), c) support for stricter measures from European incumbent lighting firms who struggled to compete with Chinese firms in the ILB-market and therefore strategically reoriented from ILBs towards CFLs and LEDs.

The rapid diffusion of tele-working was due to the COVID-lockdowns, which abruptly destabilised existing work practices and forced managers and organisations to overcome their hesitations regarding tele-working (such as less oversight and control). The diffusion was also enabled by learning processes and technical developments that had improved tele-working's feasibility in preceding years. The forced diffusion also stimulated further learning processes that are likely to lead to some degree of permanent change (for some occupations), as workers can gain time (and money) by not having to commute and organisations realise they may be able to cut costs by reducing office space.

London's rapid modal shift from cars to other transport modes since the early 2000s was driven by the following momentum-enhancing mechanisms: a) political leadership from successive London mayors to reconfigure local transport systems, b) investments, improvements, and expansions of alternative transport modes (bus, tube, rail, cycling), c) improved coordination and alignment of transport modes by TfL and through integrated payment mechanisms (such as the 2003 Oyster Card) that expedited intermodal transport and kept public transport costs relatively low.

The following 'lock-in weakening mechanisms' also created opportunity structures for the modal shift: a) steadily increasing congestion, parking problems, and parking restrictions, which reduced the convenience and practicality of car use in London, b) negative public debates about worsening problems (e.g., congestion, parking, local air pollution) that created pressure on local policymakers to enact reforms, c) the 2003 London congestion charge, which (through successive increases) made car use more expensive.

The rapid increase in cycling in London since the early 2000s was driven by the following momentum-enhancing mechanisms: a) substantial infrastructure investments in dedicated cycling infrastructure, b) political leadership from successive London mayors that provided drive and visibility for cycling plans, targets, and campaigns as part of wider local transport reconfigurations, c) stronger governance and implementation capacities (including financial and policy responsibilities) through the creation of Transport for London (TfL) in 2000, d) local NGO campaigns and positive public debates about cycling and further safety improvements, e) increased cycling uptake by segments of the population and the articulation of new identities, f) company interest and consumer uptake of bicycle sharing.

The cycling expansion also benefited from the same opportunity structures that stimulated London's modal shift, as discussed previously.

In terms of general lessons, these brief summary analyses confirm that acceleration does not result from a single driver but from multiple interacting mechanisms across techno-economic, social, and political dimensions, which together drive innovations and system destabilisation across tipping points where the speed of change alters from slow to rapid.

Cost decreases were important in several cases, but this in itself was both a consequence and a driver of increasing deployment by consumers, investments by companies, and stronger policy support in the context of public debates. Carbon pricing, which economists have peddled for decades as the most important instrument, was only marginally important in one case (RET diffusion), suggesting that carbon pricing is neither necessary nor sufficient to accelerate low-carbon transitions. Purchase subsidies, R&D subsidies, and public infrastructure investment were clearly much more important financial instruments. Regulatory instruments, including bans and phase-out policies, were also important in most cases, as well as the creation of dedicated governance units or ministries to plan, coordinate, and drive implementation. Most cases also showed the importance of public support and positive discourses, and the influence on these from NGOs and political leaders.

Most of these acceleration cases, except London's cycling and modal shift, represented modular incremental or modular substitution options rather than whole system reconfigurations, and in that sense had relatively limited scope. This suggests there may be trade-offs between the speed and scope of system reconfiguration, as others have also found in historical transitions (Wilson, 2012). The reason for this is that it is easier and quicker to substitute particular components in an existing system than to change or build a whole new system, which usually requires more coordination, investment, and stakeholder engagement. In tightly coupled systems there may be limits to this logic, because modular component substitutions may have knock-on effects that require adjustments in other parts of the system (Berkers and Geels, 2011). This did indeed occur in the electricity system, where generation and use of electricity need to be closely linked to avoid blackouts. Increasing amounts of intermittent renewables thus required adjustments in the electricity grid and consumption sub-systems, which is why the scope of electricity system reconfiguration has increased with RET-diffusion.

Generalising from these findings, we can formulate the following propositions:

> **P7: Faster reconfiguration results from combinations of increasing momentum of (niche)innovations and changing opportunity structures resulting from weakening system lock-in mechanisms.**

P8: Innovations that are framed as addressing multiple issues tend to have greater legitimacy, and political and corporate backing, resulting in higher momentum.

P9: Shocks can decisively accelerate reconfigurations if they occur when emerging innovations have stabilised and acquired some endogenous momentum.

P10: There are likely trade-offs between speed and scope of system reconfiguration, since focused activities and interventions on a few elements (e.g., modular substitutions) are easier and quicker to implement than altering entire systems.

7.2 Cross-Cutting Themes

Drawing on the rich material in our empirical chapters, this section highlights several findings and insights with regard to relevant cross-cutting themes in socio-technical transitions research.

7.2.1 Incumbent Firms

The socio-technical transitions literature traditionally emphasises the role of new entrants in pioneering radical niche-innovations (Kemp et al., 1998; Schot and Geels, 2007). Our empirical chapters did indeed show instances of this pattern, for example, community energy groups deploying wind turbines, Tesla developing electric vehicles, new companies pioneering ride-hailing (e.g., Uber) and car sharing, ICT companies (co)developing driverless cars, new companies (e.g., Tesla, Powervault, Moixa, Sonnen) developing home batteries, and environmentally conscious architects, self-builders, and organisations such as the Passivhaus Trust pioneering low-energy house designs.

In all three systems, however, incumbent firms have been the dominant actors in enacting the low-carbon reconfigurations. In the electricity generation and consumption sub-systems, these included electric utilities, energy companies, project developers, appliance manufacturers, and lighting firms, which all reoriented their views and strategies to accommodate low-carbon transitions. In the automobility system, incumbent automakers have started to reorient towards electric vehicles, while also co-developing driverless cars (often with ICT companies) and dipping a toe in the car sharing market. Incumbent firms did not reorient much in the railway, bus, and cycling systems, but they also faced limited challenges from new entrants.

The continued dominance of incumbents in these systems further explains the prevalence of 'modular incremental' and 'modular substitution' options in the associated system reconfigurations. While incumbents may be willing to act at speed (if they are sufficiently incentivised or become convinced about economic opportunities), they are likely to prefer narrow transition paths that maintain or protect their core business models. This has, so far, limited the scope of UK low-carbon transitions and marginalised more radical niche-innovations that would entail 'architectural reshaping' or deeper social and institutional innovation.

In the buildings system, the role of incumbent firms has been more pernicious as they actively resisted low-carbon transitions and successfully lobbied for the removal of the zero-carbon homes policy in the mid-2010s. In the heating system, gas supply and boiler companies continue to defend and improve existing heating technologies and gas grids, while also advancing visions of hydrogen or biomethane dissemination through adjusted gas pipelines and engaging in some demonstration projects. At present, we interpret this mostly as a strategic response aimed at reducing policy commitments to heat pumps and district heating, although we acknowledge the future possibility of stronger company commitments to hydrogen or biomethane, which would avoid the gas grid becoming a stranded asset.

The continued dominance of incumbent firms also relates to characteristics of the UK governance style and is thus not an inevitable characteristic of low-carbon reconfigurations.

7.2.2 Governance Style and Politics

Although the UK governance style and policy instruments have changed substantially for some low-carbon innovations, as discussed earlier, there are some recurring features across the three systems that politically privilege incumbent interests and particular kinds of transitions pathways. First, the UK has a highly centralised style of policymaking, the so-called Westminster model (Lijphart, 2012), and close-knit policy networks that provide some access to big incumbent firms but remain closed for new entrants and start-ups (Bailey, 2007). In all three systems, this has reproduced a 'working with incumbents' governance style (Geels et al., 2016b) that tailors policy support to the interests and concerns of electric utilities, automakers, volume housebuilders, and gas industry companies.

Second, the UK's liberal market economy (Hall and Soskice, 2001) has for the past few decades been characterised by a neo-liberal political ideology, which explains the policy preference for market-based policy instruments (e.g., auctions in low-carbon electricity) that tend to favour incumbents over new entrants. In the 1980s and 1990s, this ideology also led to privatisation and liberalisation in the electricity, gas, railways, and bus systems, which produced fragmented industry structures with multiple roles and interfaces between organisations with different incentives. Despite the economic rhetoric that increased competition would improve services, increase efficiency, and lower costs, these promises did not always materialise (especially not in rail and bus transport). It also led to significant worsening on non-monetised public goods dimensions such as accessibility, fairness, and energy poverty, which departs significantly from many other countries that approach access to energy or mobility as fundamental rights. The fragmented industry structures also created coordination problems and incentives that generated misalignments with the need for low-carbon transitions (because they hampered collaboration and knowledge flows between actors). In line with neo-liberal ideology, policymakers also adopted a hands-off policy style in the electricity, gas, railways, and bus systems, leading to the disbanding of the Department of Energy in 1992 and the creation of independent regulators to oversee the efficient functioning of market forces (e.g., the Office of Gas and Electricity Markets; the Office of Rail Regulation).

Third, the UK has a technocratic, top-down governance style that pays limited attention to engaging a wider set of stakeholders: 'The government in the UK is still meant to govern – full stop. ... The government of the day acts. Others react. ... Reforms ... are not negotiated painstakingly with stakeholders. They are handed down from above by governments' (King, 2015: 283). In the electricity system, this 'bulldozer' style (Geels et al., 2016b) has generated social acceptance problems for onshore wind, biomass combustion in converted coal-fired plants, and shale gas, which were pushed through with limited consultation of citizens and societal actors, who then mobilised to express their concerns through protests. In the building system, this top-down style led to the sudden introduction of the 2006 Zero-Carbon Homes policy, against which housebuilders successfully lobbied, leading to its equally sudden removal in 2015. Policymakers also seem to think they can drive change by top-down formulation of demanding targets, without paying much attention to complementary policies that address skills, supply chains, public acceptance, or industry support. This has led to a recurring pattern of over-promising and under-delivering (e.g., with CCS, nuclear power, zero-carbon homes), which may well be repeated with heat pumps and hydrogen.

Fourth, the high degree of centralisation also implies limited autonomy for cities and local policymakers (McCann, 2016), who often do not have the regulatory or financial responsibilities to address allocated tasks, especially with regard to local transport systems, as discussed in Chapter 5. Rather than giving local policymakers sufficient long-term budgets, the Treasury requires them to apply for central funding for specific projects (e.g., new cycle paths or cleaner buses), which often leads to on-the-ground fragmentation because cities may win money in one allocation round but not in another, leading to stop-start dynamics. Eadson (2016) characterises the resulting implementation style as 'disordered, syndromic experimentation and government-by-project rather than any systematic programme of government'. The recent bus and cycling strategies aim to reconfigure local transport governance but still represent a top-down push that may encounter local implementation problems.

In all instances of more rapid reconfiguration, discussed earlier, one or more of the aforementioned characteristics were changed, leading to more interventionist policies that mobilised multiple kinds of instruments (financial, regulatory, infrastructure investment) that were strengthened over time to drive change that was overseen or coordinated by dedicated new agencies (TfL, OLEV) or ministries (DECC, BEIS) with sufficient budgets and responsibilities.

In the systems with less transitional progress, most of those governance characteristics are still pertinent, leading to a reliance on single market-based instruments (e.g., the RHI, the Green Deal), selective infrastructure investments (e.g., 'flashy' rail projects), and frequent chopping and changing of policies such as the sudden scrapping of CCS-support (in 2016), the unforeseen downscaling (in 2015) and removal (in 2019) of Feed-in-Tariffs, the unexpected closure of Renewables Obligations for small-scale renewables (in 2016), the removal of the ZCH-policy (in 2015) and the Green Deal (in 2015). The CCC (2020: 99) characterises these changes as 'shortcomings', noting that 'frequent changing of policy should be avoided' because it 'can damage faith in Government policy and reduce business willingness to invest'.

7.2.3 Users

The role of users has been relatively muted in the low-carbon system reconfigurations we analysed. Despite calls by critical theorists and sustainable consumption scholars for deep value changes and a move away from consumer societies, our analyses of the three systems shows little evidence of widespread behavioural or cultural revolutions, except perhaps for tele-working, which is likely to permanently alter work practices (in some professions).

Our analyses do, however, show instances of more limited change such as purchasing greener products or behavioural adjustments that are not particularly demanding or disruptive. For instance, some segments of the population have been willing to buy electric vehicles, LEDs, energy-efficient appliances, wood burners, or electric heat pumps, which suggests they have some willingness to pay if product performance has sufficiently improved and additional costs are not too high. We also found that people are doing more cycling, more train travel, and more intermodal transport in London, but these changes are relatively easy and sometimes enacted in response to changing incentives (e.g., London's congestion charge). Increasing levels of public concern about climate change thus do not seem to have translated into widespread deep behaviour change in the direction of frugality, sufficiency, or down-scaling.

These findings resonate with Dubois et al. (2019), whose large-scale survey research of users in Germany, France, Sweden, and Norway found that most consumers are willing to implement relatively simple low-carbon changes, such as enhanced waste recycling or buying energy-efficient appliances, but are unwilling to make more drastic changes such as abandoning private cars or flying (unless forced by external shocks such as the COVID-pandemic). These findings imply that the limited scope of unfolding low-carbon reconfigurations is not only due to incumbent interests and governance styles but also to limited mainstream user interest in more radical options such as passive house, whole house retrofit, modal shift or intermodality (except for London), car sharing, mobility services, or electricity prosumption.

One possible reason for this limited user interest is that policymakers, opinion leaders, and other relevant actors have not yet sufficiently tried to shape consumer preferences and social practices in low-carbon directions (through information campaigns, consumer training, or infrastructure provision), because they have, so far, mostly focused on technological changes. Another possible reason is that mainstream users, despite expressing high concerns about climate change, are not really willing to sacrifice convenience, comfort, or practicality, nor pay high switching costs, when considering low-carbon options for heating, lighting, appliances, or mobility. This explanation partly relates to the value-action gap literature (Flynn et al., 2009), which has found substantial discrepancies between what people say they care about and what they actually do. On the other hand, our summary-analysis in Section 7.1.6 of London's modal shift from cars to other transport modes suggests that interventionist policies, public debates, and investments in alternatives can indeed shape user practices and preferences to some degree. Open questions for the coming years are how far such user practices can be modulated and how strongly policymakers are willing to push for demand-side change.

7.2.4 Wider Publics and Civil Society Organisations

The socio-technical transitions literature suggests that public debates about issues are important because they shape cultural meanings and legitimacy, which can influence consumer preferences, create a sense of urgency, and exert credibility pressure on policymakers (Hermwille, 2016; Roberts, 2017; Roberts and Geels, 2018; Rosenbloom et al., 2016; Sovacool, 2019). Civil society and social movement organisations can shape these public debates (Benford and Snow, 2000) and also nurture grassroots innovations to develop alternative options themselves

(Feola and Nunes, 2014; Smith and Seyfang, 2013). Our findings confirm many of these influences but in varying degrees.

At a *general level*, the fluctuating level of public attention and public debate about climate change (Figure 1.1) was important to create conditions for the strengthening of climate policies. The public attention increase in the mid-2000s was an important driver for the pathbreaking 2008 Climate Change Act and subsequent translation policies. Weakening public attention in the years after the 2007/8 financial crisis then created conditions for the political weakening of climate mitigation policies, for instance in the electricity and heating systems. The resurging attention in the late 2010s, and the emergence of new framings such as 'climate emergency', then helped to create the conditions for net-zero commitments and stronger climate policies by the UK government and other countries, although many of these have not yet been translated into substantial changes.

Civil society and social movement organisations helped to shape these general climate change debates in various ways. Environmental NGOs, for instance, organised the 2006 Big Ask campaign, which mobilised wider constituencies (including the National Federation of Women's Institutes, Christian Aid, the National Trust, Oxfam, UNISON, and the RSPB[13]) and the wider public to pressure policymakers for a climate law with ambitious GHG reduction targets (Carter and Jacobs, 2014). And in 2019, public protests by school children and civil society organisations (e.g., Extinction Rebellion, Climate Justice movement) helped to push climate change and framings high onto public and political agendas.

For *specific technologies*, public debates and civil society activities also shaped socio-cultural meanings that, in turn, affected public policies. We found multiple instances of *negative* debates and campaigns that eroded legitimacy and created pressure for stronger policies. In the late 1990s and early 2000s, environmental NGOs (e.g., WWF, Greenpeace) campaigned at the EU-level against the inefficiency of ILBs, framing these in terms of 'energy waste', which helped to create the conditions for the 2009 ILB phase-out ban (Franceschini and Alkemade, 2016). In the late 2000s, campaigns by activist groups (e.g., Climate Camp) and public debates succeeded in halting plans to expand coal-fired power stations, and created negative discourses that later contributed to coal phase-out policies. In the early 2010s, environmental NGO campaigns and public debates about the sustainability of imported biomass pellets led to the creation of sustainability standards, although they did not succeed in halting the expansion of industrial-scale biomass combustion in converted coal-fired power plants. Throughout the

[13] Royal Society for the Protection of Birds.

2010s, environmental NGOs and local communities contested plans for fracking and shale gas technologies, which in 2019 resulted in termination, even though the government for many years had dismissed the protests. The public outrage following the 2015 Dieselgate scandal is another example of how negative public debates can erode legitimacy and prepare the ground for stricter policy action, in this case a sales ban by 2030.

Positive public debates can also stimulate reconfigurations by shaping consumer preferences and creating legitimacy for stronger policy support. Positive public debates about electric vehicles (which increasingly align multiple issues, including cleaner air, climate mitigation, cost reductions, new industry and job creation) underpinned and legitimated stronger policy support. Debates about onshore wind have been more mixed, with some groups emphasising positive issues (like clean energy, cost reductions, jobs) and other groups underlining negative issues (e.g., visual burdens, noise, shadow flicker, industrialisation of the countryside, bird fatalities, intermittency), which, in conjunction with other issues, led to fluctuating policy support (including a moratorium on new-built turbines in 2016, which was overturned again in 2020). Debates on offshore wind have become increasingly positive, as costs fell and industrial prospects became clearer, which underpinned an escalation of future targets to 40GW installed capacity by 2030.

Civil society and social movement organisations also helped to develop grassroots innovations, but in all three systems these struggled to diffuse and scale-up because of unfavourable contexts and policies. There are thousands of UK energy community projects, often organised as small wind parks, but their cumulative electricity generation has remained small, leading Strachan et al. (2015: 105) to conclude that 'community renewables remain weakly developed in the UK'. They suggested that 'this can be attributed to the persistence of key features of the socio-technical regime for electricity provision, which continues to favour large corporations and major facilities. Indeed, key structuring elements – in systems of market support and planning policy in the UK – have arguably become more supportive of hard energy paths in the years since 2000, not less' (p. 106). This resonates with our more up-to-date findings that UK policies systematically and increasingly favour large-scale electricity generation technologies over small-scale ones, including community energy projects.[14]

Civil society and social movement organisations also pioneered radical innovations such as low-energy house designs in the building sector. Since the 1970s, a sustainable building movement, including pioneering architects,

[14] Recent examples include the 2016 closure of Renewables Obligations for small-scale renewables, the scrapping of Feed-in-Tariffs in 2019, and the inclusion of large-scale solar-PV and onshore wind in the fourth round of the Contracts-for-Difference (CfD) auction in 2021.

specialised suppliers, and self-builders, developed new knowledge and construction principles in dedicated niches. Although some of these insights and principles were selectively adopted by volume housebuilders (Smith, 2007), the grassroots eco-housing niche has remained very small in the UK, as shown in Chapter 6.

In the mobility sector, environmentally concerned citizens pioneered car sharing in the late 1980s and early 1990s at the local community level (Ornetzeder and Rohracher, 2013). But the wider diffusion of this new practice involved professionalisation and the creation of commercial business models (Truffer, 2003). Although car sharing has deep transformative potential, the niche has remained relatively small, confined to a few big cities (especially London) and user segments (such as young urban professionals without children).

In sum, wider publics and civil society organisations have been more influential in shaping UK low-carbon reconfigurations through public debates and campaigns than through grassroots innovations. Even with regard to public debates, however, the effects on low-carbon reconfigurations have been mitigated by public concerns about other issues such as rising prices (for electricity, gas, railways, buses, petrol/diesel), fuel poverty, energy security, housing shortage, road congestion, jobs, growth, industrial opportunities, overcrowding (in trains), reliability, and punctuality (trains, buses). Particularly in the heat and mobility systems, these other public concerns have often been more important than climate change mitigation, which has had constraining effects on low-carbon transitions.

7.2.5 Landscape Developments

The socio-technical transitions literature also suggests that exogenous 'landscape' developments play important roles because they may help to open up, orient, or accelerate opportunities for fundamental change. Several exogenous shocks and landscape developments influenced the UK low-carbon reconfigurations in various ways. The rapid rise of oil prices, which reached $140 per barrel in 2008 and remained high for several subsequent years, made car-driving more expensive and, combined with the 2007/8 financial-economic crisis, reduced automobility, car-related GHG emissions, and car sales in the 2008–2012 period. These effects were short-lived, however, and trends rebounded as oil prices decreased after 2012 and economic growth picked up again. In the electricity sector, rising oil prices increased gas prices (because both are linked), which led to increased coal use and GHG emissions between 2010 and 2012. This temporary landscape effect, however, did not overturn the post-2007 trend of declining coal use.

While the oil price rise mostly had short-term effects, the 2007/8 financial-economic crisis and the subsequent recession and austerity politics, had longer-lasting effects. They changed political priorities and perceptions, leading to greater emphasis on renewed economic growth and less political emphasis on climate change and low-carbon transitions, which the Conservative Prime Minister David Cameron in 2013 reportedly referred to as 'green crap' that raised energy bills and created red tape for industries. Since the early 2010s, these changing political priorities and perceptions, which assumed opposition and trade-offs between low-carbon transitions and economic growth, resulted in the weakening of several policy instruments, including weaker financial support for onshore wind, bio-power, and solar-PV (particularly through the 2015 energy reset), and the watering down and scrapping of the zero-carbon homes policy (2015) and the Green Deal (2015).

It was not until the late 2010s that the effects of the financial crisis began to fade and low-carbon transition policies strengthened again in the context of the 2017 *Industrial Strategy* and the 2017 *Clean Growth Strategy*, which both reduced the presumed opposition with economic growth. The stronger policies did, however, concur with stronger alignment of climate change and industrial policies, which in many instances led to a focus on large-scale technological mitigation options and job-related issues.

The 2016 Brexit referendum decision was a landscape development that especially affected the UK car industry, which was closely entwined with Europe through supply chains and exports. The prolonged Brexit negotiations, which were not resolved until 2020, created deep uncertainties about trade barriers and tariffs, which led automakers to reduce investments in UK manufacturing plants by about 75% for several years. Uncertainties and slow economic growth also reduced UK car sales and car manufacturing in subsequent years, creating challenging circumstances for the industry. The re-appearance of an explicit industrial strategy, including for low-carbon transitions, was thus also partly a response to the economic effects of Brexit, driven by new priorities such as increased national independence, revived manufacturing, and job creation.

Another consequence of Brexit was the delayed publication of major policy documents (such as the Energy White Paper, and the Buildings and Heat Strategy), because the prolonged Brexit negotiations not only created much uncertainty but also reduced the political bandwidth for other policy discussions. Further long-term Brexit consequences are still uncertain, for instance how it may affect interconnectors and electricity flows between the UK and European countries, or in which countries automakers will decide to build new electric vehicle manufacturing plants.

The COVID-19 pandemic, which led to a series of lockdowns and significantly reduced economic activity in 2020, is still ongoing at the time of writing. It is too early to definitively appraise its effects because long-term consequences are still in the making and because the emergence of new variants means the pandemic may

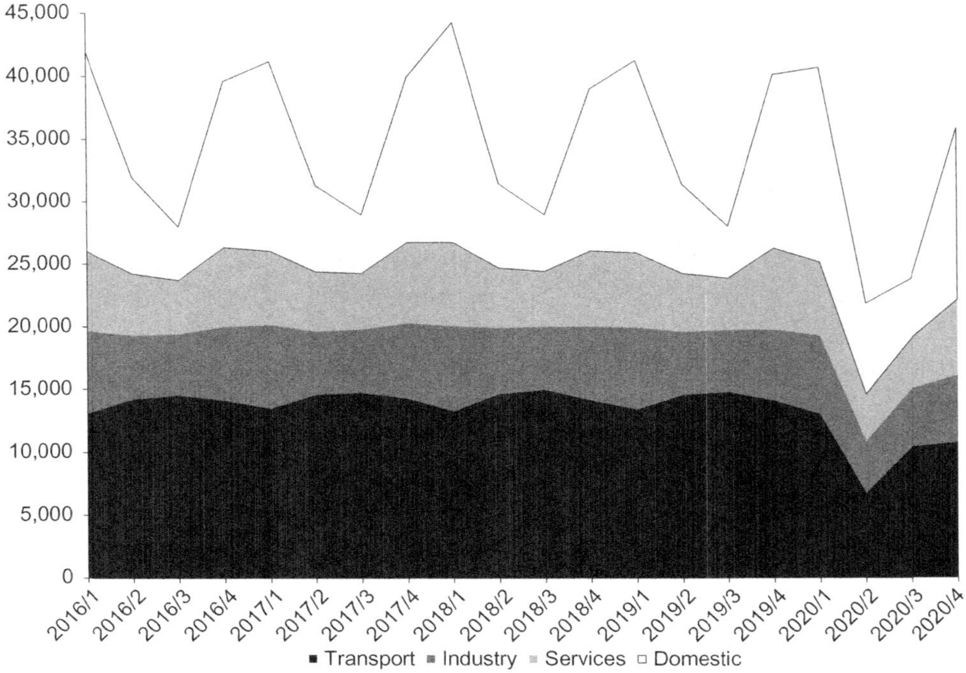

Figure 7.4 Final energy consumption (in million tonnes of oil equivalent) by user category (constructed using data from DUKES Energy Trends March 2021; supply and use of fuels)

not yet be over. Nevertheless, we can make some tentative evaluations. With regard to our three domains, COVID-lockdowns had the largest effects on mobility, where lockdowns reduced transport-related energy use by 46% in the second quarter of 2020 (Figure 7.4). For the whole of 2020, transport-related energy use and GHG emissions were 29% lower than in 2019 (CCC, 2021), which is likely a temporary effect.

Although automobility decreased substantially in the second quarter of 2020, it has since then almost entirely rebounded (Figure 5.4), which suggests that COVID was a temporary shock with limited structural effects on automobility behaviours. Bus and rail travel also plummeted in 2020 but have not rebounded as much because people have remained cautious about traveling with others in confined spaces. COVID could thus have longer-term effects on public transport, and potentially even lead to a modal shift towards personal cars. Cycling increased in 2020, but subsequently returned to pre-pandemic levels. Tele-working rapidly increased and is likely to have long-term structural effects for many skilled professions. Lockdowns and tele-working also increased home energy use, which raised emissions from residential buildings by 2% in 2020 (CCC, 2021). There are also indications that COVID may have affected people's housing preferences, notably a desire for larger suburban houses with gardens, which would also increase heat demand and residential GHG emissions. Longer-term COVID consequences could also include greater acceptance of a more interventionist role of the state, which might enable stronger governance of low-carbon transitions. They could also include empty office blocks (if people work more from home) and closures of high street shops (if people shop more online), which may both seriously affect city centres, with unclear effects on low-carbon transitions.

7.3 Future Low-Carbon Transitions and Policy Recommendations

Although GHG emissions decreased by 41% between 1990 and 2019, the UK is not on track to reach net-zero by 2050. The reason is that the bulk of emission reductions have so far come from low-carbon transitions in the electricity system and from industrial offshoring (which is not captured by our evaluation) and efficiency improvements. Low-carbon transitions in heat and mobility are not yet progressing with sufficient speed and scope to reach the 2050 targets.

Current trends, policies, future targets, and actor commitments suggest that low-carbon transitions in electricity are likely to continue in the coming years, although this will probably require deeper reconfigurations in the electricity grid sub-system, especially related to battery storage, capacity markets, smart grids, and perhaps also demand-side response.

In the automobility system, current trends, policies, future targets, and actor commitments suggest that a low-carbon transition to electric vehicles is likely to continue, although reaching the 2030 targets will be challenging and require faster creation of rapid battery-recharging facilities as well as upgrading of local electricity grids (because increased home-charging will intensify local electricity flows). In cities, there are also opportunities to reconfigure local transport systems more deeply by increasing public transport, cycling, car sharing, mobility services, and ride-hailing, and simultaneously reducing car travel through traffic restrictions (e.g., one-way streets, closing city centre access) and pricing mechanisms (e.g., congestion charging, emission zones). London has led the way in this regard, and the recent bus and cycling strategies and the 2021 Transport Decarbonisation Plan aim to stimulate similar changes in other cities, but success is not guaranteed due to structural problems (e.g., car dependence, lack of capabilities, local opposition). The railway system has expanded since the mid-1990s, but the pandemic (and a shift to tele-working) has substantially reduced rail travel, so a future transition is uncertain, which is compounded by the recent institutional reforms that aim to reduce the need for public subsidies. The bus system outside London has been declining for decades, and it remains to be seen if the new bus strategy can reverse this, especially in the COVID context.

In the heat system, current trends, policies, and actor commitments suggest that a low-carbon transition is not underway. Policy terminations, policy reversals, and weakened governance in the past decade have slowed low-carbon progress and weakened actor commitments. There has been some progress in the deployment of low-carbon heating devices, but more radical heat and building options have remained relatively small. The government's green recovery plan, launched in 2020, announced a £1.5 billion Green Homes Grant to stimulate energy-efficiency upgrades and the ambition to roll out 600,000 heat pumps per year by 2028. But the Green Homes Grant experienced implementation problems and was scrapped in March 2021, while the absence of clear implementation policies make the heat pump target rather unrealistic, reinforcing the 'overpromising and underdelivering' pattern.

In light of this evaluation, reaching the 2050 net-zero target will require significantly stronger low-carbon governance and more comprehensive policies in heat and passenger mobility systems, as well as in agri-food, waste, and industrial production systems, which the book did not analyse. Our analyses show that this not only involves tinkering with market incentives but also a shift from a hands-off governance style to an interventionist approach that actively seeks to shape markets and technologies as well as develop skills and infrastructures. Financial instruments (such as R&D subsidies, capital grants, purchase subsidies, feed-in-tariffs, contracts-for-difference, congestion charges, emission zones, and public infrastructure investments) are important in this regard, but so are regulatory instruments (such as energy efficiency regulations, building standards, emission standards, supplier obligations, phase-out bans) and instruments that enable learning-by-doing and skills development (such as demonstration projects, apprenticeships, retraining schemes, accreditation). Stronger governance and effective policy implementation may also require the creation of dedicated agencies or bodies with sufficient budgets, oversight, and responsibilities, which is noticeably missing in the heat system, as we noted previously. To avoid recurrent and detrimental policy reversals, it is also important to nurture processes that help the ratcheting up of low-carbon policy commitments and implementation stringency, as we further discuss later.

Policymakers cannot change governance styles and policy frameworks at will because they are themselves locked-in by procedures, capabilities, and policy networks (Pierson, 2000). Instead of voluntarist appeals to 'political will', we need to better understand the political conditions within which major policy change can occur (Roberts and Geels, 2019a, 2019b). These conditions include increased pressure from interest groups, citizens, and public opinion, as Hall (1993: 287) also notes: 'It is not civil servants or policy experts, but politicians and the media who play preeminent roles in third-order policy change. . . . The struggle to replace one policy paradigm with another is a society-wide affair, mediated by the press, deeply imbricated with electoral competition, and fought in the public arena.'

Additionally, policymakers can nurture the creation of conditions for deeper policy change by setting targets and using critical evaluations of initial policy instruments to implement stronger ones. This policy-sequencing and ratcheting pattern happened with energy-efficient lighting, where policymakers started with voluntary instruments (e.g., labelling and give-away programmes), then increased energy-efficiency standards for light bulbs, and then introduced an ILB-ban when alternatives were sufficiently developed. Renewable electricity policies also strengthened through policy-sequencing, starting with the auction-based Non-Fossil Fuels Obligation (1990), then moving to the non-technology specific Renewables Obligation (2002), and then to the technology-specific Renewables Obligation (2009), which was followed by a raft of other instruments.

Conditions for deeper policy change can also be nurtured through programmes of on-the-ground demonstration projects, which help demonstrate the practical feasibility of particular options and build the confidence and validated knowledge base required for upscaling (Turnheim and Geels, 2019), and through industrial policy targeted at supporting new firms that can lobby for policy change (Meckling et al., 2017, 2015; Roberts and Geels, 2019a).

7.4 Future Research

Research on socio-technical sustainability transitions has progressed rapidly in the past two decades (Geels, 2019; Köhler et al., 2019). The expansion of the research field has also given rise to increasing specialisation, often along disciplinary lines, leading to more research on the roles of particular actors (e.g., users, business, policymakers, intermediary actors) or dimensions (e.g., economic, political, cultural, consumption). While this specialisation has deepened our understandings in some respects, it also carries the potential risk of disciplinary fragmentation and a loss of attention for multi-actor interactions and co-evolution between multiple dimensions, which has characterised socio-technical transitions research from the start and contributed to the field's intellectual excitement.

While the tendency to 'zoom' in and focus on the roles of particular actors or dimensions is understandable, this book has hopefully demonstrated the importance and fruitfulness of 'zooming out' and investigating 'whole system' transitions, which was arguably the original motivation of the socio-technical transitions literature (Elzen et al., 2004; Geels, 2005). We hope that future research will further explore this topic, which our book has only begun to open up.

Such socio-technical whole system research is important to complement, and critically engage with, the engineering and modelling research communities that have long made 'whole system' analyses of low-carbon transitions, which are, however, based on rather restrictive assumptions that oversimplify social realities and pay little attention to actors and behaviours, including politics, power struggles, beliefs, meanings, strategies (Turnheim et al., 2015; Turnheim and Nykvist, 2019). As policymakers shift attention from abstract modelling to real-world implementation of low-carbon transition pathways, there will be an increasing demand for better understandings and analyses of real-world dynamics. Socio-technical research on whole system reconfiguration, as advanced in this book, is well suited to address this policy demand.

We also hope that future research on socio-technical system reconfiguration will be interdisciplinary, as this is the only way to investigate the co-evolutionary interactions and feedbacks that are essential in long-term, large-scale transformations, as Section 3.1 explained. This is not easy because 'the social sciences are highly balkanized and tribal' (Pierson, 2004: 7). Many *disciplinary* analyses of low-carbon transitions are therefore characterised by pleas for deeper investigations of political, economic, or cultural dimensions, while downplaying the importance of other dimensions. While interesting in some respects, this approach is ultimately reductionist and does not generate comprehensive multi-dimensional understandings of low-carbon transitions. Our book has tried to alleviate this problem by explicitly addressing techno-economic, social, political, and cultural processes and the interactions between them. As a trade-off, the analyses of these processes are probably not deep enough for disciplinary scholars, so we should anticipate and accept criticisms along these lines. Nevertheless, we hope that future research will continue along the path of making inter-disciplinary co-evolutionary analyses of unfolding low-carbon system reconfigurations.

Comparative research between socio-technical systems is also a fruitful avenue for future research. Although comparative socio-technical research between countries (for the same innovation or socio-technical system) is becoming more common, there is almost no research that compares between systems. But because socio-technical transitions research has over the past two decades produced a wealth of studies on single innovations, particular actors, and specific systems, it is now becoming more feasible to make comparative analyses of whole system reconfigurations: 'There is an unmistakable drive towards systematic comparison and theory-building from cases' (Köhler et al., 2019: 19). Our book has only started to explore this comparative opportunity, so there is much more that can be done on this topic.

Our analyses mostly consisted of systematic empirical comparisons across several conceptual dimensions. Future research could try to abstract further and articulate theoretical patterns or typologies of whole system reconfiguration pathways. Future comparative research could also aim to reflect deeper on the morphological differences between socio-technical systems and the implications of different system architectures for the dynamics of change. We identified some structural characteristics of the electricity system that enabled substantial changes in the generation sub-system (such as the separation of supply and demand by the grid and the passing on of costs through electricity bills, which creates 'indirect' or 'involuntary' market demand). We suspect that further structural differences can be identified for the other systems. The agri-food system, for instance, consists of hundreds of parallel commodity chains that are increasingly coordinated by supermarkets, which probably shapes reconfiguration dynamics in important ways (Mylan et al., 2019, 2015). The cultural meaning of food and the roles of diets, civil society organisations, and agri-food governance styles also make the food system structurally different in some ways from electricity, heat, and mobility, which likely has implications for the dynamics of change.

The broader implication of this point is that there may be limitations to the degree of possible theoretical generalisation about low-carbon system reconfiguration dynamics. If structural differences between socio-technical systems do indeed have substantial implications for reconfiguration dynamics, then insights from electricity system analyses may not entirely translate to mobility, heating, agri-food, or industrial systems. Instead of a general parsimonious theory of complex longitudinal research topics such as socio-technical transitions, it may therefore be more realistic to generalise in terms of middle-range analytical frameworks (such as the Multi-Level Perspective) and a toolbox of more differentiated theories about causal processes and mechanisms. For such research topics, Little (2016: xvi) notes that the 'hope for a comprehensive theory of social change is chimera; it does not correspond to the nature of the social world. It does not sufficiently recognize several crucial features of social phenomena: heterogeneity, causal complexity, contingency, path dependency and plasticity.' Instead, he suggests that: 'A "mature" social science would involve not a deductive theory with a few high-level generalizations and laws, but rather an eclectic ensemble of theories of mid-level processes and mechanisms' (Little, 2016: 260). Elsewhere, he made a thoughtful comment about large-scale historical research, which also holds relevance for low-carbon system reconfigurations:

> The best hope we have for generalizations about large historical processes . . . is not at the level of wholes . . .. Rather, what we can hope to do is to discover a number of recurring mid-level processes and mechanisms (political, demographic, technology, institutional, and economic) that can be identified and studied in multiple historical cases. (Little, 2010: 89)

Although there are clear differences between the three systems we analysed, there are also similarities, particularly with regard to the governance style and the role of incumbent firms, as we discussed earlier. Since country contexts matter, future research could also fruitfully make comparative analyses of system reconfigurations in other countries that differ from the UK in one or more dimensions, for instance countries with a stronger role for the state (e.g., France, India, Japan) or a decentralised state (e.g., Germany), countries with a less market-driven or neo-liberal forms of capitalism, countries with a more mission-driven governance style (e.g., Costa Rica, Austria), or countries with significant manufacturing capacity (e.g., China). It would be interesting to investigate how these country differences affect system reconfiguration dynamics.

Another interesting topic for future research concerns the conditions and drivers of accelerated low-carbon transitions. Section 7.1.6 already identified several conditions and drivers and illustrated these with multiple examples. However, more can be done on this topic, which is likely to become more important in the coming years. A related topic are the trade-offs and tensions between different aspects of transitions. Section 7.1.6 already identified potential trade-offs between speed and scope of low-carbon system reconfigurations. Scholars also identified trade-offs between speed and inclusive deliberation (Ciplet and Harrison, 2020; Skjølsvold and Coenen, 2021). Investigating such trade-offs and tensions, and how they can best be navigated, is interesting and important, and arguably more realistic than assuming that multiple social problems (e.g., climate change, democratic deficits, poverty, inequality, exclusion) will be solved in one great or deep transition.

To conclude, low-carbon system reconfiguration is an interesting, exciting, and ambitious topic for future socio-technical transitions research. This topic opens up many new questions and research puzzles that build on but go beyond existing transitions research. It enables (and requires) new conceptual contributions and new empirical research templates, which support systematic multi-dimensional comparisons between systems. We hope that our book has not only produced novel insights but also opened up fruitful research avenues that other transition scholars may want to further elaborate. Considering that socio-technical transition research has come a long way in the past two decades, it will be interesting to see what the next two decades will bring.

References

Abbott, A., 2004. *Methods of Discovery: Heuristics for the Social Sciences*. W. W. Norton, New York.

Abbott, A., 2001. *Time Matters: On Theory and Method*. University of Chicago Press, Chicago.

Acemoglu, D., Akcigit, U., Hanley, D., and Kerr, W., 2016. Transition to clean technology. *Journal of Political Economy* 124, 52–104.

ADBA, 2020. *Biomethane: The Pathway to 2030*. Anaerobic Digestion and Bioresources Association, London.

Adelle, C. and Russel, D., 2013. Climate policy integration: A case of déjà vu? *Environmental Policy and Governance* 23, 1–12.

AEA, 2012. RHI Phase II – Technology assumptions: Key technical assumptions for selected technologies. Report for DECC, AEA/R/ED57097, Issue 3.

AECOM, 2011. Study on public transport smartcard. Final report for DG Move at the European Commission.

Akyelken, N., Banister, D., and Givoni, M., 2018. The sustainability of shared mobility in London: The dilemma for governance. *Sustainability* 10, 420.

Alcott, B., 2007. The sufficiency strategy: Would rich-world frugality lower environmental impact? *Ecological Economics* 64, 770–786.

Aldred, R., 2012. Governing transport from welfare state to hollow state: The case of cycling in the UK. *Transport Policy* 23, 95–102.

Aldred, R. and Jungnickel, K., 2014. Why culture matters for transport policy: The case of cycling in the UK. *Journal of Transport Geography* 34, 78–87.

Alexander, B., Dijst, M., and Ettema, D., 2010. Working from 9 to 6? An analysis of in-home and out-of-home working schedules. *Transportation* 37, 505–523.

Alexander, J. C., 2003. *The Meanings of Social Life: A Cultural Sociology*. Oxford University Press, New York.

Aman, M. M., Jasmon, G. B., Mokhlis, H., and Bakar, A. H. A., 2013. Analysis of the performance of domestic lighting lamps. *Energy Policy* 52, 482–500.

Ambrose, J. and Vinter, R., 2021. Nissan sets out plans for £1bn electric car hub in Sunderland. *The Guardian*. www.theguardian.com/business/2021/jul/01/nissan-sets-out-plans-for-1bn-electric-car-hub-in-sunderland

Ambrosio-Albalá, P., Upham, P., and Bale, C. S. E., 2019. Purely ornamental? Public perceptions of distributed energy storage in the United Kingdom. *Energy Research & Social Science* 48, 139–150.

AMDEA, 2014. *Promoting Highly Efficient Electrical Appliances*. Association of Manufacturers of Domestic Appliances, London.

Aminzade, R., 1992. Historical sociology and time. *Sociological Methods and Research* 20, 456–480.

Andersen, A. D., 2014. No transition without transmission: HVDC electricity infrastructure as an enabler for renewable energy? *Environmental Innovation and Societal Transition* 13, 75–95.

Andersen, A. D., Steen, M., Mäkitie, T., Hanson, J., Thune, T. M., and Soppe, B., 2020. The role of inter-sectoral dynamics in sustainability transitions: A comment on the transitions research agenda. *Environmental Innovation and Societal Transitions* 34, 348–351.

Anderson, P. and Tushman, M. L., 1990. Technological discontinuities and dominant designs: A cyclical model of technological change. *Administrative Science Quarterly* 35, 604–633.

Arapostathis, S., Carlsson-Hyslop, A., Pearson, P. J. G., Thornton, J., Gradillas, M., Laczay, S., et al., 2013. Governing transitions: Cases and insights from two periods in the history of the UK gas industry. *Energy Policy* 52, 25–44. https://doi.org/10.1016/j.enpol.2012.08.016

Arapostathis, S., Pearson, P. J. G., and Foxon, T. J., 2014. UK natural gas system integration in the making, 1960–2010: Complexity, transitional uncertainties and uncertain transitions. *Environmental Innovation and Societal Transitions* 11, 87–102. https://doi.org/10.1016/j.eist.2014.01.004

Arbib, J. and Sebat, T., 2017. Rethinking Transportation 2020–2030: The Disruption of Transportation and the Collapse of the Internal Combustion Vehicle and Oil Industries. RethinkX.

Archer, T. and Cole, I., 2014. Still not plannable? Housing supply and the changing structure of the housebuilding industry in the UK in 'austere' times. *People, Place and Policy* 8, 97–112. https://doi.org/10.3351/ppp.0008.0002.0002

Arthur, W. B., 1989. Competing technologies, increasing returns, and lock-in by historical events. *The Economic Journal* 99, 116–131.

Arthur, W. B., 1988. Competing technologies: An overview, in: Dosi, G., Freeman, C., Nelson, R., Silverberg, G., and Soete, L. (Eds.), *Technical Change and Economic Theory*. Pinter, London, pp. 590–607.

Audenaert, A., De Clyn, S. H., and Vankerckhove, B., 2008. Economic analysis of passive houses and low-energy house compared with standard houses. *Energy Policy* 36, 47–55.

Bailey, I., 2007. Market environmentalism, new environmental policy instruments, and climate change policy in the United Kingdom and Germany. *Annals of the Association of American Geographers* 97, 530–550.

Balcombe, P., Rigby, D., and Azapagic, A., 2014. Investigating the importance of motivations and barriers related to microgeneration uptake in the UK. *Applied Energy* 130, 403–418.

Baldwin, C. Y. and Clark, K. B., 1997. Managing in an age of modularity. *Harvard Business Review*, September, pp. 84–93. https://hbr.org/1997/09/managing-in-an-age-of-modularity

Barlow, J., 1999. From craft production to mass customisation: Innovation requirements for the UK housebuilding industry. *Housing Studies* 14, 23–42.

Barnes, W., Gartland, M., and Stack, M., 2004. Old habits die hard: Path dependency and behavioral lock-in. *Journal of Economic Issues* 38, 371–377.

Barney, J. B., 1991. Firm resources and sustained competitive advantage. *Journal of Management* 17, 99–120.

Barnham, K., 2014. *The Burning Answer: A User's Guide to the Solar Revolution*. Orion Publishing Group, London.
Bastian, A., Börjesson, M., and Eliasson, J., 2017. Response to Wadud and Baierl: 'Explaining "peak car" with economic variables: An observation'. *Transportation Research Part A: Policy and Practice* 95, 386–389.
Bastian, A., Börjesson, M., and Eliasson, J., 2016. Explaining 'peak car' with economic variables. *Transportation Research Part A: Policy and Practice* 88, 236–250.
Batel, S., 2018. A critical discussion of research on the social acceptance of renewable energy generation and associated infrastructures and an agenda for the future. *Journal of Environmental Policy and Planning* 20, 356–369.
BBC, 2019. Smart meter roll-out delayed for four years. www.bbc.co.uk/news/business-49721436
BEIS, 2020a. *Digest of UK Energy Statistics (DUKES)*. Department for Business, Energy and Industrial Strategy, London.
BEIS, 2020b. *Future Support for Low Carbon Heat: Impact Assessment*. Department for Business, Energy and Industrial Strategy, London.
BEIS, 2020c. *Future Support for Low Carbon Heat: Consultation*. Department for Business, Energy and Industrial Strategy, London.
BEIS, 2020d. *Heat Networks: Building a Market Framework*. Department for Business, Energy & Industrial Strategy, London.
BEIS, 2019. *Energy Consumption in the UK*. Department of Business, Energy and Industrial Strategy, London.
BEIS, 2018a. *Clean Growth: Transforming Heating*. Department of Business, Energy and Industrial Strategy, London.
BEIS, 2018b. *Digest of UK Energy Statistics (DUKES)*. Department of Business, Energy and Industrial Strategy, London.
BEIS, 2017a. *The Clean Growth Strategy: Leading the Way to a Low Carbon Future*. Department of Business, Energy and Industrial Strategy, London.
BEIS, 2017b. *Energy Consumption in the UK*. Department of Business, Energy and Industrial Strategy, London.
Benford, R. D. and Snow, D. A., 2000. Framing processes and social movements: An overview and assessment. *Annual Review of Sociology* 26, 611–639.
Bergek, A., Berggren, C., Magnusson, T., and Hobday, M., 2013. Technological discontinuities and the challenge for incumbent firms: Destruction, disruption or creative accumulation? *Research Policy* 42, 1210–1224.
Berggren, C., Magnusson, T., and Sushandoyo, D., 2015. Transition pathways revisited: Established firms as multi-level actors in the heavy vehicle industry. *Research Policy* 44, 1017–1028. https://doi.org/10.1016/j.respol.2014.11.009
Bergman, N., Schwanen, T., and Sovacool, B. K., 2017. Imagined people, behaviour and future mobility: Insights from visions of electric vehicles and car clubs in the United Kingdom. *Transport Policy* 59, 165–173.
Berkers, E. and Geels, F. W., 2011. System innovation through stepwise reconfiguration: The case of technological transitions in Dutch greenhouse horticulture (1930–1980). *Technology Analysis & Strategic Management* 23, 227–247.
Berkhout, F., Smith, A., and Stirling, A., 2004. Socio-technological regimes and transition contexts, in: Elzen, B., Geels, F. W., and Green, K. (Eds.), *System Innovation and the Transition to Sustainability: Theory, Evidence and Policy*. Edward Elgar, Cheltenham, pp. 48–75.
BERR, 2007. *Appraisal of Costs & Benefits of Smart Meter Roll Out Options*. Department for Business, Enterprise and Regulatory Reform, Brighton.

Beynon, H., Cam, S., Fairbrother, P., and Nichols, T., 2003. The Rise and Transformation of the UK Domestic Appliances Industry. Cardiff University School of Social Sciences, Working Paper 42.

Bidmon, C. M. and Knab, S. F., 2018. The three roles of business models in societal transitions: New linkages between business model and transition research. *Journal of Cleaner Production* 178, 903–916.

Blythe, P., 2004. Improving public transport ticketing through smart cards. *Municipal Engineer* 157, 47–54.

BNEF, 2020. Battery Price Survey. Bloomberg New Energy Finance.

Boardman, B., 2007. *Home Truths: A Low-Carbon Strategy to Reduce UK Housing Emissions by 80% by 2015*. Environmental Change Institute, Oxford.

Boardman, B., 2004. Achieving energy efficiency through product policy: The UK experience. *Environmental Science and Policy* 7, 165–176.

Bobrova, Y., 2020. The adoption process of low-carbon home retrofits among UK homeowners: A socio-technical perspective and system dynamics model. University College London.

Bohnsack, R., Kolk, A., Pinkse, J., and Bidmon, C., 2020. Driving the electric bandwagon: The dynamics of incumbents' sustainable product innovation. *Business Strategy and the Environment* 29, 727–743.

Boltanski, L. and Thévenot, L., 2006. *On Justification: Economies of Worth*. Princeton University Press, Princeton, NJ.

Bolton, R. and Foxon, T. J., 2015. Infrastructure transformation as a socio-technical process: Implications for the governance of energy distribution networks in the UK. *Technological Forecasting and Social Change* 90, 538–550. https://doi.org/10.1016/j.techfore.2014.02.017

Bonfield, P., 2016. Each Home Counts: An Independent Review of Consumer Advice, Protection, Standards and Enforcement for Energy Efficiency and Renewable Energy. Department for Business, Energy & Industrial Strategy and the Department for Communities and Local Government, London.

Borup, M., Brown, N., Konrad, K., and van Lente, H., 2006. The sociology of expectations in science and technology. *Technology Analysis & Strategic Management* 18, 285–298.

Bourdieu, P., 1977. *Outline of a Theory of Practice*. Cambridge University Press, London.

Bowman, A., Folkman, P., Froud, J., Johal, S., Law, J., Leaver, A., et al., 2013. The Great Train Robbery: The Economic and Political Consequences of Rail Privatisation, Public Interest Report. Centre for Research on Socio-Cultural Change, Manchester University.

Bradshaw, M., 2018. Future UK Gas Security: Midstream Infrastructure.

Bradshaw, M., Bridge, G., Bouzarovski, S., Watson, J., and Dutton, J., 2014. *The UK's Global Gas Challenge*. Research Report. London.

Brand, C., Anable, J., Ketsopoulou, I., and Watson, J., 2020. Road to zero or road to nowhere? Disrupting transport and energy in a zero carbon world. *Energy Policy* 139, 111334.

Brand, C., Anable, J., and Tran, M., 2013. Accelerating the transformation to a low carbon passenger transport system: The role of car purchase taxes, feebates, road taxes and scrappage incentives in the UK. *Transportation Research Part A* 49, 132–148.

Broad, O., Hawker, G., and Dodds, P. E., 2020. Decarbonising the UK residential sector: The dependence of national abatement on flexible and local views of the future. *Energy Policy* 140, 111321. https://doi.org/10.1016/j.enpol.2020.111321

Brown, D., 2018. Whole-house retrofit: The role of new business models, finance mechanisms, and their implications for policy. University of Sussex.

Brown, D., Kivimaa, P., and Sorrell, S., 2019. An energy leap? Business model innovation and intermediation in the 'Energiesprong' retrofit initiative. *Energy Research and Social Science* 58, 101253. https://doi.org/10.1016/j.erss.2019.101253

Butler, S., 2021. Uber drivers entitled to workers' rights, UK supreme court rules. *The Guardian.* www.theguardian.com/technology/2021/feb/19/uber-drivers-workers-uk-supreme-court-rules-rights

Caird, S., Roy, R., and Potter, S., 2012. Domestic heat pumps in the UK: User behaviour, satisfaction and performance. *Energy Efficiency* 5, 283–301. https://doi.org/10.1007/s12053–012-9146-x

Campbell, P. and Inagaki, K., 2021. UK carmakers after Brexit: A race to attract battery production. *Financial Times.*

Campbell, P., Pfeifer, S., and Pooler, M., 2019. How the UK is falling behind in the global electric car race. *Financial Times.* www.ft.com/content/0e097085-5b80-4bc1-8c34-ec5df5be0acc

Carter, N. and Jacobs, M., 2014. Explaining radical policy change: The case of climate change and energy policy under the British Labour Government 2006–10. *Public Administration* 92, 125–141.

Cary, R., 2010. *Future Proof: An Electricity Network for the 21st Century.* Green Alliance, London.

Cary, R. and Benton, D., 2012. *Creating a Market for Electricity Savings: Paying for Energy Efficiency through the Energy Bill.* Green Alliance, London.

CCC, 2021. *Progress in Reducing Emissions: 2021 Report to Parliament.* Committee on Climate Change, London.

CCC, 2020. *Reducing UK Emissions: 2020 Progress Report to Parliament.* Committee on Climate Change, London.

CCC, 2019a. *Reducing UK Emissions: 2019 Progress Report to Parliament.* Committee on Climate Change, London.

CCC, 2019b. *Net Zero: The UK's Contribution to Stopping Global Warming.* Committee on Climate Change, London.

CCC, 2019c. *UK Housing: Fit for the Future?* Committee on Climate Change, London.

CCC, 2018a. *Reducing UK Emissions 2018: Progress Report to Parliament.* Committee on Climate Change, London.

CCC, 2018b. *Biomass in a Low-Carbon Economy.* Committee on Climate Change, London.

CCC, 2018c. *Hydrogen in a Low-Carbon Economy.* Committee on Climate Change, London.

CCC, 2016. *Next Steps for UK Heat Policy.* Committee on Climate Change, London.

CCC, 2015. *Meeting Carbon Budgets: Progress in Reducing the UK's Emissions 2015 Report to Parliament.* Committee on Climate Change, London.

CCC, 2013. *Meeting Carbon Budgets: 2013 Progress Report to Parliament.* Committee on Climate Change, London.

CCC, 2011. *The Renewable Energy Review.* Committee on Climate Change, London.

CCC, 2010. *The Fourth Carbon Budget: Reducing Emissions through the 2020s.* Committee on Climate Change, London.

Chandler, A. D., 2001. *Inventing the Electronic Century: The Epic Story of the Consumer Electronics and Computer Industries.* The Free Press, New York.

Chappells, H. and Shove, E., 2005. Debating the future of comfort: Environmental sustainability, energy consumption and the indoor environment. *Building Research and Information* 33, 32–40. https://doi.org/10.1080/0961321042000322762

Chaudry, M., Abeysekera, M., Hosseini, S. H. R., Jenkins, N., and Wu, J., 2015. Uncertainties in decarbonising heat in the UK. *Energy Policy* 87, 623–640. https://doi.org/10.1016/j.enpol.2015.07.019

Chenari, B., Dias Carrilho, J., and Gameiro da Silva, M., 2016. Towards sustainable, energy-efficient and healthy ventilation strategies in buildings: A review. *Renewable and Sustainable Energy Reviews* 59, 1426–1447.

Cherp, A., Vinichenko, V., Jewell, J., Brutschin, E., and Sovacool, B., 2018. Integrating techno-economic, socio-technical and political perspectives on national energy transitions: A meta-theoretical framework. *Energy Research and Social Science* 37, 175–190. https://doi.org/10.1016/j.erss.2017.09.015

Christensen, C., 1997. *The Innovator's Dilemma: When New Technologies Cause Great Firms to Fail*. Harvard Business School Press, Boston.

Christensen, C. M., Kaufman, S. P., and Shih, W. C., 2008. Innovation killers: How financial tools destroy our capacity to do new things. *Harvard Business Review* 86, 98–105.

Christoff, P., 1996. Ecological modernisation, ecological modernities. *Environmental Politics* 5, 476–500.

Ciplet, D. and Harrison, J. L., 2020. Transition tensions: Mapping conflicts in movements for a just and sustainable transition. *Environmental Politics* 29, 435–456.

CIVITAS, 2016. *Mobility-as-a-Service: A New Transport Model*. Civitas: The Institute for the Study of Civil Society, London.

Coenen, L., Benneworth, P., and Truffer, B., 2012. Toward a spatial perspective on sustainability transitions. *Research Policy* 41, 968–979. https://doi.org/10.1016/j.respol.2012.02.014

Cohen-Blankshtain, G. and Rotem-Mindali, O., 2016. Key research themes on ICT and sustainable urban mobility. *International Journal of Sustainable Transportation* 10, 9–17.

Colli, E., 2020. Towards a mobility transition? Understanding the environmental impact of Millennials and Baby Boomers in Europe. *Travel Behaviour and Society* 20, 273–289.

Conebi, 2017. European Bicycle Market. Confederation of the European Bicycle Industry. Brussels. https://issuu.com/conebi/docs/20170713_european_bicyle_industry_a

Connor, P. M., Axon, C. J., Xenias, N., and Balta-Ozkan, N., 2018. Sources of risk and uncertainty in UK smart grid deployment: An expert stakeholder analysis. *Energy* 161, 1–9.

Connor, P. M., Xie, L., Lowes, R., Britton, J., and Richardson, T., 2015. The development of renewable heating policy in the United Kingdom. *Renewable Energy* 75, 733–744. https://doi.org/10.1016/j.renene.2014.10.056

Cotton, M. and Devine-Wright, P., 2013. Putting pylons into place: A UK case study of public perspectives on the impacts of high voltage overhead transmission lines. *Journal of Environmental Planning and Management* 56, 1225–1245.

Cotton, M. and Devine-Wright, P., 2012. Making electricity networks 'visible': Industry actor representations of 'publics' and public engagement in infrastructure planning. *Public Understanding of Science* 21, 17–35.

Cowie, J., 2002. Acquisition, efficiency and scale economies: An analysis of the British bus industry. *Transport Reviews* 22, 147–157.

Cramer, J. and Krueger, A. B., 2016. Disruptive change in the taxi business: The case of Uber. *American Economic Review* 106, 177–182.

Crosbie, T., 2008. Household energy consumption and consumer electronics: The case of television. *Energy Policy* 36, 2191–2199.

Cullen, J. M. and Allwood, J. M., 2010. The efficient use of energy: Tracing the global flow of energy from fuel to service. *Energy Policy* 38, 75–81. https://doi.org/10.1016/j.enpol.2009.08.054

Currie, G. and Wallis, I., 2008. Effective ways to grow urban bus markets: A synthesis of evidence. *Journal of Transport Geography* 16, 419–429.

Cyert, R. and March, J., 1963. *A Behavioral Theory of the Firm*. Prentice Hall, Englewood Cliffs, NJ.

Dangerman, A. J. and Schellnhuber, H. J., 2013. Energy systems transformation. *Proceedings of the National Academy of Sciences of the United States of America* 110, 549–558.

Darby, S., 2006. The effectiveness of feedback on energy consumption: A review for DEFRA of the literature on metering, billing and direct displays. Oxford.

Darby, S. J. and McKenna, E., 2012. Social implications of residential demand response in cool temperate climates. *Energy Policy* 49, 759–769.

Darnhofer, I., D'Amico, S., and Fouilleux, E., 2019. A relational perspective on the dynamics of the organic sector in Austria, Italy, and France. *Journal of Rural Studies*. https://doi.org/10.1016/j.jrurstud.2018.12.002

Daunton, M. J., 1987. *A Property-Owning Democracy? Housing in Britain*. Faber & Faber, London.

Davis, G. F. and Marquis, D., 2005. Prospects for organization theory in the early twenty-first century: Institutional fields and mechanisms. *Organization Science* 16, 332–343.

DCLG, 2017. *Fixing our Broken Housing Market*. Department for Communities and Local Government, London.

De Boer, M. A. H. and Caprotti, F., 2017. Getting Londoners on two wheels: A comparative approach analysing London's potential pathways to a cycling transition. *Sustainable Cities and Society* 32, 613–626.

DECC, 2016. *Energy Consumption in the UK*. Department of Energy and Climate Change, London.

DECC, 2014a. *Delivering UK Energy Investment*. Department of Energy and Climate Change, London.

DECC, 2014b. *Life Cycle Impacts of Biomass Electricity in 2020*. Department of Energy and Climate Change, London.

DECC, 2014c. *UK Solar PV Strategy Part 2: Delivering a Brighter Future*. Department of Energy and Climate Change, London.

DECC, 2014d. *Impact Assessment of Smart Meter Roll-Out for the Domestic and Small and Medium Non-Domestic Sectors*. Department of Energy and Climate Change, London.

DECC, 2014e. *Community Energy Strategy: People Powering Change*. Department of Energy and Climate Change, London.

DECC, 2013a. *UK Solar PV Strategy Part 1: Roadmap to a Brighter Future*. Department of Energy and Climate Change, London.

DECC, 2013b. *The Future of Heating: Meeting the Challenge*. Department of Energy and Climate Change, London.

DECC, 2013c. *United Kingdom Housing Energy Fact File*. Department of Energy and Climate Change, London.

DECC, 2013d. *Summary Evidence on District Heating Networks in the UK*. Department of Energy and Climate Change, London.

DECC, 2013e. *Research into Barriers to Deployment of District Heating Networks*. Department of Energy and Climate Change, London.

DECC, 2012a. *The Energy Efficiency Strategy: The Energy Efficiency Opportunity in the UK: Strategy and Annexes*. Department of Energy and Climate Change, London.

DECC, 2012b. *UK Bioenergy Strategy*. Department of Energy and Climate Change, London.

DECC, 2012c. *United Kingdom Housing Energy Fact File*. Department of Energy and Climate Change, London.

DECC, 2012d. *The Future of Heating: A Strategic Framework for Low Carbon Heat in the UK*. Department of Energy and Climate Change, London.
DECC, 2011a. *UK Renewable Energy Roadmap*. Department of Energy and Climate Change, London.
DECC, 2011b. *Carbon Plan*. Department of Energy and Climate Change, London.
DECC, 2010. *2050 Pathway Analysis*. Department of Energy and Climate Change, London.
DECC, 2009. *The UK Low Carbon Transition Plan: National Strategy for Climate and Energy*. Department of Energy and Climate Change, London.
DEFRA, 2020. *Crops Grown for Bioenergy in the UK: 2019*. Department for Environment, Food and Rural Affairs, London.
DEFRA, 2009. *Saving Energy through Better Products and Appliances: A Report on Analysis, Aims and Indicative Standards for Energy Efficient Products 2009–2030*. Department for Environment, Food and Rural Affairs, London.
Delbosc, A., 2016. Delay or forgo? A closer look at youth driver licensing trends in the United States and Australia. Transportation 1–8. https://doi.org/10.1007/s11116-016-9685-7
DfT, 2021a. *Decarbonising Transport: A Better, Greener Britain*. Department for Transport, London.
DfT, 2021b. *Great British Railways: Williams-Shapps Plan for Rail*. Department for Transport, London.
DfT, 2021c. *Bus Back Better: National Bus Strategy for England*. Department for Transport, London.
DfT, 2021d. *Road Traffic Estimates Great Britain 2020*. Department for Transport, London.
DfT, 2021e. *Gear Change: One-Year-On Review*. Department for Transport, London.
DfT, 2020a. *Decarbonising Transport: Setting the Challenge*. Department for Transport, London.
DfT, 2020b. *Gear Change: A Bold Vision for Cycling and Walking*. Department for Transport, London.
DfT, 2020c. *Reported Road Casualties in Great Britain: 2019 Annual Report*. Department for Transport, London.
DfT, 2020d. *Renewable Fuel Statistics 2019 Final Report*. Department for Transport, London.
DfT, 2019a. *Rail Factsheet*. Department for Transport, London.
DfT, 2019b. *Code of Practice: Automated Vehicle Trialing*. Department for Transport, London.
DfT, 2019c. *Transport Statistics Great Britain 2019: Moving Britain Ahead*. Department for Transport, London.
DfT, 2018a. *Analyses from the National Travel Survey*. Department for Transport, London.
DfT, 2018b. *The Road to Zero: Next Steps towards Cleaner Road Transport and Delivering Our Industrial Strategy*. Department for Transport, London.
DfT, 2017. *Transport Statistics Great Britain 2017*. Department for Transport, London.
DfT, 2015a. *The Pathway to Driverless Cars: Summary Report and Action Plan*. Department for Transport, London.
DfT, 2015b. *The Pathway to Driverless Cars: A Code of Practice for Testing*. Department for Transport, London.
DfT, 2014. *National Travel Survey 2013*, England. Department for Transport, London.
DfT, 2013. *Door to Door: A Strategy for Improving Sustainable Transport Integration*. Department for Transport, London.

DfT, 2009. *Smart and Integrated Ticketing Strategy*. Department for Transport, London.
Di Gregorio, M., Nurrochmat, D. R., Paavola, J., Sari, I. M., Fatorelli, L., Pramova, E., et al., 2017. Climate policy integration in the land use sector: Mitigation, adaptation and sustainable development linkages. *Environmental Science & Policy* 67, 35–43.
Diercks, G., 2018. Lost in translation: How legacy limits the OECD in promoting new policy mixes for sustainability transitions. *Research Policy* 48. https://doi.org/10.1016/j.respol.2018.09.002
Dijk, M., Orsato, R. J., and Kemp, R., 2013. The emergence of an electric mobility trajectory. *Energy Policy* 52, 135–145. https://doi.org/10.1016/j.enpol.2012.04.024
Dijk, M., Wells, P., and Kemp, R., 2016. Will the momentum of the electric car last? Testing an hypothesis on disruptive innovation. *Technological Forecasting and Social Change* 105, 77–88. https://doi.org/10.1016/j.techfore.2016.01.013
Dosi, G., 1982. Technological paradigms and technological trajectories: A suggested interpretation of the determinants and directions of technical change. *Research Policy* 11, 147–162. https://doi.org/10.1016/0048-7333(82)90016-6
Doty, D. H., Glick, W. H., and Huber, G. P., 1993. Fit, equifinality, and organizational effectiveness: A test of two configurational theories. *Academy of Management Journal* 36, 1196–1250.
Dowson, M., Poole, A., Harrison, D., and Susman, G., 2012. Domestic UK retrofit challenge: Barriers, incentives and current performance leading into the Green Deal. *Energy Policy* 50, 294–305. https://doi.org/10.1016/j.enpol.2012.07.019
Dubois, G., Sovacool, B., Aall, C., Nilsson, M., Barbier, C., Herrmann, A., et al., 2019. It starts at home? Climate policies targeting household consumption and behavioral decisions are key to low-carbon futures. *Energy Research and Social Science* 52, 144–158. https://doi.org/10.1016/j.erss.2019.02.001
Dupont, C., 2016. *Climate Policy Integration into EU Energy Policy: Progress and Prospects*. Routledge, London.
Eadson, W., 2016. State enrolment and energy-carbon transitions: Syndromic experimentation and atomisation in England. *Environment and Planning C: Government and Policy* 34, 1612–1631.
EC, 2019. The European Green Deal, Communication from the Commission to the European Parliament, the European Council, the Council, the European Economic and Social Committee and the Committee of the Regions, COM(2019) 640 final, Brussels, 11.12.2019.
EC, 2018. A Clean Planet for all: A European strategic long-term vision for a prosperous, modern, competitive and climate neutral economy, COM(2018) 773 final. Brussels.
Ecofys, 2014. Overview of UK Biofuel Producers: Input to Post-Implementation Review, Report for Department for Transport.
Edmondson, D. L., Rogge, K. S., and Kern, F., 2020. Zero carbon homes in the UK? Analysing the co-evolution of policy mix and socio-technical system. *Environmental Innovation and Societal Transitions* 35, 135–161.
EEA, 2019a. *Sustainability Transitions: Policy and Practice*. European Environment Agency, Copenhagen. https://doi.org/10.2800/641030
EEA, 2019b. *The European Environment – State and Outlook 2020: Knowledge for Transition to a Sustainable Europe*. European Environment Agency, Copenhagen.
EEA, 2018. *Perspectives on Transitions to Sustainability, EEA Report 25/2017*. European Environment Agency, Copenhagen.
EEA, 2017. *Food in a Green Light: A Systems Approach to Sustainable Food*. European Environment Agency, Copenhagen. https://doi.org/10.2800/884986

EEA, 2016. *Sustainability Transitions: Now for the Long Term.* European Environment Agency, Copenhagen.

EHPA, 2018. Heat pumps: Integrating technologies to decarbonise heating and cooling. European Heat Pump Association.

Ellis, G., Cowell, R., Warren, C., Strachan, P., and Szarka, J., 2009. Expanding wind power: A problem of planning, or of perception? *Planning Theory & Practice* 10, 521–547.

Elzen, B., Geels, F. W., and Green, K., 2004. *System Innovation and the Transition to Sustainability: Theory, Evidence and Policy.* Edward Elgar, Cheltenham.

ENA, 2020a. Gas Goes Green: Delivering the Pathway to Net Zero. Energy Networks Association.

ENA, 2020b. Gas Goes Green: Britain's Hydrogen Network Plan. Energy Networks Association.

ENA, 2018. Secure, affordable, low carbon: Gas in our future energy system. Energy Networks Association.

Energiesprong UK, 2018. Performance Specification for Demonstrators v3.03.

Energy Transitions Commission, 2017. Better Energy, Greater Prosperity: Achievable Pathways to Low-Carbon Energy Systems.

Energy Transitions Commission, 2016. Shaping Energy Transitions: Position Paper of the Energy Transitions Commission.

ENSG, 2012. *Our Electricity Transmission Network: A Vision for 2020.* Electricity Networks Strategy Group, London.

ENSG, 2010. *A Smart Grid Routemap.* Electricity Networks Strategy Group, London.

ENSG, 2009. *A Smart Grid Vision.* Electricity Networks Strategy Group, London.

Environmental Audit Committee, 2021. Energy Efficiency of Existing Homes, Fourth Report of Session 2019–2021. House of Commons Environmental Audit Committee.

EST, 2011. *The Elephant in the Living Room: How Our Appliances and Gadgets Are Trampling the Green Dream.* Energy Saving Trust, London.

ETI, 2016. Housing retrofits: A new start. Energy Technologies Institute.

Evans, D. M. and Mylan, J., 2019. Market coordination and the making of conventions: Qualities, consumption and sustainability in the agro-food industry. *Economy and Society* 48, 426–449.

Eyre, N. and Baruah, P., 2015. Uncertainties in future energy demand in UK residential heating. *Energy Policy*, 1–13. https://doi.org/10.1016/j.enpol.2014.12.030

Fagella, D., 2020. Self-driving car timeline from 11 top automakers [WWW Document]. VentureBeat website. URL https://emerj.com/ai-adoption-timelines/self-driving-car-timeline-themselves-top-11-automakers/ (accessed 14 March 2020).

Fagnant, D. J. and Kockelman, K., 2015. Preparing a nation for autonomous vehicles: Opportunities, barriers and policy recommendations. *Transportation Research Part A: Policy and Practice* 77, 167–181.

Fairclough, N., 1995. *Critical Discourse Analysis: The Critical Study of Language.* Longman, London.

Faisal, A., Yigitcanlar, T., Kamruzzaman, M., and Currie, G., 2019. Understanding autonomous vehicles: A systematic literature review on capability, impact, planning and policy. *Journal of Transport and Land Use* 12, 45–72.

Falleti, T. G., 2016. Process tracing of extensive and intensive processes. *New Political Economy* 21, 455–462. https://doi.org/10.1080/13563467.2015.1135550

Felstead, A., 2012. Rapid change or slow evolution? Changing places of work and their consequences in the UK. *Journal of Transport Geography* 21, 31–38.

Feola, G., 2015. Societal transformation in response to global environmental change: A review of emerging concepts. *Ambio* 44, 376–390. https://doi.org/10.1007/s13280–014-0582-z

Feola, G. and Nunes, R., 2014. Success and failure of grassroots innovations for addressing climate change: The case of the transition movement. *Global Environmental Change* 24, 232–250. https://doi.org/10.1016/j.gloenvcha.2013.11.011

Ferguson, J., 2010. The uses of neoliberalism. *Antipode* 41, 166–184.

Finnish Transport Agency, 2015. Maas Services and business opportunities, Research report of the Finnish Transport Agency 56/2015. Helsinki.

Fiorentini, M., 2013. *Solar Thermal UK 2013 A Multi-Client Study*. BSRIA Limited Consultancy, Preston.

Firestone, J., Bates, A. W., and Prefer, A., 2018. Power transmission: Where the offshore wind energy comes home. *Environmental Innovation and Societal Transitions* 29, 90–99.

Fiss, P. C., Marx, A., and Cambre, B. (Eds.), 2013. *Configurational Theory and Methods in Organizational Research*. Emerald Group Publishing Limited, Bingley.

Fligstein, N. and McAdam, D., 2012. *A Theory of Fields*. Oxford University Press, Oxford.

Flynn, R., Bellaby, P., and Ricci, M., 2009. The 'value-action gap' in public attitudes towards sustainable energy: The case of hydrogen energy. *The Sociological Review* 57, 159–180.

Fontaras, G., Zacharof, N.-G., and Ciuffo, B., 2017. Fuel consumption and CO2 emissions from passenger cars in Europe Laboratory versus real-world emissions. *Progress in Energy and Combustion Science* 60, 97–131.

Foxon, T. J. and Pearson, P. J. G., 2007. Towards improved policy processes for promoting innovation in renewable electricity technologies in the UK. *Energy Policy* 35, 1539–1550.

Franceschini, S. and Alkemade, F., 2016. Non-disruptive regime changes: The case of competing energy efficient lighting trajectories. *Environmental Innovation and Societal Transitions* 21, 56–68.

Franceschini, S., Borup, M., and Rosales-Carreón, J., 2018. Future indoor light and associated energy consumption based on professionals' visions: A practice- and network-oriented analysis. *Technological Forecasting and Social Change* 129, 1–11.

Fridahl, M. and Lehtveer, M., 2018. Bioenergy with carbon capture and storage (BECCS): Global potential, investment preferences, and deployment barriers. *Energy Research & Social Science* 42, 155–165.

Fu, M., Kelly, J. A., Clinch, J. P., and King, F., 2012. Environmental policy implications of working from home: Modelling the impacts of land-use, infrastructure and socio-demographics. *Energy Policy* 47, 416–423.

Fuchs, D. and Lorek, S., 2005. Sustainable consumption governance: A history of promises and failures. *Journal of Consumer Policy* 28, 261–288.

Fuenfschilling, L. and Truffer, B., 2014. The structuration of socio-technical regimes: Conceptual foundations from institutional theory. *Research Policy* 43, 772–791.

Gardiner, D., Schmidt, O., Heptonstall, P., Gross, R., and Staffel, I., 2020. Quantifying the impact of policy on the investment case for residential electricity storage in the UK. *Journal of Energy Storage* 27, 101140.

Garikapati, V. M., Pendyala, R. M., Morris, E. A., and Mokhtarian, P. L., 2016. Activity patterns, time use, and travel of millennials: A generation in transition? *Transport Reviews* 36, 558–584.

Gas Safe Register, 2017. The decade review: The UK gas industry considers its past, present and future.

Gavetti, G., 2005. Cognition and hierarchy: Rethinking the microfoundations of capabilities' development. *Organization Science* 16, 599–617.

Gavetti, G. and Rivkin, J. W., 2007. On the origin of strategy: Action and cognition over time. *Organization Science* 18, 420–439.

Geels, F. W., 2021. From leadership to followership: A suggestion for interdisciplinary theorising of mainstream actor reorientation in sustainability transitions. *Environmental Innovation and Societal Transitions* 41, 45–48.

Geels, F. W., 2020a. Transformative innovation and socio-technical transitions to address grand challenges, in: *Science, Research and Innovation Performance of the EU 2020: A Fair, Green and Digital Europe. DG Research and Innovation*. European Commission, Brussels, pp. 572–607.

Geels, F. W., 2020b. Micro-foundations of the Multi-Level Perspective on socio-technical transitions: Developing a multi-dimensional model of agency through crossovers between social constructivism, evolutionary economics and neo-institutional theory. *Technological Forecasting and Social Change*, 152.

Geels, F. W., 2019. Socio-technical transitions to sustainability: A review of criticisms and elaborations of the Multi-Level Perspective. *Current Opinion in Environmental Sustainability* 39, 187–201. https://doi.org/10.1016/j.cosust.2019.06.009

Geels, F. W., 2018a. Disruption and low-carbon system transformation: Progress and new challenges in socio-technical transitions research and the Multi-Level Perspective. *Energy Research and Social Science* 37, 224–231. https://doi.org/10.1016/j.erss.2017.10.010

Geels, F. W., 2018b. Low-carbon transition via system reconfiguration? A socio-technical whole system analysis of passenger mobility in Great Britain (1990–2016). *Energy Research and Social Science* 46, 86–102. https://doi.org/10.1016/j.erss.2018.07.008

Geels, F. W., 2014. Regime resistance against low-carbon transitions: Introducing politics and power into the Multi-Level Perspective. *Theory, Culture & Society* 31, 21–40. https://doi.org/10.1177/0263276414531627

Geels, F. W., 2012. A socio-technical analysis of low-carbon transitions: Introducing the multi-level perspective into transport studies. *Journal of Transport Geography* 24, 471–482. https://doi.org/10.1016/j.jtrangeo.2012.01.021

Geels, F. W., 2007a. Analysing the breakthrough of rock ‘n’ roll (1930–1970): Multi-regime interaction and reconfiguration in the multi-level perspective. *Technological Forecasting and Social Change* 74, 1411–1431.

Geels, F. W., 2007b. Transformations of large technical systems: A multi-level analysis of the Dutch highway system (1950–2000). *Science Technology & Human Values* 32, 123–149.

Geels, F. W., 2005. *Technological Transitions and System Innovations: A Co-evolutionary and Socio-Technical Analysis*. Edward Elgar, Cheltenham.

Geels, F. W., 2004. From sectoral systems of innovation to socio-technical systems: Insights about dynamics and change from sociology and institutional theory. *Research Policy* 33, 897–920. https://doi.org/10.1016/j.respol.2004.01.015

Geels, F. W., 2002. Technological transitions as evolutionary reconfiguration processes: A multi-level perspective and a case-study. *Research Policy* 31, 1257–1274. https://doi.org/10.1016/S0048–7333(02)00062-8

Geels, F. W., Berkhout, F., and van Vuuren, D. P., 2016a. Bridging analytical approaches for low-carbon transitions. *Nature Climate Change* 6, 576–583. https://doi.org/10.1038/nclimate2980

Geels, F. W., and Deuten, J. J., 2006. Local and global dynamics in technological development: A socio-cognitive perspective on knowledge flows and lessons from reinforced concrete. *Science and Public Policy* 33, 265–275.

Geels, F. W., Kern, F., Fuchs, G., Hinderer, N., Kungl, G., Mylan, J., et al., 2016b. The enactment of socio-technical transition pathways: A reformulated typology and a comparative multi-level analysis of the German and UK low-carbon electricity transitions (1990–2014). *Research Policy* 45, 896–913. https://doi.org/10.1016/j.respol.2016.01.015

Geels, F. W., McMeekin, A., Mylan, J., and Southerton, D., 2015. A critical appraisal of Sustainable Consumption and Production research: The reformist, revolutionary and reconfiguration positions. *Global Environmental Change* 34, 1–12. https://doi.org/10.1016/j.gloenvcha.2015.04.013

Geels, F. W. and Penna, C. C. R., 2015. Societal problems and industry reorientation: Elaborating the Dialectic Issue LifeCycle (DILC) model and a case study of car safety in the USA (1900–1995). *Research Policy* 44, 67–82. https://doi.org/10.1016/j.respol.2014.09.006

Geels, F. W., Pieters, T., and Snelders, S., 2007. Cultural enthusiasm, resistance and the societal embedding of new technologies: Psychotropic drugs in the 20th century. *Technology Analysis & Strategic Management* 19, 145–165.

Geels, F. W., Sareen, S., Hook, A., and Sovacool, B. K., 2021. Navigating implementation dilemmas in technology-forcing policies: Insights from a comparative analysis of accelerated smart meter diffusion in the Netherlands, UK, Norway, and Portugal (2000–2019). *Research Policy* 50, 104–272.

Geels, F. W. and Schot, J., 2007. Typology of sociotechnical transition pathways. *Research Policy* 36, 399–417.

Geels, F. W. and Schot, J. W., 2010. Reflections: Process theory, causality and narrative explanation, in: Grin, J., Rotmans, J., Schot, J., Geels, F. W., and Loorbach, D. (Eds.), *Transitions to Sustainable Development: New Directions in the Study of Long Term Transformative Change*. Routledge, pp. 93–102.

Geels, F. W. and Smit, W. A., 2000. Failed technology futures: Pitfalls and lessons from a historical survey. *Futures* 32, 867–885.

Geels, F. W., Sovacool, B. K., Schwanen, T., and Sorrell, S., 2017. Sociotechnical transitions for deep decarbonization. *Science* 357, 1242–1244.

Geels, F. W. and Verhees, B., 2011. Cultural legitimacy and framing struggles in innovation journeys: A cultural-performative perspective and a case study of Dutch nuclear energy (1945–1986). *Technological Forecasting and Social Change* 78, 910–930. https://doi.org/10.1016/j.techfore.2010.12.004

George, A. L. and Bennett, A., 2004. *Case Studies and Theory Development in the Social Sciences*. MIT Press, Cambridge, MA.

GGCS, 2020. GGCS 2019 Annual Report. Green Gas Certification Scheme.

Gibbs, D. and O'Neill, K., 2015. Building a green economy? Sustainability transitions in the UK building sector. *Geoforum* 59, 133–141. https://doi.org/10.1016/j.geoforum.2014.12.004

Gillard, R., 2016. Unravelling the United Kingdom's climate policy consensus: The power of ideas, discourse and institutions. *Global Environmental Change* 40, 26–36.

GLA, 2014. *District Heating Manual for London*. Greater London Authority, London.

Global Alliance for the Future of Food, 2019. Beacons of Hope: Accelerating Transformations to Sustainable Food Systems.

Godoe, H., 2000. Innovation regimes, R&D and radical innovations in telecommunications. *Research Policy* 29, 1033–1046.

Goodwin, P. and van Dender, K., 2013. Peak car: Themes and issues. *Transport Reviews* 33, 243–254.

Göpel, M., 2016. *The Great Mindshift: How a New Economic Paradigm and Sustainability Transformations go Hand in Hand*. Springer, Berlin.

Gough, I., 2010. Economic crisis, climate change and the future of the welfare state. *Twenty-First Century Society* 5, 51–64.

Goulden, M., Ryley, T., and Dingwall, R., 2014. Beyond 'predict and provide': UK transport, the growth paradigm and climate change. *Transport Policy* 32, 139–147.

Government Office for Science, 2018. Mobility as a Service (MaaS) in the UK: Change and its implications, Report commissioned as part of the UK government's Foresight Future of Mobility project. London.

Green Alliance, 2007. A manifesto for sustainable heat.

Greening, B. and Azapagic, A., 2014. Domestic solar thermal water heating: A sustainable option for the UK? *Renewable Energy* 63, 23–36. https://doi.org/10.1016/j.renene.2013.07.048

Greening, B. and Azapagic, A., 2012. Domestic heat pumps: Life cycle environmental impacts and potential implications for the UK. *Energy* 39, 205–217. https://doi.org/10.1016/j.energy.2012.01.028

Greenwood, R. and Hinings, C. R., 1996. Understanding radical organizational change: Bringing together the old and the new institutionalism. *The Academy of Management Review* 21, 1022–1054.

Greenwood, R. and Hinings, C. R., 1993. Understanding strategic change: The contribution of archetypes. *Academy of Management Journal* 36, 1052–1081.

Gross, R. and Hanna, R., 2019. Path dependency in provision of domestic heating. *Nature Energy* 4, 358–364. https://doi.org/10.1038/s41560–019-0383-5

Gross, R., Heptonstall, P., Greenacre, P., Candelise, C., Jones, F., and Castillo, A. C., 2013. Presenting the Future: An assessment of future costs estimation methodologies in the electricity generation sector, UKERC report. UK Energy Research Centre.

Grubb, M., 2014. *Planetary Economics: Energy, Climate Change and the Three Domains of Sustainable Development*. Routledge, Oxford.

Grubler, A., Wilson, C., and Nemet, G., 2016. Apples, oranges, and consistent comparisons of the temporal dynamics of energy transitions. *Energy Research & Social Science* 22, 18–25.

Grünewald, P. and Torriti, J., 2013. Demand response from the non-domestic sector: Early UK experiences and future opportunities. *Energy Policy* 61, 423–429.

Haddad, H., Lyons, G., and Chatterjee, K., 2009. An examination of determinants influencing the desire for and frequency of part-day and whole-day homeworking. *Journal of Transport Geography* 17, 124–133.

Hajer, M. and Versteeg, W., 2005. A decade of discourse analysis of environmental politics: Achievements, challenges, perspectives. *Journal of Environmental Policy and Planning* 7, 175–184. https://doi.org/10.1080/15239080500339646

Hall, P., 2003. Aligning ontology and methodology in comparative research, in: Mahoney, J. and Rueschemeyer, D. (Eds.), *Comparative Historical Analysis in the Social Sciences*. Cambridge University Press, Cambridge, pp. 373–404.

Hall, P. A., 1993. Policy paradigms, social learning, and the state: The case of economic policymaking in Britain. *Comparative Politics*, 275–296.

Hall, P. A. and Soskice, D. (Eds.), 2001. *Varieties of Capitalism: The Institutional Foundations of Comparative Advantage*. Oxford University Press, New York.

Hand, M. and Shove, E., 2007. Condensing practices: Ways of living with a freezer. *Journal of Consumer Culture* 7, 79–104.

Hanmer, C. and Abram, S., 2017. Actors, networks, and translation hubs: Gas central heating as a rapid socio-technical transition in the United Kingdom. *Energy Research & Social Science* 34, 176–183. https://doi.org/10.1016/j.erss.2017.03.017

Hanna, R., Leach, M., and Torriti, J., 2018. Microgeneration: The installer perspective. *Renewable Energy* 116, 458–469. https://doi.org/10.1016/j.renene.2017.09.023

Hannon, M. J., 2015. Raising the temperature of the UK heat pump market: Learning lessons from Finland. *Energy Policy* 85, 369–375. https://doi.org/10.1016/j.enpol.2015.06.016
Hardt, L., Barrett, J., Brockway, P., Foxon, T. J., Heun, M. K., Owen, A., et al., 2017. Outsourcing or efficiency? Investigating the decline in final energy consumption in the UK productive sectors. *Energy Procedia* 142, 2409–2414.
Hardt, L., Owen, A., Brockway, P., Heun, M. K., Barrett, J., Taylor, P. G., et al., 2018. Untangling the drivers of energy reduction in the UK productive sectors: Efficiency or offshoring? *Applied Energy* 223, 124–133.
Hargreaves, T., Hielscher, S., Seyfang, G., and Smith, A., 2013. Grassroots innovations in community energy: The role of intermediaries in niche development. *Global Environmental Change* 23, 868–880. https://doi.org/10.1016/j.gloenvcha.2013.02.008
Hargreaves, T., Longhurst, N., and Seyfang, G., 2013. Up, down, round and round: Connecting regimes and practices in innovation for sustainability. *Environment and Planning A* 45, 402–420.
Harvey, M. and Pilgrim, S., 2013. Rudderless in a sea of yellow: The European political economy impasse for renewable transport energy. *New Political Economy* 18, 364–390.
Haubrich, D., 2001. UK rail privatisation five years down the line: An evaluation of nine policy objectives. *Policy & Politics* 29, 317–336.
Hawken, P., 2017. Drawdown: The Most Comprehensive Plan Ever Proposed to Reverse Global Warming. Tantor Media Inc.
Hawkey, D. and Webb, J., 2014. District energy development in liberalised markets: Situating UK heat network development in comparison with Dutch and Norwegian case studies. *Technology Analysis & Strategic Management* 26, 1228–1241.
Haywood, R., 2007. Britain's national railway network: Fit for purpose in the 21st century? *Journal of Transport Geography* 15, 198–216.
HBF, 2018. The Economic Footprint of House Building in England and Wales. Home Builders Federation.
Heat Pump Association, 2019. Delivering net zero: A roadmap for the role of heat pumps.
Heffernan, E., Pan, W., Liang, X., and de Wilde, P., 2015. Zero carbon homes: Perceptions from the UK construction industry. *Energy Policy* 79, 23–36. https://doi.org/10.1016/j.enpol.2015.01.005
Henderson, R. M. and Clark, K. B., 1990. Architectural innovation: The reconfiguration of existing product technologies and the failure of established firms. *Administrative Science Quarterly* 35, 9–30.
Heptonstall, P., Gross, R., Greenacre, P., and Cockerill, T., 2012. The cost of offshore wind: Understanding the past and projecting the future. *Energy Policy* 4, 815–821.
Heracleous, L. and Marshak, R. J., 2004. Conceptualizing organizational discourse as situated symbolic action. *Human Relations* 57, 1285–1312.
Héretier, A., 2008. Causal explanation, in: Della Porta, D. M. and Keating, M. (Eds.), *Approaches and Methodologies in the Social Sciences: A Pluralist Perspective*. Cambridge University Press, Cambridge, pp. 61–79.
Hermwille, L., 2016. The role of narratives in socio-technical transitions: Fukushima and the energy regimes of Japan, Germany, and the United Kingdom. *Energy Research and Social Science* 11, 237–246. https://doi.org/10.1016/j.erss.2015.11.001
Hess, D.J., 2014. Sustainability transitions: A political coalition perspective. *Research Policy* 43, 278–283. https://doi.org/10.1016/j.respol.2013.10.008
Hewlett, J. G., 2005. De-regulated electric power markets and operating nuclear power plants: The case of British energy. *Energy Policy* 33, 2293–2297.

Hielscher, S. and Sovacool, B., 2018. Contested smart and low-carbon energy futures: Media discourses of smart meters in the United Kingdom. *Journal of Cleaner Production* 195, 978–990.

Hirsch, P. M. and Lounsbury, M., 1997. Ending the family quarrel: Toward a reconciliation of "old" and "new" institutionalisms. *American Behavioral Scientist* 40, 406–418.

Hirschhorn, F., Paulsson, A., Sorensen, C. H., and Veeneman, W., 2019. Public transport regimes and mobility as a service: Governance approaches in Amsterdam, Birmingham, and Helsinki. *Transportation Research Part A: Policy and Practice* 130, 178–191.

Hiteva, R. and Watson, K., 2019. Governance of interactions between infrastructure sectors: The making of smart grids in the UK. *Environmental Innovation and Societal Transitions* 32, 140–152.

HM Government, 2020a. The Energy White Paper: Powering our Net Zero Future, CP 337.

HM Government, 2020b. The Ten Point Plan for a Green Industrial Revolution.

HM Government, 2018. Growing the Bioeconomy. Improving lives and strengthening our economy: A national bioeconomy strategy to 2030.

Hoffman, A. J., 1999. Institutional evolution and change: Environmentalism and the US chemical industry. *Academy of Management Journal* 42, 351–371.

Holden, E., Banister, D., Gössling, S., Gilpin, G., and Linnerud, K., 2020. Grand Narratives for sustainable mobility: A conceptual review. *Energy Research & Social Science* 65, 10145.

Hommels, A., 2005. Studying obduracy in the city: Toward a productive fusion between technology studies and urban studies. *Science Technology and Human Values* 30, 323–351. https://doi.org/10.1177/0162243904271759

HoP, 2017. Decarbonising the Gas Network. POSTNote N°565. Houses of Parliament Parliamentary Office of Science and Technology.

Hope, A., Roberts, T., and Walker, I., 2018. Consumer engagement in low-carbon home energy in the United Kingdom: Implications for future energy system decentralization. *Energy Research & Social Science* 44, 362–370. https://doi.org/10.1016/j.erss.2018.05.032

Hopkins, D. and Schwanen, T., 2018. Automated mobility transitions: Governing processes in the UK. *Sustainability* 10, 956.

House of Lords Science and Technology Select Committee, 2017. *Connected and Autonomous Vehicles: The Future? 2nd Report of Sessions 2016–17*. HM Government, London.

Howarth, N. A. A. and Rosenow, J., 2014. Decision-making, institutional evolution and the phased ban of incandescent light bulbs. *Energy Policy* 67, 737–746.

Howlett, M. and Rayner, J., 2013. Patching vs packaging in policy formulation: Assessing policy portfolio design. *Politics and Governance* 1, 170–182.

Hughes, T. P., 1994. Technological momentum, in: Smith, M. R. and Marx, L. (Eds.), *Does Technology Drive History?* The Dilemma of Technological Determinism. MIT Press, Cambridge, MA, pp. 101–113.

Hynes, M., 2016. Developing (tele)work? A multi-level sociotechnical perspective of telework in Ireland. *Research in Transportation Economics* 57, 21–31.

IEA, IRENA, 2017. Perspectives for the Energy Transition: Investment Needs for a Low-Carbon Energy System. International Energy Agency and International Renewable Energy Agency.

IET, 2020. Scaling up retrofit 2050. The Institution of Engineering and Technology.

Institute for Government, 2021. Decarbonising heating at home: Learning from past successes and failures to improve energy policy making.

IoD, 2015. *Not Too Clever: Will Smart Meters Be the Next Government IT Disaster?* IoD policy document, Institute of Directors, London.

IPCC, 2021. Climate Change 2021 The Physical Science Basis: Working Group 1 Contribution to the Sixth Assessment Report of the Intergovernmental Panel on Climate Change. Intergovernmental Panel on Climate Change.

IPCC, 2018. Global warming of 1.5°C. An IPCC Special Report on the impacts of global warming of 1.5°C above pre-industrial levels and related global greenhouse gas emission pathways, in the context of strengthening the global response to the threat of climate change. Intergovernmental Panel on Climate Change.

IPPR, 2013. *Sustainable Consumption in the UK: A Selection of Case Studies*. Institute for Public Policy Research, London.

IRENA, 2021. *Renewable Power Generation Costs in 2020*. International Renewable Energy Agency, Abu Dhabi.

Isoaho, K. and Markard, J., 2020. The politics of technology decline: Discursive struggles over coal phase-out in the UK. *Review of Policy Research* 37, 342–368. https://doi.org/10.1111/ropr.12370

Jackson, T. and Victor, P., 2011. Productivity and work in the 'green economy': Some theoretical reflections and empirical tests. *Environmental Innovation and Societal Transitions* 1, 101–108.

Jamasb, T. and Pollitt, M., 2008. Liberalisation and R&D in network industries: The case of the electricity industry. *Research Policy* 37, 995–1008.

Jamasb, T. and Pollitt, M., 2007. Incentive regulation of electricity distribution networks: Lessons of experience from Britain. *Energy Policy* 35, 6163–6187.

Jamasb, T. and Pollitt, M. G., 2015. Why and how to subsidise energy R+D: Lessons from the collapse and recovery of electricity innovation in the UK. *Energy Policy* 83, 197–205.

Jenkins, N., Long, C., and Wu, J., 2015. An overview of the smart grid in Great Britain. *Engineering* 1, 413–421.

Jeswani, H. K., Whiting, A., and Azapagic, A., 2019. Environmental and Economic Sustainability of Biomass Heat in the UK. *Energy Technology* 1901044. https://doi.org/10.1002/ente.201901044

Johnstone, P. and Stirling, A., 2020. Comparing nuclear trajectories in Germany and the United Kingdom: From regimes to democracies in sociotechnical transitions and discontinuities. *Energy Research and Social Science* 59, 101–245. https://doi.org/10.1016/j.erss.2019.101245

Jolly, J., 2021a. UK slashes grants for electric car buyers while retaining petrol vehicle support. The Guardian. https://www.theguardian.com/environment/2021/mar/18/uk-slashes-grants-for-electric-car-buyers-while-increasing-petrol-vehicle-support

Jolly, J., 2021b. JLR to make Jaguar brand electric-only by 2025. *The Guardian*. www.theguardian.com/business/2021/feb/15/jlr-to-make-jaguar-brand-electric-only-by-2025

Jolly, J., 2020a. Polluting vehicles could be pulled from UK sale, say carmakers. *The Guardian*. www.theguardian.com/business/2020/jan/13/polluting-vehicles-could-be-pulled-from-uk-sale-say-carmakers#:~:text=Under%20new%20EU%20rules%2C%20average,designed%20to%20meet%20the%20goal

Jolly, J., 2020b. Electric cars "as cheap to manufacture" as regular models by 2024. *The Guardian*. www.theguardian.com/environment/2020/oct/21/electric-cars-as-cheap-to-manufacture-as-regular-models-by-2024

Jolly, J., 2020c. Toyota will not invest in electric cars in UK until after 2027. *The Guardian*. www.theguardian.com/business/2020/dec/07/toyota-will-not-invest-in-electric-cars-in-uk-until-at-least-2034

Jolly, S., Spodniak, P., and Raven, R. P. J. M., 2016. Institutional entrepreneurship in transforming energy systems towards sustainability: Wind energy in Finland and India. *Energy Research and Social Science* 17, 102–118. https://doi.org/10.1016/j.erss.2016.04.002

Jordan, A. and Lenschow, A., 2010. Environmental policy integration: A state of the art review. *Environmental Policy and Governance* 20, 147–158.

Jørgensen, U., 2012. Mapping and navigating transitions: The multi-level perspective compared with arenas of development. *Research Policy* 41, 996–1010. https://doi.org/10.1016/j.respol.2012.03.001

JRC, 2020. *Projecting Opportunities for INdustrial Transitions (POINT): Concepts, Rationales and Methodological Guidelines for Territorial Reviews of Industrial Transition*. Joint Research Centre, European Commission, Ispra, Italy.

JRC, 2017. *Smart Grid Projects Outlook 2017: Facts, Figures and Trends in Europe, EUR 28614 EN*. Joint Research Centre, European Commission, Brussels.

Judson, E., Bell, S., Bulkeley, H., Powells, G., and Lyon, S., 2015. The co-construction of energy provision and everyday practice: Integrating heat pumps in social housing in England. *Science and Technology Studies* 28, 26–53.

Jupe, R., 2013. New development: Going off the rails? Rail franchising after the cancellation of the West Coast franchise competition. *Public Money & Management* 33, 337–341.

Kallis, G., 2011. In defense of degrowth. *Ecological Economics* 70, 873–880.

Kamargianni, M., Li, W., Matyas, M., and Schafer, A., 2016. A critical review of new mobility services for urban transport. *Transportation Research Procedia* 14, 3294–3303.

Kanger, L., Geels, F. W., Sovacool, B., and Schot, J., 2019. Technological diffusion as a process of societal embedding: Lessons from historical automobile transitions for future electric mobility. *Transportation Research Part D: Transport and Environment* 71, 47–66. https://doi.org/10.1016/j.trd.2018.11.012

Kannan, R. and Strachan, N., 2009. Modelling the UK residential energy sector under longterm decarbonisation scenarios: Comparison between energy systems and sectoral modelling approaches. *Applied Energy* 86, 416–428.

Karlsson, I. C. M., Mukhtar-Landgren, D., Smith, G., Koglin, T., Kronsell, A., Lund, E., et al., 2020. Development and implementation of Mobility-as-a-Service: A qualitative study of barriers and enabling factors. *Transportation Research Part A: Policy and Practice* 131, 283–295.

Karvonen, A. and Guy, S., 2018. Urban energy landscapes and the rise of heat networks in the United Kingdom. *Journal of Urban Technology* 25, 19–38.

Kay, D., Hill, N., and Newman, D., 2013. *Powering Ahead: The Future of Low-Carbon Cars and Fuels*. RAC Foundation.

Keay, M., 2016. UK energy policy: Stuck in ideological limbo? *Energy Policy* 94, 247–252. https://doi.org/10.1016/j.enpol.2016.04.022

Kemp, R. and Van Lente, H., 2011. The dual challenge of sustainability transitions. *Environmental Innovation and Societal Transitions* 1, 121–124.

Kemp, R., Schot, J., and Hoogma, R., 1998. Regime shifts to sustainability through processes of niche formation: The approach of strategic niche management. *Technology Analysis & Strategic Management* 10, 175–198.

Kern, F., 2011. Ideas, institutions, and interests: Explaining policy divergence in fostering "system innovations" towards sustainability. *Environment and Planning C: Government and Policy* 29, 1116–1134. https://doi.org/10.1068/c1142

Kern, F., Gaede, J., Meadowcroft, J., and Watson, J., 2016. The political economy of carbon capture and storage: An analysis of two demonstration projects. *Technological Forecasting and Social Change* 102, 250–260.

Kern, F. and Howlett, M., 2009. Implementing transition management as policy reforms: A case study of the Dutch energy sector. *Policy Sciences* 42, 391–408.

Kern, F., Kivimaa, P., and Martiskainen, M., 2017. Policy packaging or policy patching? The development of complex energy efficiency policy mixes. *Energy Research & Social Science* 23, 11–25. https://doi.org/10.1016/j.erss.2016.11.002

Kern, F., Kuzemko, C., and Mitchell, C., 2014a. Measuring and explaining policy paradigm change: The case of UK energy policy. *Policy & Politics* 42, 513–530.

Kern, F., Rogge, K. S., and Howlett, M., 2019. Policy mixes for sustainability transitions: New approaches and insights through bridging innovation and policy studies. *Research Policy* 48, 103832. https://doi.org/10.1016/j.respol.2019.103832

Kern, F., Smith, A., Shaw, C., Raven, R., and Verhees, B., 2014b. From laggard to leader: Explaining offshore wind developments in the UK. *Energy Policy* 69, 635–646. https://doi.org/10.1016/j.enpol.2014.02.031

Kern, F., Verhees, B., Raven, R., and Smith, A., 2015. Empowering sustainable niches: Comparing UK and Dutch offshore wind developments. *Technological Forecasting and Social Change* 100, 344–355.

Kerr, N. and Winskel, M., 2020. Household investment in home energy retrofit: A review of the evidence on effective public policy design for privately owned homes. *Renewable and Sustainable Energy Reviews* 123, 109778. https://doi.org/10.1016/j.rser.2020.109778

Ketchen, D. J., Thomas, J. B., and Snow, C. C., 1993. Organizational configurations and performance: A comparison of theoretical approaches. *Academy of Management Journal* 36, 1278–1313.

Khan, N. and Abas, N., 2011. Comparative study of energy saving light sources. *Renewable and Sustainable Energy Reviews* 15, 296–309.

King, A., 2015. *Who Governs Britain?* Penguin Random House, London.

Kivimaa, P. and Martiskainen, M., 2018. Dynamics of policy change and intermediation: The arduous transition towards low-energy homes in the United Kingdom. *Energy Research and Social Science* 44, 83–99. https://doi.org/10.1016/j.erss.2018.04.032

Knowles, R. and Abrantes, P., 2008. Buses and light rail: Stalled en route?, in: Docherty, I. and Shaw, J. (Eds.), *Traffic Jam: Ten Years of Sustainable Transport in the UK*. Policy Press, Bristol, pp. 97–116.

Köhler, J., Geels, F. W., Kern, F., Markard, J., Wieczorek, A., Alkemade, F., et al., 2019. An agenda for sustainability transitions research: State of the art and future directions. *Environmental Innovation and Societal Transitions* 31, 1–32. https://doi.org/10.1016/j.eist.2019.01.004

Kolk, A. and Pinkse, J., 2007. Multinationals' political activities on climate change. *Business & Society* 46, 201–228.

Konrad, K., Truffer, B., and Voss, J., 2008. Multi-regime dynamics in the analysis of sectoral transformation potentials: Evidence from German utility sectors. *Journal of Cleaner Production* 16, 1190–1202.

Kopp, S. D., 2015. The UK Gas Market, in: *Politics, Markets and EU Gas Supply Security. Case Studies of the UK and Germany*. Springer VS, Wiesbaden, pp. 91–180.

KPMG, 2016. 2050 Energy Scenarios: The UK Gas Networks role in a 2050 whole energy system, for Energy Networks Association.

KPMG, 2012. *Offshore Transmission: An Investor Perspective*. KPMG Consultancy, London.

Kuzemko, C., 2016. Energy depoliticisation in the UK: Depoliticisation destroying political energy in the UK. *The British Journal of Politics and International Relations* 18, 107–124. https://doi.org/10.1111/1467-856X.12068

Kuzemko, C., 2014. Politicising UK energy: What "speaking energy security" can do. *Policy & Politics* 42, 259–274. https://doi.org/10.1332/030557312X655990

Laganakou, G., 2019. What feasible macro-scale interventions could stimulate a sustainable growth in the UK housing retrofit industry? An examination of the potential impact from supply-chain innovations on low energy retrofit of pre-1919 housing. University of Northampton.

Laker, L., 2021. Big rise in UK weekend cycling amid calls for more investment. *The Guardian.* www.theguardian.com/lifeandstyle/2021/jul/22/big-rise-in-uk-weekend-cycling-amid-calls-for-more-investment

Laker, L., 2020. Are the Tories really serious about supporting cycling? *The Guardian.* www.theguardian.com/environment/bike-blog/2020/may/11/this-could-be-the-time-to-usher-in-a-golden-age-for-cycling-in-britain#comment-140570965

Langendahl, P.-A., Roy, H., Potter, S., and Cook, M., 2019. Smoothing peaks and troughs: Intermediary practices to promote demand side response in smart grids. *Energy Research & Social Science* 58, 101277.

Langley, A., 1999. Strategies for theorizing from process data. *Academy of Management Review* 24, 691–710.

Langley, A., Smallman, C., Tsoukas, H., and van de Ven, A. H., 2013. Process studies of change in organization and management: Unveiling temporality, activity, and flow. *Academy of Management Journal* 56, 1–13.

Langridge, R. and Sealey, R., 2000. The demand for public transport: The effects of fares, quality of service, income and car ownership. *Transport Policy* 7, 105–115.

Lauber, V. and Jacobsson, S., 2016. The politics and economics of constructing, contesting and restricting socio-political space for renewables: The German Renewable Energy Act. *Environmental Innovations and Societal Transitions* 18, 147–163.

Laville, S., 2021. Less than 5% of green homes grant budget paid out, Labour reveals. *The Guardian.* www.theguardian.com/environment/2021/feb/06/government-green-homes-grant-budget-labour

Lawrence, T. B. and Phillips, N., 2004. From Moby Dick to Free Willy: Macro-cultural discourse and institutional entrepreneurship in emerging institutional fields. *Organization* 11, 689–711.

Le Quéré, C., Korsbakken, J. I., Wilson, C., Tosun, J., Andrew, R., Andres, R. J., et al., 2019. Drivers of declining CO2 emissions in 18 developed economies. *Nature Climate Change* 9, 213–217. https://doi.org/10.1038/s41558–019-0419-7

Leadbeater, C. and Winhall, J., 2020. *Building Better Systems: A Green Paper on System Innovation*. Rockwool Foundation, Copenhagen.

Lees, T. and Sexton, M., 2014. An evolutionary innovation perspective on the selection of low and zero-carbon technologies in new housing. *Building Research and Information* 42, 276–287.

Leveque, F. and Robertson, A., 2014. Future Heat Series. Part 1. Pathways for Heat: Low Carbon Heat for Buildings, A report by Carbon Connect.

Levinthal, D. A., 1998. The slow pace of rapid technological change: Gradualism and punctuation in technological change. *Industrial and Corporate Change* 7, 217–247.

Li, F. G. N. and Pye, S., 2018. Uncertainty, politics, and technology: Expert perceptions on energy transitions in the United Kingdom. *Energy Research and Social Science* 37, 122–132. https://doi.org/10.1016/j.erss.2017.10.003

Lie, M. and Sørensen, K. H. (Eds.), 1996. *Making Technology Our Own: Domesticating Technology into Everyday Life*. Scandinavian University Press, Oslo.

Lijphart, A., 2012. *Patterns of Democracy: Government Forms and Performance in Thirty-Six Countries*. Yale University Press, New Haven, CT.

Lindblom, C. E., 1979. Still muddling, not yet through. *Public Administration Review* 39, 517–526.

Little, D., 2016. *New Directions in the Philosophy of Social Science*. Rowman & Littlefield International, Lanham, MD.

Little, D., 2010. *New Contributions to the Philosophy of History*. Springer, New York.

Lockwood, M., 2017. The Development of the Capacity Market for Electricity in Great Britain. EPG Working Paper 1702.

Lockwood, M., 2016. Creating protective space for innovation in electricity distribution networks in Great Britain: The politics of institutional change. *Environmental Innovation and Societal Transitions* 18, 111–127. https://doi.org/10.1016/j.eist.2015.05.007

Lockwood, M., 2013. The political sustainability of climate policy: The case of the UK Climate Change Act. *Global Environmental Change* 23, 1339–1348.

Lockwood, M., Kuzemko, C., Mitchell, C., and Hoggett, R., 2017. Historical institutionalism and the politics of sustainable energy transitions: A research agenda. *Environment and Planning C* 35, 312–333.

Lockwood, M., Mitchell, C., and Hoggett, R., 2020. Incumbent lobbying as a barrier to forward-looking regulation: The case of demand-side response in the GB capacity market for electricity. *Energy Policy* 140, 111426.

Lovell, H., 2008. Discourse and innovation journeys: The case of low energy housing in the UK. *Technology Analysis & Strategic Management* 20, 613–632.

Lovio, R. and Kivimaa, P., 2012. Comparing alternative path creation frameworks in the context of emerging biofuel fields in the Netherlands, Sweden and Finland. *European Planning Studies* 20, 773–790. https://doi.org/10.1080/09654313.2012.667925

Lowes, R. and Woodman, B., 2020. Disruptive and uncertain: Policy makers' perceptions on UK heat decarbonisation. *Energy Policy* 142, 111494.

Lowes, R., Woodman, B., and Clark, M., 2018. A transformation to sustainable heating in the UK : Risks and opportunities for UK heat sector businesses. Exeter.

Lowes, R., Woodman, B., and Fitch-Roy, O., 2019. Policy change, power and the development of Great Britain's Renewable Heat Incentive. *Energy Policy* 131, 410–421. https://doi.org/10.1016/j.enpol.2019.04.041

Lowes, R., Woodman, B., and Speirs, J., 2020. Heating in Great Britain: An incumbent discourse coalition resists an electrifying future. *Environmental Innovation and Societal Transitions* 37, 1–17. https://doi.org/10.1016/j.eist.2020.07.007

Lynch, H., 2014. Passivhaus in the UK: The challenges of an emerging market. A case study of innovation using mixed methods research. University College London.

Mahoney, J., 2000. Path dependence in historical sociology. *Theory and Society* 29, 507–548. https://doi.org/10.1023/A:1007113830879

Mallaburn, P. S. and Eyre, N., 2014. Lessons from energy efficiency policy and programmes in the UK from 1973 to 2013. *Energy Efficiency* 7, 23–41.

Maréchal, K., 2010. Not irrational but habitual: The importance of "behavioural lock-in" in energy consumption. *Ecological Economics* 69, 1104–1114.

Markard, J., 2018. The next phase of the energy transition and its implications for research and policy. *Nature Energy*. https://doi.org/10.1038/s41560–018-0171-7

Markard, J. and Hoffmann, V. H., 2016. Analysis of complementarities: Framework and examples from the energy transition. *Technological Forecasting & Social Change* 111, 63–75. https://doi.org/10.1016/j.techfore.2016.06.008

Markard, J., Suter, M., and Ingold, K., 2016. Socio-technical transitions and policy change: Advocacy coalitions in Swiss energy policy. *Environmental Innovation and Societal Transitions* 18, 215–237. https://doi.org/10.1016/j.eist.2015.05.003

Markard, J., Wirth, S., and Truffer, B., 2016. Institutional dynamics and technology legitimacy: A framework and a case study on biogas technology. *Research Policy* 45, 330–344.

Marletto, G., 2014. Car and the city: Socio-technical transition pathways to 2030. *Technological Forecasting and Social Change* 87, 164–178. https://doi.org/10.1016/j.techfore.2013.12.013

Marsden, G., Ferreira, A., Bache, I., Flinders, M., and Bartle, I., 2014. Muddling through with climate change targets: A multi-level governance perspective on the transport sector. *Climate Policy* 14, 617–636.

Martínez Arranz, A., 2017. Lessons from the past for sustainability transitions? A meta-analysis of socio-technical studies. *Global Environmental Change* 44, 125–143. https://doi.org/10.1016/j.gloenvcha.2017.03.007

Martinot, E. and Borg, N., 1998. Energy-efficient lighting programs: Experience and lessons from eight countries. *Energy Policy* 26, 1071–1081.

Martiskainen, M. and Kivimaa, P., 2019. Role of knowledge and policies as drivers for low-energy housing: Case studies from the United Kingdom. *Journal of Cleaner Production* 215, 1402–1414. https://doi.org/10.1016/j.jclepro.2019.01.104

Martiskainen, M. and Kivimaa, P., 2018. Creating innovative zero carbon homes in the United Kingdom: Intermediaries and champions in building projects. *Environmental Innovation and Societal Transitions* 26, 15–31. https://doi.org/10.1016/j.eist.2017.08.002

Mayer, A., 2018. A just transition for coal miners? Accountability frames, community economic identity, and just transition policy support among local policy actors. *Environmental Innovation and Societal Transition* 28, 1–13.

Mayntz, R., 2004. Mechanisms in the analysis of social macro-phenomena. *Philosophy of the Social Sciences* 34, 237–259.

Mazur, C., Contestabile, M., Offer, G. J., and Brandon, N. P., 2015. Assessing and comparing German and UK transition policies for electric mobility. *Environmental Innovation and Societal Transitions* 14, 84–100. https://doi.org/10.1016/j.eist.2014.04.005

Mazur, C., Offer, G. J., Contestabile, M., and Brandon, N. B., 2018. Comparing the effects of vehicle automation, policy-making and changed user preferences on the uptake of electric cars and emissions from transport. *Sustainability (Switzerland)* 10, 4–6. https://doi.org/10.3390/su10030676

McAdam, D. and Scott, W. R., 2005. Organizations and movements, in: Davis, G. F., McAdam, D., Scott, W. R., and Zald, M. N. (Eds.), *Social Movements and Organization Theory*. Cambridge University Press, Cambridge, pp. 4–40.

McCann, P., 2016. *The UK Regional (and National) Economic Problem: Geography, Globalisation and Governance*. Routledge, London.

McDonald, R. C., 2015. Are millennials really the "Go-Nowhere" generation? *Journal of the American Planning Association* 81, 90–103.

McGee, M. C., 1980. The 'ideograph': A link between rhetoric and ideology. *The Quarterly Journal of Speech* 66, 1–16.

McGlade, C., Pye, S., Ekins, P., Bradshaw, M., and Watson, J., 2018. The future role of natural gas in the UK: A bridge to nowhere? *Energy Policy* 113, 454–465. https://doi.org/10.1016/j.enpol.2017.11.022

McKinsey Global Institute, 2013. Disruptive technologies: Advances that will transform life, *Business and the Global Economy*. McKinsey Global Institute, New York.

McMeekin, A., Geels, F. W., and Hodson, M., 2019. Mapping the winds of whole system reconfiguration: Analysing low-carbon transformations across production, distribution and consumption in the UK electricity system (1990–2016). *Research Policy*. https://doi.org/10.1016/j.respol.2018.12.007

McTigue, C., Monios, J., and Rye, T., 2018. Identifying barriers to implementation of local transport policy: An analysis of bus policy in Great Britain. *Utilities Policy* 50, 133–143.

McVeigh, K., 2017. Solar thermal in the UK: Opportunities and developments. *Energy World*, 30–31.

Meadowcroft, J., 2009. What about the politics? Sustainable development, transition management, and long term energy transitions. *Policy Sciences* 42, 323–340. https://doi.org/10.1007/s11077–009-9097-z

Meadowcroft, K., Stephens, J. C., Wilson, E. J., and Rowlands, I. H., 2018. Social dimensions of smart grid: Regional analysis in Canada and the United States. Introduction to special issue. *Renewable and Sustainable Energy Reviews* 82, 1909–1912.

Meadows, 2018. More than one million smart meters currently in "dumb" mode. *The Telegraph*. www.telegraph.co.uk/bills-and-utilities/gas-electric/one-million-smart-meters-currently-dumb-mode/

Meckling, J. and Allan, B. B., 2020. The evolution of ideas in global climate policy. *Nature Climate Change* 10, 434–438.

Meckling, J., Kelsey, N., Biber, E., and Zysman, J., 2015. Winning coalitions for climate policy. *Science* 349, 1170–1171.

Meckling, J., Sterner, T., and Wagner, G., 2017. Policy sequencing toward decarbonization. *Nature Energy* 2, 918–922.

Menanteau, P. and Lefebvre, H., 2000. Competing technologies and the diffusion of innovations: The emergence of energy-efficient lamps in the residential sector. *Research Policy* 29, 375–389.

Metz, D., 2013. Peak car and beyond: The fourth era of travel. *Transport Reviews* 33, 255–270.

Metz, D., 2010. Saturation of demand for daily travel. *Transport Reviews* 30, 659–674.

Meyer, A. D., Tsui, A., and Hinings, C., 1993. Configurational approaches to organizational analysis. *Academy of Management Journal* 36, 1175–1195.

Decorte, M., Tessens, S., Francisco, F., Repullo, D., McCarthy, P., Oriordan, B., et al., 2020. D6.1: Mapping the state of play of renewable gases in Europe. REGATRACE project.

Millar, M.-A., Burnside, N. M., and Yu, Z., 2019. District heating challenges for the UK. *Energies* 12, 310.

Millard-Ball, A. and Schipper, L., 2011. Are we reaching peak travel? Trends in passenger transport in eight industrialized countries. *Transport Reviews* 31, 357–378.

Miller, D., 1996. Configurations revisited. *Strategic Management Journal* 17, 505–512.

Miller, D., 1990. Organizational configurations: Cohesion, change and prediction. *Human Relations* 43, 771–789.

Miller, J., 2019. Global car market shrinking at fastest rate since financial crisis. *Financial Times*. www.ft.com/content/b38adcac-169f-11ea-9ee4-11f260415385

Milovanoff, A., Posen, I. D., and MacLean, H. L., 2020. Electrification of light-duty vehicle fleet alone will not meet mitigation targets. *Nature Climate Change* 10, 1102–1107.

Mirzania, P., Ford, A., Andrews, D., Ofori, G., and Maidment, G., 2019. The impact of policy changes: The opportunities of Community Renewable Energy projects in the UK and the barriers they face. *Energy Policy* 129, 1282–1296.

Mitchell, C. and Connor, P. M., 2004. Renewable energy policy in the UK 1990–2003. *Energy Policy* 32, 1935–1947.

Mladenović, M. N., Stead, D., Milakis, D., Pangbourne, K., and Givoni, M., 2020. Governance cultures and sociotechnical imaginaries of self-driving vehicle technology: Comparative analysis of Finland, UK and Germany. *Advances in Transport Policy and Planning* 5, 235–262.

Mlecnik, 2013. *Innovation Development for Highly Energy-Efficient Housing: Opportunities and Challenges Related to the Adoption of Passive Houses*. Delft University Press, Delft.

Mohr, L. B., 1982. *Explaining Organizational Behavior*. Jossey-Bass, San Francisco.

Monreal, A. C., McMeekin, A., and Southerton, D., 2016. Beyond acquisition: Exploring energy consumption through the appreciation and appropriation of domestic lighting in the UK. *Sustainable Production and Consumption* 7, 37–48.

Mora, L., Wu, X., and Panori, A., 2020. Mind the gap: Developments in autonomous driving research and the sustainability challenge. *Journal of Cleaner Production* 275, 124087.

Moran, M., 2003. *The British Regulatory State: High Modernism and Hyper-Innovation*. Oxford University Press, Oxford.

Murray, J., 2021. Zero-carbon hydrogen injected into gas grid for first time in groundbreaking UK trial. *The Guardian*. www.theguardian.com/environment/2020/jan/24/hydrogen-uk-gas-grid-keele-university

Mylan, J., 2017. The business of 'behaviour change': Analysing the consumer-oriented corporate sustainability journey of low-temperature laundry. *Organization & Environment* 30, 283–303.

Mylan, J., 2016. The directionality of desire in the economy of qualities: The case of retailers, refrigeration and re-constituted orange juice, in: Bulkeley, H., Paterson, M., and Stripple, J. (Eds.), *Towards a Cultural Politics of Climate Change: Devices, Desires and Dissent*. Cambridge University Press, Cambridge, pp. 142–159.

Mylan, J., Geels, F. W., McMeekin, A., Gee, S., and Foster, C., 2015. Eco-innovation and retailers in UK milk, beef and bread chains: Enriching environmental supply chain management with insights from innovation studies. *Journal of Cleaner Production* 107, 20–30.

Mylan, J., Morris, C., Beech, E., and Geels, F. W., 2019. Rage against the regime: Niche-regime interactions in the societal embedding of plant-based milk. *Environmental Innovation and Societal Transitions* 31, 233–247. https://doi.org/10.1016/j.eist.2018.11.001

Mylan, J. and Southerton, D., 2018. The social ordering of an everyday practice: The case of laundry. *Sociology* 52, 1134–1151.

NAO, 2018. Low-carbon heating of homes and businesses and the Renewable Heat Incentive.

National Grid, 2019. Enabling the Gas Markets Plan 2019/2020. National Grid Plc.

Nelson, R. R. and Winter, S. G., 1982. *An Evolutionary Theory of Economic Change*. The Belknap Press of Harvard University Press, Cambridge, MA; London.

Nemet, G., 2019. How Solar Energy Became Cheap: A Model for Low Carbon Innovation. Earthscan.

NESTA, 2013. Systems Innovation. National Endowment for Science, Technology and the Arts, London.

Newell, P., 2021. *Power Shift: The Global Political Economy of Energy Transitions*. Cambridge University Press, Cambridge.

Newell, P. and Paterson, M., 1998. Climate for business: Global warming, the State, and capital. *Review of International Political Economy* 5, 679–704.

Newman, P. and Kenworthy, J., 2011. 'Peak car use': Understanding the demise of automobile dependence. *World Transport Policy and Practice* 17, 31–42.

Newson, C. and Sloman, L., 2018. *The Value of the Cycling Sector to the British Economy: A Scoping Study*, Transport for Quality of Life.

NIC, 2020. *Rail Needs Assessments for the Midlands and the North: Final Report*. National Infrastructure Commission, London.

Nilsson, M. and Nykvist, B., 2016. Governing the electric vehicle transition: Near term interventions to support a green energy economy. *Applied Energy* 179, 1360–1371. https://doi.org/10.1016/j.apenergy.2016.03.056

Nilsson, M. and Persson, A., 2017. Policy note: Lessons from environmental policy integration for the implementation of the 2030 Agenda. *Environmental Science and Policy* 78, 36–39.

North, D. C., 1990. *Institutions, Institutional Change and Economic Performance*. Cambridge University Press, Cambridge.

Nykvist, B., Sprei, F., and Nilsson, M., 2019. Assessing the progress toward lower priced long range battery electric vehicles. *Energy Policy* 124, 144–155.

Nykvist, B. and Whitmarsh, L., 2008. A multi-level analysis of sustainable mobility transitions: Niche development in the UK and Sweden. *Technological Forecasting and Social Change* 75, 1373–1387. https://doi.org/10.1016/j.techfore.2008.05.006

O'Brien, K. and Signa, L., 2018. Transformations in socio-ecological systems, in: EEA (Ed.), *Perspectives on Transitions to Sustainability, EEA Report 25/2017*. European Environment Agency, Copenhagen, pp. 29–45.

O'Neill, K. and Gibbs, D., 2020. Sustainability transitions and policy dismantling: Zero carbon housing in the UK. *Geoforum* 108, 119–129. https://doi.org/10.1016/j.geoforum.2019.11.011

O'Neill, K. and Gibbs, D., 2014. Towards a sustainable economy? Socio-technical transitions in the green building sector. *Local Environment* 19, 572–590.

OECD, 2018. *Financing Climate Futures: Rethinking Infrastructure*. Organisation for Economic Co-operation and Development, Paris.

OECD, 2015. System Innovation: Synthesis Report.

Ofgem, 2020. Domestic RHI tariffs table (Q1 – 2021/22). Office of Gas and Electricity Markets.

Ofgem, 2019. *State of the Energy Market 2019*. Office of Gas and Electricity Markets, London.

Ofgem, 2017. *Update on Extending Competition in Transmission, Letter to interested parties*. Office of Gas and Electricity Markets, London.

Ofgem, 2013. *Strategy Decision for the RIIO-ED1 Electricity Distribution Price Control*. Office of Gas and Electricity Markets, London.

OLEV, 2020. Electric Vehicle Homecharge Scheme: Guidance for Consumers. Office for Low Emission Vehicles.

Olleros, F.-J., 1986. Emerging industries and the burnout of pioneers. *Journal of Product Innovation Management* 3, 5–18.

Ornetzeder, M. and Rohracher, H., 2013. Of solar collectors, wind power, and car sharing: Comparing and understanding successful cases of grassroots innovations. *Global Environmental Change* 23, 856–867. https://doi.org/10.1016/j.gloenvcha.2012.12.007

ORR, 2020. *Rail Industry Finance (UK) 2019–20*. Office of Rail and Road, London.

Owaineh, A., Leach, M., Guest, P., and Wehrmeyer, W., 2015. Policy, niches and diffusion in UK smart grid innovation. Centre for Environmental Strategy Working Paper 01/15, University of Surrey.

Owens, S., 1995. From 'predict and provide' to 'predict and prevent'? Pricing and planning in transport policy. *Transport Policy* 2, 43–49.

Pangbourne, K., Mladenovic, M. N., Stead, D., and Milakis, D., 2020. Questioning mobility as a service: Unanticipated implications for society and governance. *Transportation Research Part A: Policy and Practice* 131, 35–49.

Papachristos, G., Sofianos, A., and Adamides, E., 2013. System interactions in socio-technical transitions: Extending the multi-level perspective. *Environmental Innovation and Societal Transitions* 7, 53–69. https://doi.org/10.1016/j.eist.2013.03.002

Park, W. P., Phadke, A., Shah, N., and Letschert, V., 2013. Efficiency improvement opportunities in TVs: Implications for market transformation programs. *Energy Policy* 59, 361–372.

Parkhurst, G. and Lyons, G., 2018. The many assumptions about self-driving cars – Where are we heading and who is in the driving seat? Paper presented at 16th Annual Transport Practitioners' Meeting, 5–6 July 2018.

Parrish, B., Gross, R., and Heptonstall, P., 2019. On demand: Can demand response live up to expectations in managing electricity systems? *Energy Research & Social Science* 51, 107–118.

Passivhaus Trust, 2019. Passivhaus Construction Costs. London.

Passivhaus Trust, 2015. Passivhaus Capital Cost Research Project. London.

Paterson, M. and P-Laberge, X., 2018. Political economies of climate change. *WIRES Climate Change* 9, e506.

Pearson, P. and Watson, J., 2012. UK Energy Policy 1980–2010: A History and Lessons to be Learnt. London.

Pearson, P. J. G. and Arapostathis, S., 2017. Two centuries of innovation, transformation and transition in the UK gas industry: Where next? *Proceedings of the Institution of Mechanical Engineers, Part A: Journal of Power and Energy* 231, 478–497. https://doi.org/10.1177/0957650917693482

Penna, C. C. R. and Geels, F. W., 2015. Climate change and the slow reorientation of the American car industry (1979–2012): An application and extension of the Dialectic Issue LifeCycle (DILC) model. *Research Policy* 44, 1029–1048. https://doi.org/10.1016/j.respol.2014.11.010

Perry, M. and Rosillo-Calle, F., 2008. Recent trends and future opportunities in UK bioenergy: Maximising biomass penetration in a centralised energy system. *Biomass and Bioenergy* 32, 688–701.

Persson, A. and Runhaar, H., 2018. Conclusion: Drawing lessons for environmental policy integration and prospects for future research. *Environmental Science and Policy* 85, 141–145.

Pettifor, A., 2019. The Case for the Green Deal. London.

PHE, 2019. Review of Interventions to Improve Outdoor Air Quality and Public Health. Public Health England.

Phillips, M. E., 1994. Industry mindsets: Exploring the cultures of two macro-organizational settings. *Organization Science* 5, 384–402.

Pierson, P., 2004. *Politics in Time: History, Institutions, and Social Analysis*. Princeton University Press, Princeton, NJ.

Pierson, P., 2000. Increasing returns, path dependence, and the study of politics. *The American Political Science Review* 94, 251–267.

Pinkse, J. and Van den Buuse, D., 2012. The development and commercialization of solar PV technology in the oil industry. *Energy Policy* 40, 11–20.

Pitts, A., 2017. Passive house and low energy buildings: Barriers and opportunities for future development within UK practice. *Sustainability* 9, 272.

Polanyi, K., 2001. *The Great Transformation: The Political and Economic Origins of Our Time [1944]*. Beacon Press, Boston.

Poole, M. S., Van de Ven, A. H., Dooley, K., and Holmes, M. E., 2000. *Organizational Change and Innovation Processes: Theory and Methods for Research*. Oxford University Press, New York.

Pooley, C., Horton, D., Scheldeman, G., Tight, M., Jones, T., Chisholm, A., et al., 2011. Household decision-making for everyday travel: A case study of walking and cycling in Lancaster (UK). *Journal of Transport Geography* 19, 1601–1607.

Powell, W. and DiMaggio, P. (Eds.), 1991. *The New Institutionalism in Organizational Analysis*. The University of Chicago Press, Chicago.

Princen, T., 2005. *The Logic of Sufficiency*. MIT Press, Cambridge, MA.

Pucher, J. and Buehler, R., 2008. Making cycling irresistible: Lessons from The Netherlands, Denmark and Germany. *Transport Reviews* 28, 495–528. https://doi.org/10.1080/01441640701806612

RAC, 2012. *On the Move: Making Sense of Car and Train Travel Trends in Britain*. Royal Automobile Club Foundation, London.

RAE, 2017. *Sustainability of Liquid Biofuels*. Royal Academy of Engineering, London.

Ragin, C., 2008. *Redesigning Social Inquiry: Fuzzy Sets and beyond*. University of Chicago Press, Chicago.

Raman, S. and Shove, E., 2000. The business of building regulation, in: Fineman, S. (Ed.), *The Business of Greening*. Routledge, London, pp. 134–150.

Raskin, P., 2016. *Journey to Earthland: The Great Transition to Planetary Civilization*. Tellus Institute, Boston.

Raven, R., 2007. Niche accumulation and hybridisation strategies in transition processes towards a sustainable energy system: An assessment of differences and pitfalls. *Energy Policy* 35, 2390–2400. https://doi.org/10.1016/j.enpol.2006.09.003

Raven, R., Heiskanen, E., Lovio, R., Hodson, M., and Brohmann, B., 2008. The contribution of local experiments and negotiation processes to field-level learning in emerging (niche) technologies: Meta-analysis of 27 new energy projects in Europe. *Bulletin of Science, Technology & Society* 28, 464–477. https://doi.org/10.1177/0270467608317523

Raven, R., Kern, F., Verhees, B., and Smith, A., 2016. Niche construction and empowerment through socio-political work: A meta-analysis of six low-carbon technology cases. *Environmental Innovation and Societal Transitions* 18, 164–180. https://doi.org/10.1016/j.eist.2015.02.002

Raven, R., Schot, J., and Berkhout, F., 2012. Space and scale in socio-Technical transitions. *Environmental Innovation and Societal Transitions* 4, 63–78. https://doi.org/10.1016/j.eist.2012.08.001

Raven, R. and Verbong, G., 2007. Multi-regime interactions in the Dutch energy sector: The case of combined heat and power technologies in the Netherlands 1970–2000. *Technology Analysis & Strategic Management* 19, 491–507. https://doi.org/10.1080/09537320701403441

Regen, 2020. The decarbonisation of heat. Exeter.

Richards, S. J. and Al Zaili, J., 2020. Contribution of encouraging the future use of biomethane to resolving sustainability and energy security challenges: The case of the UK. *Energy for Sustainable Development* 55, 48–55. https://doi.org/10.1016/j.esd.2019.12.003

Rip, A. and Kemp, R., 1998. Technological change, in: Rayner, S. and Malone, L. (Eds.), *Human Choice and Climate Change*, Vol 2 Resources and Technology. Batelle Press, Washington, DC, pp. 327–399.

Roberts, C. and Geels, F. W., 2019a. Conditions for politically accelerated transitions: Historical institutionalism, the multi-level perspective, and two historical case studies in transport and agriculture. *Technological Forecasting and Social Change* 140, 221–240. https://doi.org/10.1016/j.techfore.2018.11.019

Roberts, C. and Geels, F. W., 2019b. Conditions and intervention strategies for the deliberate acceleration of socio-technical transitions: Lessons from a comparative multi-level analysis of two historical case studies in Dutch and Danish heating. *Technology Analysis & Strategic Management* 31, 1081–1103.

Roberts, C. and Geels, F. W., 2018. Public storylines in the British transition from rail to road transport (1896–2000): Discursive struggles in the multi-level perspective. *Science as Culture* 27, 513–542. https://doi.org/10.1080/09505431.2018.1519532

Roberts, E., 2020. Warming with wood: Exploring the everyday heating practices of rural off-gas households in Wales. *Energy Policy* 142, 111451. https://doi.org/10.1016/j.enpol.2020.111451

Roberts, J. C. D., 2017. Discursive destabilisation of socio-technical regimes: Negative storylines and the discursive vulnerability of historical American railroads. *Energy Research and Social Science* 31, 86–99. https://doi.org/10.1016/j.erss.2017.05.031

Robertson, P. L., 1992. Networks and innovation in a modular system: Lessons from the microcomputer and stereo component industries. *Research Policy* 21, 297–313.

Rockström, J., Gaffney, O., Rogelj, J., Meinshausen, M., Nakicenovic, N., and Schellnhuber, H. J., 2017. A roadmap for rapid decarbonisation. *Science* 355, 1269–1271.

Rogge, K. S. and Reichardt, K., 2016. Policy mixes for sustainability transitions: An extended concept and framework for analysis. *Research Policy* 45, 1620–1635.

Røpke, I., Christensen, T. H., and Jensen, J. O., 2010. Information and communication technologies: A new round of household electrification. *Energy Policy* 38, 1764–1773.

Rosenbloom, D., 2020. Engaging with multi-system interactions in sustainability transitions: A comment on the transitions research agenda. *Environmental Innovation and Societal Transitions* 34, 336–340.

Rosenbloom, D., 2017. Pathways: An emerging concept for the theory and governance of low-carbon transitions. *Global Environmental Change* 43, 37–50. https://doi.org/10.1016/j.gloenvcha.2016.12.011

Rosenbloom, D., Berton, H., and Meadowcroft, J., 2016. Framing the sun: A discursive approach to understanding multi-dimensional interactions within socio-technical transitions through the case of solar electricity in Ontario, Canada. *Research Policy* 45, 1275–1290. https://doi.org/10.1016/j.respol.2016.03.012

Rosenbloom, D., Markard, J., Geels, F. W., and Fuenfschilling, L., 2020. Why carbon pricing is not sufficient – and how a "sustainability transition policy" can help mitigate climate change. *Proceedings of the National Academy of Sciences* 117, 8664–8668.

Rosenow, J., 2012. Energy savings obligations in the UK: A history of change. *Energy Policy* 49, 373–382.

Rosenow, J. and Eyre, N., 2016. A post mortem of the Green Deal: Austerity, energy efficiency, and failure in British energy policy. *Energy Research & Social Science* 21, 141–144. https://doi.org/10.1016/j.erss.2016.07.005

Rosenow, J., Guertler, P., Sorrell, S., and Eyre, N., 2018. The remaining potential for energy savings in UK households. *Energy Policy* 121, 542–552. https://doi.org/10.1016/j.enpol.2018.06.033

Rosenow, J., Kern, F., and Rogge, K. S., 2017. The need for comprehensive and well targeted instrument mixes to stimulate energy transitions: The case of energy efficiency policy. *Energy Research & Social Science* 33, 95–104.

Rosenow, J. and Thomas, S., 2020. Net zero is nowhere in sight for UK clean heat policy, Green Alliance blog. https://greenallianceblog.org.uk/2020/05/13/net-zero-is-nowhere-in-sight-for-uk-clean-heat-policy/.

Ryan-Collins, J., Lloyd, T., and Macfarlane, L., 2017. *Rethinking the Economics of Land and Housing*. Zed Books, London.

Rycroft, R. W. and Kash, D. E., 2002. Path dependence in the innovation of complex technologies. *Technology Analysis & Strategic Management* 14, 21–35.

Ryghaug, M. and Toftaker, M., 2014. A transformative practice? Meaning, competence, and material aspects of driving electric cars in Norway. *Nature and Culture* 9, 146–163.

SAE, 2016. Taxonomy and definitions for terms related to driving automation systems for on-road motor vehicles. Society of Automotive Engineers.

Sandén, B. A. and Hillman, K. M., 2011. A framework for analysis of multi-mode interaction among technologies with examples from the history of alternative transport fuels in Sweden. *Research Policy* 40, 403–414. https://doi.org/10.1016/j.respol.2010.12.005

Sanderson, S. W. and Simons, K. L., 2014. Light emitting diodes and the lighting revolution: The emergence of a solid-state lighting industry. *Research Policy* 43, 1730–1746.

Schmidt, N. M. and Fleig, A., 2018. Global patterns of national climate policies: Analyzing 171 country portfolios on climate policy integration. *Environmental Science & Policy* 84, 177–185.

Schmidt, T. S., Matsuo, T., and Michaelowa, A., 2017. Renewable energy policy as an enabler of fossil fuel subsidy reform? Applying a socio-technical perspective to the cases of South Africa and Tunisia. *Global Environmental Change* 45, 99–110. https://doi.org/10.1016/j.gloenvcha.2017.05.004

Schmidt, T. S. and Sewerin, S., 2019. Measuring the temporal dynamics of policy mixes: An empirical analysis of renewable energy policy mixes' balance and design features in nine countries. *Research Policy* 48, 103557.

Schnaiberg, A., 1980. *The Environment: From Surplus to Scarcity*. Oxford University Press, New York.

Schor, J. B., 2014. Climate discourse and economic downturns: The case of the United States, 2008–2013. *Environmental Innovation and Societal Transitions* 13, 6–20. https://doi.org/10.1016/j.eist.2014.04.006

Schot, J. and Geels, F. W., 2008. Strategic niche management and sustainable innovation journeys: Theory, findings, research agenda, and policy. *Technology Analysis & Strategic Management* 20, 537–554. https://doi.org/10.1080/09537320802292651

Schot, J. and Steinmueller, W. E., 2018. Three frames for innovation policy: R&D, systems of innovation and transformative change. *Research Policy* 47, 1554–1567. https://doi.org/10.1016/j.respol.2018.08.011

Schot, J. W. and Geels, F. W., 2007. Niches in evolutionary theories of technical change: A critical survey of the literature. *Journal of Evolutionary Economics* 17, 605–622.

Schumpeter, J. A., 1927. The explanation of the business cycle. *Economica* 21, 286–311.

Schwanen, T., 2015. The bumpy road toward low-energy urban mobility: Case studies from two UK cities. *Sustainability* 7, 7086–7111.
Scott, W. R., 2008. Approaching adulthood: The maturing of institutional theory. *Theory and Society* 37, 427–442.
Scott, W. R., 1995. *Institutions and Organizations*. Sage Publications, Thousand Oaks, CA.
Scott, W. R., Ruef, M., Mendel, P. J., and Caronna, C., 2000. *Institutional Change and Healthcare Organizations: From Professional Dominance to Managed Care*. Chicago University Press, Chicago.
SGF, 2014. Smart Grid Vision and Roadmap. London.
Shaheen, S. and Chan, N., 2016. Mobility and the sharing economy: Potential to facilitate the first- and last-mile public transit connections. *Built Environment* 42, 573–588.
Shaw, J. and Docherty, I., 2014. *The Transport Debate*. Policy Press, Bristol.
Shaw, R., Attree, M., and Jackson, T., 2010. Developing electricity distribution networks and their regulation to support sustainable energy. *Energy Policy* 38, 5927–5937.
Sheller, M., 2012. The emergence of new cultures of mobility: Stability, openings and prospects, in: Geels, F. W., Kemp, R., Dudley, G., and Lyons, G. (Eds.), *Automobility in Transition? A Socio-Technical Analysis of Sustainable Transport*. Routledge, New York, pp. 180–202.
Sheller, M., 2004. Automotive emotions: Feeling the car. *Theory, Culture & Society* 21, 221–242.
Shipworth, D., Fell, M. J., and Elam, S., 2019. Response to "Vulnerability and resistance in the United Kingdom's smart meter transition". *Energy Policy* 124, 418–420.
Shove, E., 2003. *Comfort, Cleanliness and Convenience: The Social Organization of Normality*. Berg, Oxford.
Shove, E., Pantzar, M., and Watson, M., 2012. *The Dynamics of Social Practice: Everyday Life and How It Changes*. Sage, London.
Shove, E. and Walker, G., 2014. What is energy for? Social practice and energy demand. *Theory, Culture & Society* 31, 41–58.
Shove, E. and Walker, G., 2010. Governing transitions in the sustainability of everyday life. *Research Policy* 39, 471–476. https://doi.org/10.1016/j.respol.2010.01.019
Simon, H. A., 1973. The organization of complex systems, in: Pattee, H. (Ed.), *Hierarchy Theory: The Challenge of Complex Systems*. George Braziller, New York, pp. 1–27.
Skea, J., van Diemen, R., Hannon, M., Gazis, E., and Rhodes, A., 2019. *Energy Innovation for the Twenty-First Century: Accelerating the Energy Revolution*. Edward Elgar, Cheltenham.
Skeete, J.-P., 2019. Concentration of power: A UK case study examining the dominance of incumbent automakers and suppliers in automotive sociotechnical transitions. *Global Transitions* 1, 93–103.
Skjølsvold, T. M. and Coenen, L., 2021. Are rapid and inclusive energy and climate transitions oxymorons? Towards principles of responsible acceleration. *Energy Research & Social Science* 79.
Slavin, T., 2014. How to build a transport system that works for the whole of the UK? *The Guardian*. www.theguardian.com/public-leaders-network/2014/sep/18/transport-infrastructure-uk-future
Sminia, H., 2009. Process research in strategy formation: Theory, methodology and relevance. *International Journal of Management Reviews* 11, 97–125.
Smink, M., Negro, S. O., Niesten, E., and Hekkert, M. P., 2015. How mismatching institutional logics hinder niche-regime interaction and how boundary spanners

intervene. *Technological Forecasting and Social Change* 100, 225–237. https://doi.org/10.1016/j.techfore.2015.07.004

Smith, A., 2007. Translating sustainabilities between green niches and socio-technical regimes. *Technology Analysis and Strategic Management* 19, 427–450. https://doi.org/10.1080/09537320701403334

Smith, A., Fressoli, M., and Thomas, H., 2014. Grassroots innovation movements: Challenges and contributions. *Journal of Cleaner Production* 64, 114–124.

Smith, A., Kern, F., Raven, R., and Verhees, B., 2013. Spaces for sustainable innovation: Solar photovoltaic electricity in the UK. *Technological Forecasting & Social Change* 81, 115–130.

Smith, A. and Raven, R., 2012. What is protective space? Reconsidering niches in transitions to sustainability. *Research Policy* 41, 1025–1036. https://doi.org/10.1016/j.respol.2011.12.012

Smith, A. and Seyfang, G., 2013. Constructing grassroots innovations for sustainability. *Global Environmental Change* 23, 827–829.

Smith, A., Stirling, A., and Berkhout, F., 2005. The governance of sustainable socio-technical transitions. *Research Policy* 34, 1491–1510. https://doi.org/10.1016/j.respol.2005.07.005

Smith, G. and Hensher, D. A., 2020. Towards a framework for mobility-as-a-service policies. *Transport Policy* 89, 54–65.

Smith, G., Sochor, J., and Karlsson, M. A., 2020. Intermediary MaaS Integrators: A case study on hopes and fears. *Transportation Research Part A: Policy and Practice* 131, 163–177.

SMMT, 2020. *Motor Car Industry Facts 2020*. Society of Motor Manufacturers and Traders, London.

SMMT, 2016. *Ultra Low Emission Vehicles Guide, 2016*. Society of Motor Manufacturers and Traders, London.

Solar Heat Europe, 2018. Solar Heat Markets in Europe.

Sørensen, C. and Gudmundsson, H., 2010. The impact of governance modes on sustainable transport – the case of bus transport in Greater Manchester, UK. World Review of Intermodal Transportation Research 3.

Soteropoulos, A., Berger, M., and Ciari, F., 2019. Impacts of automated vehicles on travel behaviour and land use: An international review of modelling studies. *Transport Reviews* 39, 29–49.

Sovacool, B. and Griffiths, S., 2020. Culture and low-carbon energy transitions. *Nature Sustainability* 3, 685–693.

Sovacool, B. K., 2019. *Visions of Energy Futures: Imagining and Innovating Low-Carbon Transitions*. Routledge, New York.

Sovacool, B. K., 2017. Experts, theories, and electric mobility transitions: Toward an integrated conceptual framework for the adoption of electric vehicles. *Energy Research and Social Science* 27, 78–95. https://doi.org/10.1016/j.erss.2017.02.014

Sovacool, B. K., 2016. How long will it take? Conceptualizing the temporal dynamics of energy transitions. *Energy Research and Social Science* 13, 202–215. https://doi.org/10.1016/j.erss.2015.12.020

Sovacool, B. K., Kivimaa, P., Hielscher, S., and Jenkins, K., 2017. Vulnerability and resistance in the United Kingdom's smart meter transition. *Energy Policy* 109, 767–781.

Sovacool, B. K., Kivimaa, P., Hielscher, S., and Jenkins, K. E., 2019. Further reflections on vulnerability and resistance in the United Kingdom's smart meter transition. *Energy Policy* 124, 411–417.

Spaargaren, G. and Cohen, M., 2009. Greening lifecycles and lifestyles: Sociotechnical innovations in consumption and production as core concerns of ecological modernization theory, in: Mol, A., Sonnenfeld, D., and Spaargaren, G. (Eds.), *The Ecological Modernization Reader: Environmental Reform in Theory and Practice*. Routledge, New York, pp. 257–275.

Speirs, J., Balcombe, P., Johnson, E., Martin, J., Brandon, N., and Hawkes, A., 2018. A greener gas grid: What are the options? *Energy Policy* 118, 291–297. https://doi.org/10.1016/j.enpol.2018.03.069

Sperling, D., 2018. *Three Revolutions: Steering Automated, Shared, and Electric Vehicles to a Better Future*. Island Press, Washington, DC.

Sprei, F., 2018. Disrupting mobility. *Energy Research and Social Science* 37, 238–242. https://doi.org/10.1016/j.erss.2017.10.029

Staffell, I., 2017. Measuring the progress and impacts of decarbonising British electricity. *Energy Policy* 102, 463–475. https://doi.org/10.1016/j.enpol.2016.12.037

Steer Davies Gleave, 2019a. *Car Club Annual Survey for London 2017/18*. Collaborative Mobility UK, Leeds.

Steer Davies Gleave, 2019b. *England & Wales Car Club Annual Survey 2017/18 (Excluding London)*. Collaborative Mobility UK, Leeds.

Steer Davies Gleave, 2019c. *Car Club Annual Survey for Scotland 2018/19*. Collaborative Mobility UK, Leeds.

Steer Davies Gleave, 2016. *Carplus Annual Survey of Car Clubs 2015/16: England & Wales (Excluding London)*. Carplus, London.

Steer Davies Gleave, 2012. *Carplus Annual Survey of Car Clubs 2011/2012: London*. Carplus, London.

Steer Davies Gleave, 2011. *Carplus Annual Survey of Car Clubs 2010/2011*. Carplus, London.

Steer Davies Gleave, 2010. *Carplus Annual Survey of Car Clubs 2009/10*. Carplus, London.

Stehlin, J., Hodson, M., and McMeekin, A., 2020. Platform mobilities and the production of urban space: Toward a typology of platformization trajectories. *Environment and Planning A* 52, 1250–1268.

Stephens, J. C., Wilson, E., Peterson, T. R., and Meadowcroft, J., 2013. Getting smart: Climate change and the electric grid. *Challenges* 4, 201–216.

Steward, F., 2018. Action-oriented perspectives on transitions and system innovation, in: EEA (Ed.), *Perspectives on Transitions to Sustainability*. Copenhagen, pp. 97–119.

Stewart, H. and Walker, P., 2020. HS2 poised to get go-ahead as £5bn pledged for bus funding. *The Guardian*. www.theguardian.com/uk-news/2020/feb/10/johnson-plans-5bn-boost-for-bus-services-and-cycle-routes

Stiglitz, J. E., Stern, N., Duan, M., Edenhofer, O., Giraud, G., Heal, G. M., et al., 2017. *Report of the High-Level Commission on Carbon Prices*. World Bank, Washington DC.

Stilgoe, J., 2018. Machine learning, social learning and the governance of self-driving cars. *Social Studies of Science* 48, 25–56.

Strachan, N., Pye, S., and Kannan, R., 2009. The iterative contribution and relevance of modelling to UK energy policy. *Energy Policy* 37, 850–860.

Strachan, P. A., Cowell, R., Ellis, G., Sherry-Brennan, F., and Toke, D., 2015. Promoting community renewable energy in a corporate energy world. *Sustainable Development* 23, 96–109.

Stradling, S., Carreno, M., Rye, T., and Noble, A., 2007. Passenger perceptions and the ideal urban bus journey experience. *Transport Policy* 14, 283–292.

Svensson, O. and Nikoleris, A., 2018. Structure reconsidered: Towards new foundations of explanatory transitions theory. *Research Policy* 47, 462–473. https://doi.org/10.1016/j.respol.2017.12.007

Taylor, I. and Sloman, L., 2013. Reunifying Britain's railways: Obstacles and opportunities. *Public Money & Management* 33, 329–336.

TCPA/CHPA, 2008. Community Energy: Urban Planning for a Low Carbon Future. Town and Country Planning Association (TCPA) and Combined Heat and Power Association (CHPA), London.

TfL, 2019. *Travel in London, Report 12*. Transport for London, London.

TfL, 2015. Attitudes towards cycling. Transport for London.

Thelen, K., 1999. Historical institutionalism in comparative politics. *Annual Review of Political Science* 2, 369–404.

Thelen, K. and Mahoney, J., 2015. Comparative-historical analysis in contemporary political science, in: Mahoney, J. and Thelen, K. (Eds.), *Advances in Comparative-Historical Analysis*. Cambridge University Press, Cambridge, pp. 3–36.

Thielemans, S., Di Zenobio, D., Touhafi, A., Lataire, P., and Steenhaut, K., 2017. DC grids for smart LED-based lighting: The EDISON solution. *Energies* 10, 1454.

Thomas, S., 2016. The Hinkley Point decision: An analysis of the policy process. *Energy Policy* 96, 421–431.

Thornton, P., Occasio, W., and Lounsbury, M., 2012. *The Institutional Logics Perspective: A New Approach to Culture, Structure and Process*. Oxford University Press, Oxford.

Tilly, C., 2008. *Explaining Social Processes*. Paradigm Publishers, Boulder, CO.

Toke, D., 2011. The UK offshore wind power programme: A sea-change in UK energy policy? *Energy Policy* 39, 526–534.

Toke, D., 2005. Are green electricity certificates the way forward for renewable energy? An evaluation of the UK's renewables obligation in the context of international comparisons. *Environment and Planning C* 23, 361–375.

Toke, D., 2000. Policy network creation: The case of energy efficiency. *Public Administration* 78, 835–854.

Toke, D. and Lauber, V., 2007. Anglo Saxon and German approaches to neo-liberalism and environmental policy: The case of financing renewable energy. *Geoforum* 38, 677–687.

Topham, G., 2021. "Peak hype": Why the driverless car revolution has stalled. *The Guardian*. www.theguardian.com/technology/2021/jan/03/peak-hype-driverless-car-revolution-uber-robotaxis-autonomous-vehicle

Transport Systems Catapult, 2016. Mobility as a Service: Exploring the opportunity for mobility as a service in the UK. Birmingham.

Truffer, B., 2003. User-led innovation processes: The development of professional carsharing by environmentally concerned citizens. *Innovation: The European Journal of Social Science Research* 16, 139–154.

TSC, 2017. *Market Forecast for Connected and Autonomous Vehicles*. Transport Systems Catapult, Milton Keynes.

Turley, J., 2019. Self-Driving Cars: What the Engineers Think. *Electronic Engineering Journal* 24 April, 5–11.

Turnheim, B., 2012. The destabilisation of existing regimes in socio-technical transitions: Theoretical explorations and in-depth case studies of the British coal industry (1880–2011). University of Sussex.

Turnheim, B., Asquith, M., and Geels, F. W., 2020. Making sustainability transitions research policy-relevant: Challenges at the science-policy interface. *Environmental*

Innovation and Societal Transitions 34, 116–120. https://doi.org/10.1016/j.eist.2019.12.009

Turnheim, B., Berkhout, F., Geels, F. W., Hof, A., McMeekin, A., Nykvist, B., et al., 2015. Evaluating sustainability transitions pathways: Bridging analytical approaches to address governance challenges. *Global Environmental Change* 35, 239–253. https://doi.org/10.1016/j.gloenvcha.2015.08.010

Turnheim, B. and Geels, F. W., 2019. Incumbent actors, guided search paths, and landmark projects in infra-system transitions: Re-thinking Strategic Niche Management with a case study of French tramway diffusion (1971–2016). *Research Policy* 48, 1412–1428. https://doi.org/10.1016/j.respol.2019.02.002

Turnheim, B. and Geels, F.W., 2013. The destabilisation of existing regimes: Confronting a multi-dimensional framework with a case study of the British coal industry (1913–1967). *Research Policy* 42, 1749–1767. https://doi.org/10.1016/j.respol.2013.04.009

Turnheim, B. and Geels, F.W., 2012. Regime destabilisation as the flipside of energy transitions: Lessons from the history of the British coal industry (1913–1997). *Energy Policy* 50, 35–49. https://doi.org/10.1016/j.enpol.2012.04.060

Turnheim, B. and Nykvist, B., 2019. Opening up the feasibility of sustainability transitions pathways (STPs): Representations, potentials, and conditions. *Research Policy* 48, 775–788. https://doi.org/10.1016/j.respol.2018.12.002

Turnheim, B. and Sovacool, B. K., 2020. Forever stuck in old ways? Pluralising incumbencies in sustainability transitions. *Environmental Innovation and Societal Transitions* 35, 180–184. https://doi.org/10.1016/j.eist.2019.10.012

Tushman, M. L. and Romanelli, E., 1985. Organizational evolution: A metamorphosis model of convergence and reorientation. *Research in Organizational Behavior* 7, 171–222.

UKERC, 2019. *Review of Energy Policy 2019*. UK Energy Research Centre, London.

UKGBC, 2021. The Retrofit Playbook: Driving retrofit of existing homes – a resource for local and combined authorities. UK Green Building Council.

Upham, P., Dütschke, E., Schneider, U., Oltra, C., Sala, R., Lores, M., et al., 2018. Agency and structure in a sociotechnical transition: Hydrogen fuel cells, conjunctural knowledge and structuration in Europe. *Energy Research and Social Science* 37, 163–174. https://doi.org/10.1016/j.erss.2017.09.040

Upham, P., Kivimaa, P., and Virkamäki, V., 2013. Path dependence and technological expectations in transport policy: The case of Finland and the UK. *Journal of Transport Geography* 32, 12–22. https://doi.org/10.1016/j.jtrangeo.2013.08.004

Upreti, B. R. and Van der Horst, D., 2004. National renewable energy policy and local opposition in the UK: The failed development of a biomass electricity plant. *Biomass and Bioenergy* 26, 61–69.

Urry, J., 2004. The "system" of automobility. *Theory, Culture & Society* 21, 25–39. https://doi.org/10.1177/0263276404046059

Utility Week, 2017. Smart metering: Challenging times lead to strange bedfellows.

Van Buskirk, R., Kantner, C., Gerke, B., and Chu, S., 2014. A retrospective investigation of energy efficiency standards: Policies may have accelerated long-term declines in appliance costs. *Environment Research Letters*, 9.

Van De Poel, I., 2000. On the role of outsiders in technical development. *Technology Analysis and Strategic Management* 12, 383–397. https://doi.org/10.1080/09537320050130615

van de Ven, A. H., 2007. *Engaged Scholarship: A Guide for Organizational and Social Research*. Oxford University Press, Oxford.

Van Driel, H. and Schot, J., 2005. Radical innovation as a multi-level process: Introducing floating grain elevators in the port of Rotterdam. *Technology and Culture* 46, 51–76.

van Lente, H., Spitters, C., and Peine, A., 2013. Comparing technological hype cycles: Towards a theory. *Technological Forecasting and Social Change* 80, 1615–1628.

Van Mierlo, B., 2019. Users empowered in smart grid development? Assumptions and up-to-date knowledge. *Applied Sciences* 9, 815.

Van Waes, A., Farla, J., Frenken, K., De Jong, J. P. J., and Raven, R., 2018. Business model innovation and socio-technical transitions. A new prospective framework with an application to bike sharing. *Journal of Cleaner Production* 195, 1300–1312.

Verbong, G. P. J., Beemsterboer, S., and Sengers, F., 2013. Smart grids or smart users? Involving users in developing a low carbon electricity economy. *Energy Policy* 52, 117–125.

Verbong, G. P. J., Geels, F. W., and Raven, R. P. J. M., 2008. Multi-niche analysis of dynamics and policies in Dutch renewable energy innovation journeys (1970–2006): Hype-cycles, closed networks and technology-focused learning. *Technology Analysis & Strategic Management* 20, 555–573.

Vergragt, P. J., 2013. A possible way out of the combined economic-sustainability crisis. *Environmental Innovation and Societal Transitions* 6, 123–125.

Verhees, B., 2012. Cultural legitimacy and innovation journeys: A new perspective applied to Dutch and British nuclear power. Eindhoven University of Technology.

Victor, D., Geels, F. W., and Sharpe, S., 2019. Accelerating the Low Carbon Transition: The Case for Stronger, More Targeted and Coordinated International Action. Commissioned by the UK Department for Business, Energy & Industrial Strategy; Supported by the Energy Transitions Commission.

Vivid Economics, 2021. Greenness of Stimulus Index, 5th edition, February 2021.

Vögele, S., Kunz, P., Rübbelke, D., and Stahlke, T., 2018. Transformation pathways of phasing out coal-fired power plants in Germany. *Energy, Sustainability and Society* 8, 1–18.

Wade, F., Hitchings, R., and Shipworth, M., 2016. Understanding the missing middlemen of domestic heating: Installers as a community of professional practice in the United Kingdom. *Energy Research & Social Science* 19, 39–47. https://doi.org/10.1016/j.erss.2016.05.007

Wadud, Z. and Baierl, M., 2017. Explaining "peak car" with economic variables: A comment. *Transportation Research Part A: Policy and Practice* 95, 381–385.

Wadud, Z., MacKenzie, D., and Leiby, P., 2016. Help or hindrance? The travel, energy and carbon impacts of highly automated vehicles. *Transportation Research Part A* 86, 1–18.

Walker, G. P. and Devine-Wright, P., 2008. Community renewable energy: What should it mean? *Energy Policy* 36, 497–500.

Walker, P., 2021. Hastily abandoned low-traffic schemes could cost councils funding. *The Guardian*. www.theguardian.com/uk-news/bike-blog/2021/jul/30/hastily-abandoned-low-traffic-schemes-could-cost-councils-funding

Walker, W., 2000. Entrapment in large technology systems: Institutional commitments and power relations. *Research Policy* 29, 833–846.

Wall, R. and Crosbie, T., 2009. Potential for reducing electricity demand for lighting in households: An exploratory socio-technical study. *Energy Policy* 37, 1021–1031.

Warde, A., 2005. Consumption and theories of practice. *Journal of Consumer Culture* 5, 131–153.

Watson, J., Gross, R., Ketsopoulou, I., and Winskel, M., 2014. UK Energy Strategies Under Uncertainty, UKERC report.

Watson, S. D., Lomas, K. J., and Buswell, R. A., 2019. Decarbonising domestic heating: What is the peak GB demand? *Energy Policy* 126, 533–544. https://doi.org/10.1016/j.enpol.2018.11.001

Webb, J. and Hawkey, D., 2017. On (not) assembling a market for sustainable energy: Heat network infrastructure and British cities. *Journal of Cultural Economy* 10, 8–20.

Weiss, A. and Woodhouse, E., 1992. Reframing incrementalism: A constructive response to the critics. *Policy Sciences* 25, 255–273.

White, P., 2010. The conflict between competition policy and the wider role of the local bus industry in Britain. *Research in Transportation Economics* 29, 152–158.

Wieser, H., 2017. Ever-faster, ever-shorter? Replacement cycles of durable goods in historical perspective, in: Bakker, C. A. and Mugge, R. (Eds.), *Volume 9: PLATE: Product Lifetimes and The Environment*. Delft University of Technology, IOS Press, pp. 426–431.

Wieser, H. and Tröger, N., 2018. Exploring the inner loops of the circular economy: Replacement, repair, and reuse of mobile phones in Austria. *Journal of Cleaner Production* 172, 3042–3055.

Wilkinson, J., 2011. Convention theory and consumption, in: Southerton, D. (Ed.), *Encyclopedia of Consumer Culture*. Sage Publications, Thousand Oaks, CA, pp. 358–362.

Williams, L. and Sovacool, B. K., 2019. The discursive politics of 'fracking': Frames, storylines, and the anticipatory contestation of shale gas development in the United Kingdom. *Global Environmental Change*, 58.

Wilson, C., 2012. Up-scaling, formative phases, and learning in the historical diffusion of energy technologies. *Energy Policy* 50, 81–94. https://doi.org/10.1016/j.enpol.2012.04.077

Wilson, C., Crane, L., and Chryssochoidis, G., 2015. Why do homeowners renovate energy efficiently? Contrasting perspectives and implications for policy. *Energy Research and Social Science* 7, 12–22.

Wilson, C. and Grubler, A., 2011. Lessons from the history of technological change for clean energy scenarios and policies. *Natural Resources Forum* 35, 165–184. https://doi.org/10.1111/j.1477-8947.2011.01386.x

Wilson, I. A. and Staffell, I., 2018. Rapid fuel switching from coal to natural gas through effective carbon pricing. *Nature Energy* 3, 365–372.

Winskel, M., 2018. Energy innovation and systems change: Narratives of disruption and continuity. *Energy Research & Social Science* 37, 232–237.

Winskel, M., 2016. From optimisation to diversity: Changing scenarios of heating for buildings in the UK, in: Hawkey, D., Webb, J., Lovell, H., McCrone, D., Tingey, M., and Winskel, M. (Eds.), *Sustainable Urban Energy Policy: Heat and the City*. Routledge, London, pp. 68–90.

Wiseman, J., Edwards, T., and Luckins, K., 2013. Post carbon pathways: A meta-analysis of 18 large-scale post carbon economy transition strategies. *Environmental Innovation and Societal Transitions* 8, 76–93. https://doi.org/10.1016/j.eist.2013.04.001

Wolmar, C., 2018. *Driverless Cars: On a Road to Nowhere*. London Publishing Partnership.

Wolmar, C., 2016. *Are Trams Socialist? Why Britain Has No Transport Policy*. London Publishing Partnership.

Wood, G. and Dow, S., 2011. What lessons have been learned in reforming the Renewables Obligation? An analysis of internal and external failures in UK renewable energy policy. *Energy Policy* 39, 2228–2244. https://doi.org/10.1016/j.enpol.2010.11.012

Woodman, B. and Mitchell, C., 2011. Learning from experience? The development of the Renewables Obligation in England and Wales 2002–2010. *Energy Policy* 39, 3914–3921.

World Bank, 2015. Decarbonizing Development: Three Steps to a Zero-Carbon Future, Climate change and development series. Washington DC.

Wulf, C., Linßen, J., and Zapp, P., 2018. Review of power-to-gas projects in Europe. *Energy Procedia* 155, 367–378. https://doi.org/10.1016/j.egypro.2018.11.041

Yeatts, D. E., Auden, D., Cooksey, C., and Chen, C. F., 2017. A systematic review of strategies for overcoming the barriers to energy-efficient technologies in buildings. *Energy Research and Social Science* 32, 76–85. https://doi.org/10.1016/j.erss.2017.03.010

Zero Carbon Hub, 2013. Zero Carbon Strategies for Tomorrow's New Homes. Milton Keynes.

Printed in Dunstable, United Kingdom

84882011R00221